Institute for Nonlinear Science

T0235524

Springer

New York
Berlin
Heidelberg
Hong Kong
London
Milan
Paris
Tokyo

Institute for Nonlinear Science

Henry D.I. Abarbanel *Analysis of Chaotic Series* (1996)

Jordi García-Ojalvo, Jose M. Sancho *Noise in Spatially Extended Systems* (1999)

Leon Glass, Peter Hunter, Andrew McCullogh (Eds.) *Theory of Heart: Biomechanics, Biophysics, and Nonlinear Dynamic of Cardiac Function* (1991)

Mark Millonas (Ed.) *Fluctuations and Order: The New Synthesis* (1996)

Linda E. Reichl *The Transition to Chaos in Conservative Classical Systems: Quantum Manifestations* (1992)

Bruce West, Mauro Bologna, Paolo Grigolini *Physics of Fractal Operators* (2003)

Bruce J. West
Mauro Bologna
Paolo Grigolini

Physics of Fractal Operators

With 23 Illustrations

Springer

Bruce J. West
Department of the Army
U.S. Army Research Laboratory
Army Research Office
P.O. Box 12211
Research Triangle Park, NC 27709-2211
USA
westb@aro.arl.army.mil

Mauro Bologna
Paolo Grigolini
Department of Physics
University of North Texas
Denton, TX 76203
USA

Library of Congress Cataloging-in-Publication Data
West, Bruce J.
    Physics of fractal operators / Bruce West, Mauro Bologna, Paolo Grigolini.
      p. cm. — (Institute for nonlinear science)
    Includes bibliographical references and index.
    ISBN 978-1-4419-3054-5   e-ISBN 978-0-387-21746-8
    1. Fractional calculus. I. Bologna, Mauro. II. Grigolini, Paolo. III. Title. IV. Institute
for nonlinear science (Springer-Verlag)
    QC20.7.F75 W47 2002
    530.15′5—dc21   2002026660

Printed on acid-free paper.

9 8 7 6 5 4 3 2 1

www.springer-ny.com

Springer-Verlag   New York Berlin Heidelberg
A member of BertelsmannSpringer Science+Business Media GmbH

Preface

In Chapter One we review the foundations of statistical physics and fractal functions. Our purpose is to demonstrate the limitations of Hamilton's equations of motion for providing a dynamical basis for the statistics of complex phenomena. The fractal functions are intended as possible models of certain complex phenomena; physical systems that have long-time memory and/or long-range spatial interactions. Since fractal functions are non-differentiable, those phenomena described by such functions do not have differential equations of motion, but may have fractional-differential equations of motion. We argue that the traditional justification of statistical mechanics relies on a separation between microscopic and macroscopic time scales. When this separation exists traditional statistical physics results. When the microscopic time scales diverge and overlap with the macroscopic time scales, classical statistical mechanics is not applicable to the phenomenon described. In fact, it is shown that rather than the stochastic differential equations of Langevin describing such things as Brownian motion, we obtain fractional differential equations driven by stochastic processes.

We explore the limitations for modeling complex phenomena using the analytic functions generally taught in mathematical physics courses in Chapter Two. The need for less familiar functions, such as fractals, for modeling complex phenomena is discussed. The definitions of fractal functions, fractal dimensions, and statistical fractals are reviewed. The generalized Weierstrass function (GWF) is argued to be a fractal function, to have self-similar scaling, and is used to motivate the definition of fractional integrals and derivatives. Some elementary concepts from the fractional calculus are introduced and the non-differentiable GWF is demonstrated to have fractional integrals and derivatives. In fact the fractional derivative (integral) of a regular function is shown to yield a fractal function.

The extension of the familiar concepts from the traditional calculus, such as the chain rule for derivatives and the Leibniz rule for the derivative of products of functions, to the fractional calculus, is made in Chapter Three. We adopt a formalism for the fractional derivative operator that is different from that discussed in Chapter Two and consequently define generalizations of the exponential function, the trigonometric functions, and the hyperbolic trigonometric functions. These new functions are very useful in generalizing such concepts as the Fourier transform for the analysis of nonanalytic functions. More important, we discuss the necessary properties for physical phenomena to be described by such functions.

We briefly review Fourier analysis in Chapter Four at a level sufficient to understand the modeling of most linear physical phenomena. For example, we look at nondispersive wave propagation and ordinary diffusion. This discussion motivates the use of the generalized exponential function to define a generalization of the Fourier transform and the Dirac delta function. We use the Weyl fractional operators to suggest a generalization of

Fourier transforms and from this generalization construct fractional wave equations and fractional diffusion equations.

In Chapter Five we briefly review Laplace transforms at a level sufficient to understand how to solve systems of linear differential equations, whether ordinary or fractional. We discuss how to use the generalized exponential functions in the Laplace transform formalism to find unfamiliar inverse Laplace transforms. We explore how to solve an initial value problem using Laplace and Mellin transforms and relate the results to viscoelastic phenomena. In addition we introduce fractional Green's functions as an alternate method for solving such equations.

The discussions in Chapter Six focus on ways to generate a stochastic process with long-term memory, that is, a random process with an inverse power-law correlation function. One way to model the non-local nature of the memory is by using the properties of fractional derivatives in time. We review ordinary random walks, continuous time random walks, and finally fractional random walks. We also investigate the properties of time series that are generated by complex phenomena and show that the inverse power-law memory is the signature of statistical fractal processes. This is shown to lead to Lévy stable processes.

We review some of the essential elements of classical rheology in Chapter Seven, in order to show that ordinary and partial differential equations are insufficient to model the stress relaxation in viscoelastic materials. The fractional calculus is used to construct a general theory of rheology and the predictions compare favorably with experiments.

In Chapter Eight the fractional calculus is used to obtain insight into the stochastic dynamical equations considered earlier. The exact solutions to these equations are obtained using the techniques of the fractional calculus developed in these lectures. The central moments are calculated exactly using the solutions to the fractional dynamical stochastic equation. We also discuss the kinds of physical systems that could give rise to such behavior.

Wave propagation and transport in heterogeneous media, such as spatial fractals, are discussed in Chapter Nine. We find that the eigenfunction expansion technique for solving partial differential equations is applicable to these propagation-transport fractional equations describing fractional wave propagation and anomalous diffusion. We demonstrate that these exact solutions share a number of properties with familiar physical systems, such as relaxation to a Boltzmann equilibrium.

A number of fractional derivatives and fractional integrals are briefly discussed in Chapter Ten, which is actually an Appendix. The special functions from mathematical physics, such as the Bessel function, the Hankel function, and so on, are shown to generalize and to have generators in terms of fractional derivatives. In addition, the details of some of the techniques used to solve differential equations in the text, including Fox functions and Mellin transforms, are included in this chapter.

Contents

Chapter 1

Nondifferentiable Processes

This Is Not a Text Book   We emphasize at the outset that this is not a traditional text book. The authors do not think that modeling of the complex physical phenomena they have in mind is sufficiently well developed to warrant such a text. On the other hand, this is also not a research monograph, since it lacks the rigor that many would insist on, in such a treatment. So the book falls somewhere in between, resting on a set of lecture notes that have been polished and extended, with the view to providing insight into a new area of investigation in science, particularly in physics. The lectures present techniques from the calculus of fractional derivatives and integrals and fractional stochastic differential equations, but are not intended to form a book about mathematics. Instead of formal mathematics, we emphasize physical interpretation and high-light how to model complex physical phenomena, such as found in the world around us. The use of fractal functions and the applications of fractal opera-tors, such as fractional derivatives and integrals applied to analytic functions, are investigated with a view towards modeling complex physical phenomena. Thus, although the material may appear formal at times, our purpose is to re-veal the mechanisms underlying the complexity rather than to obscure them. Therefore we touch lightly on history and philosophy, in addition to physics and mathematics, where we think they can contribute to the discussion.

Linear Physics   Leibniz, who invented the calculus cotemporaneously with Newton, in 1695 speculated on how one might take the 1/2-power derivative of a monomial. He considered this problem in response to a query by L'Hospital and successfully obtained the right answer. However, in physics such things as fractional derivatives and integrals have been brushed aside as amusing, but not particularly important curiosities, in favor of the study of the analytic functions taught in most mathematical physics courses. It was believed, since before the time of Lagrange, and probably since Galileo, that physical phenomena can, by and large, be represented by analytic functions and the dynamics of physical phenomena can be represented by equations of motion involving such functions. The truth of these assumptions became almost unassailable with the successes of Sturm-Liouville theory in formalizing acoustics, electromagnetic theory, and heat transport in the 19th century, along with diffusion and quantum mechanics in the 20th century. Even when it became clear that such phenomena as phase transitions, turbulence, and the rheology of polymeric materials could not be explained using such an approach, it was held that more detailed analysis of the same kind would eventually unveil the secrets of even these phenomena. However, the evidence has continued to mount that the analytic functions of the 19th century do not tell the whole story.

Nonlinear Physics   Consider, for example, the findings of Poincaré, that celestial orbits need not be predictable, but may in fact be so complicated that it is nearly impossible to visualize, much less predict, their behavior. This was true for nearly everyone except Poincaré, himself. This situation persisted until the introduction of the computer into physics. The desktop computer enabled researchers with limited mathematical background to explore the behavior of systems of nonlinear ordinary differential equations describing physical phenomena. This technological breakthrough led to the widespread application of, and at least partial understanding of, chaotic solutions to nonlinear dynamical systems, and the random behavior of deterministic dynamical processes. But more about that later.

Linear but Nonanalytic   The complex physical phenomena we have mentioned require a kind of analysis that can systematically explore nonanalytic functions and their evolution. In these lectures we develop a general strategy, using fractional operators, to model the dynamics of certain kinds of complex phenomena. But we caution the reader that these lectures are better suited to the purposes of physicists than to those of mathematicians. In this regard we spend a significant amount of time and effort in working to understand the properties that physical phenomena must have in order that their changes in space and time can be properly describable by a fractional operator. Contrariwise we spend very little time on mathematical proofs.

A New Perspective   Significant changes in our knowledge of how to analyze nonlinear static and dynamical phenomena in the physical and biological sciences have occurred over the past 30 years [28, 29, 30, 35, 41]. In the physical sciences the methodologies have moved away from complete reliance on the tools of linear, analytic, quantitative mathematical physics towards a combination of nonlinear, numerical, and qualitative techniques [30]. Not only have many of the linear analytical approaches often proved to be inadequate, but the entrenched geometry of Euclid, classically used to describe natural phenomena, has not always been adequate to the task. In fact the geometry of Euclid does not describe any naturally occurring phenomenon, but is better suited to the artifacts of humans, the products of the machine age. However, when Newton formulated his laws of mechanics he chose to discuss the motion of bodies using geometrical arguments, even though he had much earlier developed the differential calculus. Nowhere in the *Principia* is the now famous equation for force, $\mathbf{F} = m\mathbf{a}$, to be found, nor is there a discussion of derivatives or solutions to differential equations. Rather, there is the geometry of Euclid and the ratios of geometric lengths that are assumed to converge to finite values as the size of the time interval goes to zero. Implicit in these arguments sprinkled throughout the *Principia* is the notion of a limit, but nowhere is the concept of limit explicitly discussed in his treatise. The dynamical arguments of Newton are based on the assumptions of absolute space and time, dimensions that are without beginning or end and continuity is everywhere. These two assumptions, unbounded continuous space and time, combined with the geometry of Euclid, almost guarantee the continuous character of such derived quantities as velocity, momentum, acceleration, and force.

Differentiability Is Questionable In a modern context we find that these assumptions regarding the absolute nature of space and time are no longer tenable, and furthermore that the geometry of Euclid has little, if anything, to do with the physical world. In his lectures on the principles of mechanics, the father of statistical physics, Ludwig Boltzmann [6], writes:

> Whoever has studied mechanics may well remember how difficult it was to understand proofs that motion could be thought of as straight and uniform, and forces as invariant, during very short time intervals. These difficulties stem simply from the fact that the proofs are incorrect. We choose the analytical functions precisely for the purpose of representing the facts of experience. Their differentiability cannot be cited as proof of the differentiability of empirically given functions, since the number of conceivable undifferentiable functions is just as infinitely large as the number of differentiable functions. Likewise, the fact that every hand-drawn or machine-drawn dash corresponds to a differentiable function is only a proof that, as far as we can tell with our present observational tools, the differentiability of the empirical functions of mechanics is something given by experience. For this reason we have, without apology, presented differentiability as an assumption that agrees with the experimental facts to date.

Fractals Arrive In the 1960s the mathematician Benoit Mandelbrot began discussing a new geometry of nature, one that embraces the irregular shapes of objects such as coastlines, lightning bolts, clouds, and molecular trajectories and that mathematicians had initiated in the late 19th century. A common feature of these objects, which Mandelbrot called fractals, is that their boundaries are so irregular that it is not easy to understand how to apply simple metrical ideas and operations to them [28, 30]. Towards this end we consider some of the metric peculiarities of unusual mathematical objects, fractal functions, and discuss the possible physical implications of their evolution in time in terms of the fractal calculus and the phenomena they are used to model.

Fractal Functions   One of the defining properties of a fractal function is that it does not possesses a characteristic scale length and consequently its derivatives diverge. The Weierstrass function was the first example of a function constructed such that it is continuous everywhere, but is nowhere differentiable. Consider a time-dependent process that consists of frequencies that increase as powers of a parameter $b > 1$, but with amplitudes that decrease inversely with powers of the parameter $a > 1$, such that

$$G\left(t\right) = \sum_{n=-\infty}^{\infty} \frac{1}{a^n} \sin\left[b^n t + \phi_n\right]. \tag{1.1}$$

Here we have also introduced a set of phases $\{\phi_n\}$ that can be chosen to be random or any other way we desire. We refer to the function $G\left(t\right)$ as a generalized Weierstrass function (GWF) and if $b > a$ the function can be characterized by a noninteger fractional dimension D that is related to the two parameters a and b as we subsequently show. The GWF is like the original function of Weierstrass in that it is a Fourier series and its derivative with respect to time diverges to infinity. Consider the time derivative of (1.1),

$$\frac{dG(t)}{dt} = \sum_{n=-\infty}^{\infty} \left(\frac{b}{a}\right)^n \cos\left[b^n t + \phi_n\right] \tag{1.2}$$

so that if $b > a$, the terms in the series diverge absolutely in the limit $n \to \infty$. We can conclude from this divergence of the derivative of the GWF that there is no way to express the evolution of this function in terms of differential equations of motion. More important there does not appear to be any way to describe any phenomenon, depicted by such a function, using Newton's force law. But we have a great deal more to say about that subsequently.

Scaling   The GWF differs from the original Weierstrass function in that it can represent a stochastic process when the phases in the series are chosen to be randomly distributed, but it can also represent a deterministic function, say, when all the phases are chosen to be zero. In either case the function is fractal, which is to say, that it is scaling. To show what we mean by scaling, consider multiplying the time scale in a function $F\left(t\right)$ by a factor a to obtain

$$F(at) = bF(t). \tag{1.3}$$

The original Weierstrass function was only approximately scaling in that it only satisfied the scaling relation (1.3) in the limit $t \to \infty$. However, if we scale the time in $G\left(t\right)$ given by (1.1), then after some rearranging of terms, we obtain the scaling relation (1.3) *exactly*. We discuss scaling further and its physical consequences in due course.

Turbulence   In 1926 Richardson [34] published his investigations on the irregular fluctuations in the velocity field of turbulent winds in the atmosphere. He found that the variations in wind are random in both magnitude and direction, with such sharp changes that the observed velocity field is nondifferentiable. Thus, he concluded that it was impossible to model the irregular flow of the wind using differential equations, such as Newton's force law, or its hydrodynamic analogue, the Navier-Stokes equations. It was specifically this property of nondifferentiability that Richardson believed captured the essential features of turbulent fluid flow. This view of turbulence, rather than becoming quaint and out of date, has been demonstrated to be quite modern [37] and in fact anticipated the introduction of fractals into the description of complex phenomena [28, 30] in general. Velocity fields of turbulent fluids at low viscosity have been shown to be multifractal [11].

Lévy Statistics During the same period when Richardson was struggling to understand turbulence, Lévy [25] was working to establish the most general properties necessary for a stochastic process to violate the then accepted form of the central limit theorem (CLT) and still converge to a limit distribution. Lévy was quite successful, establishing the class of infinitely divisible distributions, which as its name implies concerns processes whose statistical properties persist at each level of aggregation or refinement of the data. A large part of the class of infinitely divisible distributions are today called α-stable Lévy processes. A deep connection between Lévy stable processes and Weierstrass functions was established using random walk concepts [1] and this connection was subsequently used to understand some of the statistical properties of turbulent fluid flow [37].

Brownian Motion The physical paradigm of statistical physics is Brownian motion, which involves diffusion, dissipation, and the fluctuation-dissipation relation tying the two together. The dynamical model of this process was provided by Langevin in 1908 using a stochastic differential equation [23]. In spite of this long history it seems apparent that the randomness of such macroscopic stochastic equations is incompatible with the continuous and differentiable character of microscopic Hamiltonian dynamics. However, it is widely believed that Brownian motion can be rigorously derived from the totally deterministic Hamiltonian models of classical mechanics. Part of the reason for this conviction has to do with the wide use made in the literature of the Markov approximation. In one form or another, the attempts made to establish a unified view of mechanics and thermodynamics can be traced back to microscopic Hamiltonian models and the Markov approximation, even though the two have been shown to be fundamentally incompatible, as we later review.

Central Limit Theorem (CLT) There is a relation between the nondifferentiability of microscopic processes, the differentiability of macroscopic processes, and the conditions of the CLT. Recall that in the CLT the quantities being added together are statistically independent, or at most weakly dependent, in order for the theorem to be applicable. When there are a large number of statistically independent, identically distributed random variables added together,

Gaussian statistics emerge for the sum variable. In a dynamical system the CLT applies if the time scales for the microscopic processes are much smaller than the time scales for the macroscopic processes. This implies that the microscopic dynamics are stable, since dynamical instabilities can have arbitrarily long time scales. Once a condition of time-scale separation between the microscopic and macroscopic is established, in the long-time limit the memory of the details of microscopic dynamics is lost, and Gaussian statistics on the macroscopic scale result. This separation of time scales also means that we can again use ordinary differential calculations on the macroscopic time scale, even if the microscopic dynamics are incompatible with the methods of ordinary calculus.

Two Time Scales   There have been two approaches to describing stochastic phenomena in statistical physics. The first uses the dynamical variables as we discussed above. The existence of a separation between the microscopic and macroscopic time scales leads to a stochastic differential equation to describe the macroscopic dynamics. This is the Heisenberg representation in which we focus on the time evolution of the physical observables. The second approach uses the Schrödinger perspective corresponding to the time evolution of the Liouville density in the phase space for the system. The usual outcome of the Heisenberg approach is the derivation of an ordinary Langevin equation from mechanics. The Langevin equation is a stochastic ordinary differential equation for the dynamical variable. The result of the Schrödinger approach, on the other hand, is a master equation. The master equation usually leads to the conventional diffusion equation, with the diffusion process described by a partial differential equation that is second-order in space and first-order in time.

Two Perspectives   The relaxation of a physical observable is usually described by an exponential function in the Heisenberg perspective. The dynamics of the physical observable is determined by averaging over an ensemble of realizations of the stochastic force. The mathematical representation of the diffusion process is given, as we have said, by a second-order spatial derivative of a probability density function in the Schrödinger perspective. Therefore, the mathematical description rests on either ordinary analytical functions (exponential functions) describing the dynamics, or on conventional differential operators (second-order derivatives) describing the phase space evolution. The approach one takes is a matter of convenience because the two techniques have been taken to be equivalent for a hundred years. But even that pillar of statistical physics has recently been challenged [5], and different solutions to the same physical problem using the two techniques have been found.

No Time-Scale Separation   This traditional approach to statistical physics is only part of the story. When a time-scale separation between the macroscopic and the microscopic levels of description does not exist, the memory of the non-differentiable nature of the phenomenon at the microscopic level of description is not suppressed. We show in these lectures that the transport equations cannot be expressed in terms of ordinary differential calculations, in this case, even if we limit our observation to the macroscopic level. This inability to use the ordinary calculus at the macroscopic level is the reason why the time derivative in the Langevin equation is replaced with a fractional time derivative, yielding a fractional stochastic equation in the physical observables. This nondifferentiability is also why the Laplacian operator of normal diffusion is replaced with a fractional Laplacian yielding a fractional diffusion equation in the phase space for the system. We develop these arguments in detail in subsequent lectures.

Fractional Equations of Evolution Finally, this failure of traditional wisdom, that being the formulation of differential transport equations, also requires that the ordinary wave equation be replaced with a fractional wave equation in certain heterogeneous media. These replacements have all too often been based on phenomenological arguments that the solution to a fractional differential equation seems to best describe the phenomenon of interest. Furthermore, although there is a parallel, but disjoint, development of the fractional calculus, there has been little effort to blend the phenomenological discussion and the mathematical formalism. A notable exception is the generalization of the exponential form of physical relaxation in viscoelastic materials recently proposed by Nonnenmacher and coworkers [13, 14, 15, 16] involving the concept of a fractional differential equation and the solution of initial value problems. We take up this discussion in due course.

Traditional Dynamics In these lectures we attempt to synthesize the phenomenological and formal approaches to describing complex physical phenomena. The phenomenological approach has its basis in the traditional arguments of statistical mechanics, and relies on taking the macroscopic limits of microscopic processes described by Hamiltonian dynamics. Therefore, we spend a certain amount of time initially reviewing Hamiltonian dynamics and statistical mechanical methods to provide a context for the arguments as to why we need a fractional calculus to describe the evolution of complex phenomena. This synthesis is concerned with the emergence of fractional operators at the macroscopic level when the traditional approximations, such as the Markov approximation, cannot be justified within the standard formalism.

1.1 Classical Mechanics

Energy Not Force We do not begin our review of mechanics with Newton's equations of motion, even though the specification of all the forces acting on a body at a given time uniquely determines the subsequent dynamical behavior of that body. Rather we follow the procedure initiated by Leibniz in which the motion of a particle is determined by its potential and kinetic energies, since these two scalar quantities contain the complete dynamics of even the most complicated material system if, as pointed out by Lanczos [22], they are used as the basis of a principle rather than an equation.

Extremal Path The principle is due to Hamilton and may be understood by thinking of a particle at point P of known position and velocity at time t_1. Then at a later time t_2 the particle is at a point Q of known position and velocity. The trajectory between the two points P and Q is unknown, but can be determined from a knowledge of the potential and kinetic energies, even when the potential energy is time dependent. The only constraints are that the trajectory be continuous and pass through the initial point P at time t_1 and the terminal point Q at time t_2. The quantity that determines the dynamical behavior is the integral of the difference between the kinetic and potential energies between (P, t_1) and (Q, t_2) and is called the action. The actual motion is determined by that value of the action that minimizes this integral. The mathematical procedure by which this integral is minimized is the calculus of variation in which the variations are the virtual displacements of the path between the two end points to determine an extremum.

Variational Procedures The variational procedure for determining the trajectory of a particle was developed by Euler and Lagrange for conservative systems, that is, systems for which the total energy is conserved. The initial virtue of this procedure lay in its ability to properly account for constraints acting on a system through the proper choice of variables, that is, through the choice of canonically conjugate variables. In the Newtonian method one would have to separately determine all the forces of constraint and add them to the equations of motion. Using canonical variables in the kinetic and potential energies automatically takes the constraints into account by properly balancing the energy along the variational path. A second advantage of the variational approach is that the action minimization is independent of the choice of coordinates. The form of the equations resulting from the variation is invariant, which is to say they have the same form for any set of canonically conjugate variables. In addition if one of the coordinates is cyclic (ignorable) then a particular integration can be immediately performed. If all the coordinates are ignorable the complete evolution of the system is determined. A systematic procedure for transforming from an arbitrary set of variables to one in which all the coordinates are ignorable was developed by Hamilton and Jacobi and now bears their names. This final set of coordinates is the action-angle representation, in which all the actions are constant and the angles increase linearly with time. Finding this representation is equivalent to solving Newton's equations of motion for the system.

1.1.1 Euler-Lagrange Equations of Motion

Forces and Energy Consider a system of N particles subject to geometrical constraints and that are otherwise influenced by forces which are functions only of the positions of the particles. The jth particle is located by the three Cartesian coordinates (x_j, y_j, z_j) and $j = 1, 2, ..., N$, with the three-component force $\mathbf{F}^{(j)}$ acting on it aside from the forces of constraint. Each force is a function of the $3N$ coordinates for the N particles in the system, presuming that each particle in the system exerts a force on each of the other particles in the system, for example, through gravitational attraction. If the potential function is denoted by V then for a conservative system the forces are given by

$$\mathbf{F}^{(j)} = -\nabla_j V \qquad (1.4)$$

for the force exerted on the jth particle by all the other particles in the system and ∇_j is the nabla operator for the jth particle

$$\nabla_j \equiv \mathbf{e}_x \frac{\partial}{\partial x} + \mathbf{e}_y \frac{\partial}{\partial y} + \mathbf{e}_z \frac{\partial}{\partial z}$$

and $\mathbf{e}_q$ is the unit vector in the q direction. In addition to the potential function there is the kinetic energy of the system of particles

$$T = \frac{1}{2} \sum_{j=1}^{N} m_j \mathbf{v}_j \cdot \mathbf{v}_j. \qquad (1.5)$$

Thus, the total energy of the system may be written as

$$E = T + V \qquad (1.6)$$

and for a conservative system the dynamics of each and every particle is determined by the interchange of energy from potential to kinetic form and back again.

Constraints Every conservative system can express the constraints imposed on the particle dynamics through functions of the form

$$\Phi_n(\mathbf{x}_1, \mathbf{x}_2, \cdots \mathbf{x}_N) = 0, \qquad (1.7)$$

$n = 1, 2, \cdots k$, where we assume the system has k constraints.. The effect of a constraint is to reduce the number of independent coordinates describing the positions of the particles. The k equations given by (1.7) reduces the number of independent coordinates from $3N$ to $3N - k$. The constraining equations can in principle be used to eliminate k variables from the description of the system dynamics. This is sometimes referred to as freezing out certain degrees of freedom in the description of the system's dynamics. For example, certain microscopic degrees of freedom can be frozen out from a macroscopic description of the motion of a material body. This, however, is not always convenient, so let us introduce a new set of independent coordinates $(q_1, q_2, \cdots, q_M)$ where $M = 3N - k$, to locate the N particles. In this way we write the equivalent system of $3N$ equations, $\mathbf{x}_j = \mathbf{x}_j(q_1, q_2, \cdots, q_M)$ for $j = 1, 2, \cdots, N$. The qs in this case are called the generalized coordinates and are always consistent with the set of constraints placed on the system. The potential energy can therefore be written in terms of the generalized coordinates as $V(\mathbf{x}_1, ..., \mathbf{x}_N) = V(q_1, ..., q_M)$. The kinetic energy can, in general, be expressed in terms of the generalized coordinates using

$$\dot{\mathbf{x}}_j = \sum_{n=1}^{M} \frac{\partial \mathbf{x}_j}{\partial q_n} \dot{q}_n \tag{1.8}$$

so that the kinetic energy becomes

$$T = \sum_{j=1}^{N} \sum_{n=1}^{M} \sum_{p=1}^{M} m_j \frac{\partial \mathbf{x}_j}{\partial q_n} \cdot \frac{\partial \mathbf{x}_j}{\partial q_p} \dot{q}_n \dot{q}_p, \tag{1.9}$$

where the constraints have not been explicitly introduced, but instead reside in the coordinates. Thus, the kinetic energy is a homogeneous function of degree two of the generalized velocity components $\left(\dot{q}_1, \dot{q}_2, \cdots, \dot{q}_M \right)$; that is, it is a quadratic form in the generalized velocity components. Note that the coefficients of the products of the generalized velocity components in (1.9) are functions of the generalized coordinates. The potential energy is a function of the generalized coordinates alone in a conservative system, whereas the kinetic energy is a function of both the generalized coordinates and the generalized velocities. It is now convenient to introduce the Lagrangian

$$L\left(\mathbf{q}, \dot{\mathbf{q}} \right) = T - V, \tag{1.10}$$

where we introduce the vector notation $\mathbf{q} = (q_1, \cdots, q_M)$ and $\dot{\mathbf{q}} = \left(\dot{q}_1, \cdots, \dot{q}_M \right)$.

Hamilton's Principle   We now assume the validity of Hamilton's principle as the physical/mathematical law describing the motion of any conservative material system. The actual motion of a system with Lagrangian (1.10) integrated from times t_1 and t_2 is such that the time integral

$$I = \int_{t_1}^{t_2} L\left[\mathbf{q}\left(t \right), \dot{\mathbf{q}}\left(t \right) \right] dt, \tag{1.11}$$

where the indicated times are arbitrary, is an extremum with respect to continuously twice-differentiable functions of $\mathbf{q}\left(t \right)$ for which $\mathbf{q}\left(t_1 \right)$ and $\mathbf{q}\left(t_2 \right)$ are prescribed. One should be cognizant of the fact that although (1.11) is taken as the starting point for the system dynamics, Newton's laws are implicitly included through the definition of mass (third law) and the inertial frame of reference (first law). Two good reasons for starting from (1.11) are that the equations of motion are invariant to changes in the generalized coordinates and the method generalizes to nonmechanical systems, for example, to the electromagnetic field.

Euler-Lagrange Equations   The extremum of I is determined by varying the integral with respect to the generalized coordinates and velocities and setting this variation to zero:

$$\delta I = 0. \tag{1.12}$$

The variation in (1.12) is done systematically using the calculus of variation, but the details are well known and so are not reproduced here. We obtain from the calculus of variation the M equations

$$\frac{d}{dt}\frac{\partial L}{\partial \dot{q}_j} - \frac{\partial L}{\partial q_j} = 0, \; j = 1, ..., M, \tag{1.13}$$

the Euler-Lagrange equations of motion. Equation (1.13) constitutes a set of second-order differential equations whose solutions yield the components of $\mathbf{q}(t)$. The $2M$ constants of integration are evaluated from the initial conditions for the generalized coordinates and generalized velocities. Once the initial state of the system is prescribed its future motion is described in detail by the solution to (1.13).

Conservation of Energy We have restricted our considerations here to a time-independent Lagrangian, and because of that we may write down the identity

$$\frac{d}{dt}\left[\sum_{j=1}^{M}\dot{q}_j \frac{\partial L}{\partial \dot{q}_j} - L\right] = -\sum_{j=1}^{M}\dot{q}_j\left(\frac{\partial L}{\partial q_j} - \frac{d}{dt}\frac{\partial L}{\partial \dot{q}_j}\right). \tag{1.14}$$

The right-hand side of (1.14) vanishes identically because of the Euler-Lagrange equations and the left-hand side can be integrated to yield

$$\sum_{j=1}^{M}\dot{q}_j \frac{\partial L}{\partial \dot{q}_j} - L = E, \tag{1.15}$$

where E is a constant. Equation (1.15) is the first integral of the motion and can be identified with the system's energy. To arrive at this interpretation we note that the potential energy is independent of the generalized velocity so that using the definition of the Lagrangian,

$$\frac{\partial L}{\partial \dot{q}_j} = \frac{\partial T}{\partial \dot{q}_j} = m_j \dot{q}_j . \tag{1.16}$$

Substituting (1.16) into (1.15) we obtain

$$2T - L = 2T - (T - V) = T + V = E. \tag{1.17}$$

Thus, we have established that the total energy is an invariant of the dynamics in a conserved system and is determined by the initial choices of the generalized coordinates and velocities.

Physics Versus Mathematics Determining the generalized variables and setting up the Euler-Lagrange equations of motion constitutes the physics in the investigation of the phenomenon of interest. Solving the equations is mathematics. Physical systems in which analogues of the Euler-Lagrange equations arise in the continuum limit include elastic fields, electromagnetic fields, and nonlinear water waves.

1.1.2   Hamilton's Equations of Motion

The Euler-Lagrange Equations   The action minimization principle has provided us with the Euler-Lagrange equations of motion for an N particle system, with k constraints, leaving M degrees of freedom. This procedure requires that the system dynamics are describable by twice-differentiable functions that are the generalized coordinates. Note that each degree of freedom has a generalized coordinate q and generalized velocity $\dot{q}$ so M degrees of freedom refer to $2M$ variables. The evolution of the system can therefore be described by an orbit or trajectory in a $2M$-dimensional phase space. A point in this phase space gives a complete description of the dynamical system. A choice of initial conditions is given by the specification of such a point at time t_0 and the resulting trajectory emanating from this initial state reveals the continuous sequence of states by the solution to the Euler-Lagrange equations of motion.

Only Partial Knowledge   It is worth emphasizing that we do not need to know all the forces acting within a system to determine its evolution. For example, if we needed to know the interparticle forces in a fluid to determine its motion then fluid dynamics would be impossibly complex. In rigid body motion, as we mentioned, one freezes out degrees of freedom in which the constraints are determined by infinitely strong potentials. It is generally sufficient to know the kinetic and potential energy for the system in terms of generalized coordinates since the potential implicitly contains information on all the forces acting within the system. We also reiterate that the generalized coordinates are arbitrary in that one set is as good as another and they can be interrelated by transformations, called canonical transformations. It is generally advantageous to choose the generalized coordinates that make use of the symmetries of the system.

What Hamilton Did   Hamilton's principle of least action takes the motion of the system as a whole between the fixed endpoints of the trajectory of interest. The basic equations of nature are therefore obtained from the variational principle:

$$\delta \int_{t_1}^{t_2} L\,dt = 0. \tag{1.18}$$

Hamilton recognized that the momenta are a more useful set of mechanical variables than are the velocities, and expressed his dynamical equations for the system in terms of positions and momenta rather than positions and velocities. The partial derivative of the Lagrangian with respect to the generalized velocity defines a set of generalized momenta $\{p_j\}$ by

$$p_j \equiv \frac{\partial L}{\partial \dot{q}_j}, \; j = 1, 2, \cdots, M. \tag{1.19}$$

The kinetic energy is a general quadratic form in the generalized velocities, so the momenta are linear functions of these velocities when the potential is velocity independent. In general, however, it is not possible to specify in advance the functional relation between the velocities and momenta. It is clear, however, that the momenta are functions of the generalized coordinates as well as the generalized velocities in general; see, for example, (1.19). Thus, Hamilton defined the function of interest from (1.15), as

$$H\left(\mathbf{q}, \mathbf{p}\right) = \sum_{j=1}^{M} p_j \dot{q}_j - L, \tag{1.20}$$

where the generalized velocities in the Lagrangian are replaced in favor of the generalized momenta and the $\dot{q}$s are expressed as explicit functions of the ps. Equation (1.20) is a Legendre' transformation from the Lagrangian function L to the Hamiltonian function H and is a traditional way to change independent variables describing a system.

Hamilton's Equations of Motion Since we have replaced the Lagrangian for the system by a Hamiltonian for the system, we also want to replace the Euler-Lagrange equations of motion in terms of the qs and $\dot{q}$s with Hamilton's equations of motion in terms of the qs and ps. Hamilton's equations of motion are given by the set of first order differential equation

$$\dot{q}_j = \frac{\partial H}{\partial p_j} \; , \; \dot{p}_j = -\frac{\partial H}{\partial q_j} \; , \; j = 1, 2, \cdots, M. \tag{1.21}$$

The dynamics of the system are given entirely by the $2M$ equations of motion (1.21). Note that the Lagrange description has M second-order differential equations whereas the Hamiltonian description has $2M$ first-order differential equations. The solutions to Hamilton's equations of motion, often called the canonical equations of motion, yield a complete knowledge of the mechanical system. Thus, the system dynamics are completely determined by the Hamiltonian and the initial state of the system. These equations first appeared in one of Lagrange's papers in (1809), but their significance was not fully appreciated until the work of Hamilton (1833).

The Hamiltonian Is Not Always the Energy It is probably worth pointing out that the Hamiltonian H is not necessarily the energy of the system, whereas

$$h\left(\mathbf{q}, \dot{\mathbf{q}}, t\right) = \sum_{j=1}^{M} \dot{q}_j \frac{\partial L}{\partial \dot{q}_j} - L \tag{1.22}$$

is the total energy function and is a function of M generalized coordinates and their time derivatives. If $h\left(\mathbf{q}, \dot{\mathbf{q}}, \mathbf{t}\right)$ is time independent then it is the total energy of the system. The Hamiltonian $H\left(\mathbf{q}, \mathbf{p}, t\right)$ have the same numerical value as the energy function, but they are functions of different variables.

KAM Theory   Classical mechanics, also called analytic dynamics, as embodied in Hamilton's equations of motion, was for a long time considered by most physicists to be an arcane branch of physics. It was thought to be a useful way to calculate the evolution of systems, but not in itself an area for active research. This view began to change in the 1980s, when the content of the Kolmogorov, Arnold, and Moser (KAM) analysis of instabilities in mechanical systems began to be more widely understood. The concept of chaos emerged in dynamical systems that are nonconservative, that is, trajectories on strange attractors, but the same concept appeared in the KAM theory of instabilities in conservative Hamiltonian systems. We have occasion to explore these concepts more fully in later sections. In particular we discuss how microscopic dynamical systems that are chaotic may lead to ill-defined limiting dynamical macroscopic systems.

1.2   Langevin Equation

Macroscopic Evolution with Uncertainty   Classical mechanics is used to describe the dynamics of highly idealized mechanical systems, using limiting concepts abstracted from the physical world[1]. Statistical mechanics was the first discipline to restore to the description of these systems some of the uncertainty observed in actual measurements. In a systematic way statistical mechanics associates measures with that uncertainty, for example, the probability density functions for the state of the system [26]. When the probability density functions and the average properties of the physical system that can be calculated from them are independent of time, the system is said to be in equilibrium or in a stationary state. Time dependences, on the other hand, indicate that the system is evolving; that is, the variables are changing over time. Macroscopic (large-scale) observables vary in time more slowly than do the microscopic (small-scale) degrees of freedom. Furthermore, the macroscopic properties of a system are typically insensitive to the detailed microstructure of its constituent parts. Many authors have pointed out that this insensitivity to small-scale features is not unconditional, and requires: (1) restrictions on the space and time scales of the system, and (2) certain moment conditions upon statistical distributions and ranges of interactive forces. But in the final analysis these macroscopic observables very often cannot be described by analytic functions, calling into question how we model the evolution of such large-scale phenomena. In particular, we are unable to restrict the scales of microscopic motion, so consequently the range of microscopic interactions become macro-

---

[1] The contents of this lecture are taken from Lindenberg and West [26] and suitably modified for the purposes of the present discussion.

scopic in an unpredictable way; see, for example, the macroscopic trajectory of a Brownian particle in Figure 1.1.

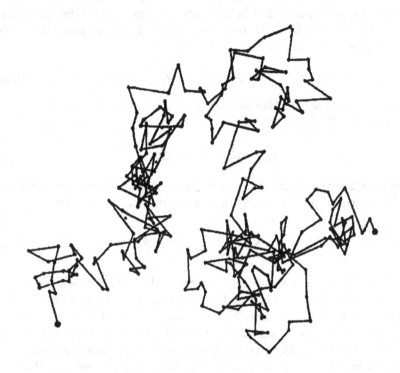

Figure 1.1: *The figure depicts an experimental trajectory of a Brownian particle being buffeted by the lighter particles of the water in which the Brownian particle is floating.*

Random Forces It is not only macroscopic phenomena that suffer from the problem of not being describable by analytic functions, however. The theory of Brownian motion as formulated in 1908 by Langevin [23] has the form

$$\frac{d\mathbf{P}(t)}{dt} = -\lambda \mathbf{P}(t) + \mathbf{f}(t),\tag{1.23}$$

where $\mathbf{P}(t)$ is the momentum of the Brownian particle, λ is the dissipation rate resulting from the Stokes drag on the Brownian particle produced by the lighter fluid particles, and $\mathbf{f}(t)$ is the random influence of the bath of lighter particles (molecular impacts) on the Brownian particle. In 1913 Perrin [33] observed the path of the Brownian particle depicted in Figure 1.1 through his microscope. Such a path can not be described by a continuous differentiable function. In spite of this non-analyticity, today we still understand this process by considering the equations of motion for the Brownian particle in contact with a heat bath, both of which are described by a Hamiltonian. Let us consider the standard reconciliation of these two points of view.

1.2.1   Hamiltonian Model

Model of System Plus Bath   A Hamiltonian description of the motion of a Brownian particle in a fluid of lighter particles may be given by segmenting the energy into that for the fluid, the energy for the Brownian particle, and the energy for their interaction [26]:

$$H = H_0 + H_B + H_I. \tag{1.24}$$

The Hamiltonian for a Brownian particle of mass M, with canonical momentum P and displacement Q in a potential $V(Q)$ is

$$H_0 = \frac{1}{2M} P^2 + V(Q). \tag{1.25}$$

The Hamiltonian for a heat bath, modeled as a system of harmonic oscillators indexed by n with canonical momenta p_n and canonical displacements q_n, is

$$H_B = \sum_{n=1}^{N} \left[\frac{p_n^2}{2m_n} + \frac{m_n \omega_n^2}{2} q_n^2 \right]. \tag{1.26}$$

Note that we have replaced the fluid background with a harmonic oscillator background; the assumption is that the replacement produces a quantitative, not a qualitative, change in describing the phenomenon of Brownian motion. This rests on the traditional assumption that the qualities of the heat bath that are important are the number of degrees of freedom and typical time scales, not the interactions of the bath particles among themselves. We see that there are N degrees of freedom in the oscillator background and ultimately we consider the limit $N \to \infty$. Finally, the simplest Hamiltonian for the interaction between the Brownian particle and the heat bath has the bilinear form

$$H_I = -\sum_{n=1}^{N} \Gamma_n q_n Q, \tag{1.27}$$

where $\{\Gamma_n\}$ is a set of constants coupling the Brownian particle to the heat bath. Note that the form of the Brownian particle is determined by the form of the potential $V(Q)$. A free particle would have $V(Q) = 0$. Of course, more realistic physical models of the interaction between a Brownian particle and the background fluid have been constructed, but it is not the specific model that is of concern to us here. Rather it is the limitation of *any* Hamiltonian model in describing a dissipative system that concerns us.

Hamiltonian Model of Brownian Motion   The equation of motion for the Brownian particle in the above model of a heavy particle coupled to a lattice of lighter harmonic oscillators where $M = 1$, is given by

$$\dot{Q} = \frac{\partial H}{\partial P} = P, \tag{1.28}$$

$$\dot{P} = -\frac{\partial H}{\partial Q} = -V'(Q) + \sum_{n=1}^{N} \Gamma_n q_n, \tag{1.29}$$

where the prime denotes a derivative with respect to Q. These dynamical equations for the Brownian particle contain the unknown time-dependent bath coordinates q_n. To eliminate these bath coordinates from the system equations we consider Hamilton's equations for the bath variables:

$$\dot{q}_n = \frac{\partial H}{\partial p_n} = p_n/m_n, \tag{1.30}$$

$$\dot{p}_n = -\frac{\partial H}{\partial q_n} = -m_n \omega_n^2 q_n + \Gamma_n Q. \tag{1.31}$$

Here we note the dependence of the bath dynamics on the system coordinate Q in (1.31). Equations (1.29) and (1.31) are formally linear in the bath variables and therefore can be integrated explicitly, using the initial state of the bath $\{p_n(0), q_n(0)\}$ to yield

$$q_n(t) = q_n(0) \cos \omega_n t + \frac{p_n(0)}{m_n \omega_n} \sin \omega_n t + \frac{\Gamma_n}{m_n \omega_n} \int_0^t d\tau \sin \omega_n (t - \tau) Q(\tau) \tag{1.32}$$

with the momentum of the bath particles given by

$$p_n(t) = m_n \dot{q}_n(t).$$

Subsequent interpretation of results is facilitated if we integrate the last term in (1.32) by parts, using the initial state of the Brownian particle $Q(0), P(0)$:

$$q_n(t) - \frac{\Gamma_n}{m_n \omega_n^2} Q(t) = \left[q_n(0) - \frac{\Gamma_n}{m_n \omega_n^2} Q(0) \right] \cos \omega_n t + \frac{p_n(0)}{m_n \omega_n} \sin \omega_n t$$
$$- \frac{\Gamma_n}{m_n \omega_n^2} \int_0^t d\tau \cos \omega_n (t - \tau) P(\tau). \tag{1.33a}$$

Substitution of the explicit solution (1.33a) into (1.29) gives the evolution equation for the Brownian particle strictly in terms of the system variables

$$\dot{P} + V_m'(Q) - \int_0^t d\tau K(t - \tau) P(\tau) = f(t), \tag{1.34}$$

where the memory kernel is

$$K(t - \tau) = \sum_{n=1}^{N} \frac{\Gamma_n}{m_n \omega_n^2} \cos \omega_n (t - \tau), \tag{1.35}$$

the forcing function is

$$f(t) = \sum_{n=1}^{N} \left[q_n(0) - \frac{\Gamma_n}{m_n \omega_n^2} Q(0) \right] \cos \omega_n t + \sum_{n=1}^{N} \frac{p_n(0)}{m_n \omega_n} \sin \omega_n t, \qquad (1.36)$$

and where we have introduced the modified system potential

$$V_m(Q) = V(Q) - \sum_{n=1}^{N} \frac{\Gamma_n^2}{2 m_n \omega_n^2} Q^2. \qquad (1.37)$$

Note that (1.34) now depends on the bath variables only through their initial values that occur in $f(t)$, but both the memory kernel and the modified potential depend on the strength of the bath coupling parameters and the spectrum of frequencies of the bath oscillators.

Generalized Langevin Equation   Equation (1.34) has the form of a generalized Langevin equation when the forcing function is given a stochastic interpretation, its autocorrelation function is related to the memory kernel, and the conditions under which the kernel is dissipative are established. The stochastic interpretation of the driving function is achieved by noting that the initial state of the bath is uncertain and is in general only determined via a distribution of initial states. The inhomogeneous term $f(t)$ is, therefore, a sum of a large (infinite) number of independent, identically distributed, random variables (initial states). If the initial configuration of the bath is an equilibrium state, then the distribution of all bath oscillators is the same and consequently, due to the central limit theorem, $f(t)$ is Gaussian. The mean value of the resulting Gaussian is determined by the particular form of the bath equilibrium at $t = 0$.

Canonical Distribution   In general, two possible physical situations might reasonable arise. In one, the bath has equilibrated at $t = 0$ in the absence of the system of interest, which is introduced at that point. The distribution function for the bath is then $W[\mathbf{q}(0), \mathbf{p}(0)]$ where

$$W[\mathbf{q}(0), \mathbf{p}(0)] = Z^{-1} e^{-H_B/kT}, \qquad (1.38)$$

Z is the bath partition function, and where H_B is given by (1.26). The mean value of the random force $f(t)$ is then

$$\begin{aligned} \langle f(t) \rangle &= \sum_{n=1}^{N} \frac{\Gamma_n^2}{m_n \omega_n^2} Q(0) \cos \omega_n t \\ &= -K(t) Q(0). \end{aligned} \qquad (1.39)$$

The effect of this dependence on the system evolution is only significant over times shorter than the memory of the kernel $K(t)$. Insofar as this time scale is usually short compared to the dominant intrinsic time scales of the isolated system (cf. below), the average value in (1.39) has a negligible effect on the long-time evolution of the system.

Modified Canonical Distribution   A second frequently assumed initial distribution is one in which the bath has equilibrated at $t = 0$ in the presence of the system. In this case the appropriate distribution of initial states is governed by a modified bath Hamiltonian

$$W\left[\mathbf{q}\left(0\right),\mathbf{p}\left(0\right)\right] = Z^{-1}e^{-H_B^{(m)}/kT}, \tag{1.40}$$

where

$$H_B^{(m)} = \sum_{n=1}^{\infty}\left[\frac{p_n^2}{2m_n} + \frac{m_n\omega_n^2}{2}\left(q_n - \frac{\Gamma_n Q}{m_n\omega_n^2}\right)^2\right]. \tag{1.41}$$

With this initial distribution, the fluctuations $f(t)$ are zero centered. This latter choice of distribution is thus consistent with the assumption usually made in phenomenological treatments of the Langevin equation. Clearly, any other reasonable initial distribution will at most introduce a nonzero average value $\langle f \rangle$ that decays on the time scale of the memory kernel.

Statistics of the Force   Having chosen a distribution for the initial state of the bath, we now turn to the calculation of the correlation function of the fluctuations. By direct calculation using either of the two initial distributions we obtain

$$\langle [f(t) - \langle f(t) \rangle] [f(\tau) - \langle f(\tau) \rangle] \rangle = kT\sum_{n=1}^{N}\frac{\Gamma_n^2}{m_n\omega_n^2}\cos\,\omega_n\left(t - \tau\right). \tag{1.42}$$

Comparing (1.42) with (1.35) enables us to identify the generalized fluctuation-dissipation relation (GFDR)

$$\langle [f(t) - \langle f(t) \rangle] [f(\tau) - \langle f(\tau) \rangle] \rangle = kTK\left(t - \tau\right). \tag{1.43}$$

Thus, we see that the physical source of the fluctuations is the same as that for the memory in the system. The original fluctuation-dissipation relation formulated by Einstein is obtained when the memory kernel is a delta function in time.

Memory Kernels   To complete the identification of (1.34) as a generalized Langevin equation we must establish the conditions on the coupling coefficients Γ_n, on the bath frequencies ω_n, and on the number of N of bath oscillators that will ensure that $K(t)$ is indeed dissipative. A sufficient condition for $K(t)$ to be dissipative is that it be positive definite and that it decrease monotonically with time. These conditions are achieved if $N \rightarrow \infty$ and if $[\Gamma_n^2/m_n\omega_n^2]$ and ω_n are sufficiently smooth functions of the index n [10]. If as the number of oscillators $(N \to \infty)$ one replaces the sum in (1.35) by an integral over the frequencies weighted by a density of states $g(\omega)$ then

$$K(t) = \int_0^\infty d\omega\, g(\omega)\, C(\omega) \cos \omega t, \qquad (1.44)$$

where

$$\Gamma_n^2/m_n\omega_n^2 \to C(\omega).$$

If, for example, we choose

$$g(\omega)\, C(\omega) = \frac{\lambda/\tau_c}{1 + \tau_c^2\omega^2}, \qquad (1.45)$$

which can be achieved by a variety of combinations of densities of states $g(\omega)$ and coupling coefficients $C(\omega)$, then from (1.44) we obtain the exponential memory kernel

$$K(t) = \frac{\lambda}{\tau_c}e^{-t/\tau_c}, \ t > 0. \qquad (1.46)$$

If the microscopic time scale vanishes, $\tau_c \to 0$, then $K(t) \to 2\lambda\delta(t)$ in (1.46) and we recover the usual Langevin equation from (1.34)

$$\dot{Q} = P \qquad (1.47)$$
$$\dot{P} = -V_m'(Q) - \lambda P + f(t) \qquad (1.48)$$

along with the fluctuation-dissipation relation from (1.43)

$$\langle f(t)\, f(\tau)\rangle = 2\lambda k T \delta(t - \tau). \qquad (1.49)$$

This, then, completes the derivation of the Langevin equation, which proves to be exact for bilinear coupling, independent of the magnitudes of the coupling coefficients. This means that the derivation does not require weak coupling in the sense of small Γ_n to ensure its validity. We subsequently examine the flaw in the above argument.

1.2.2 Stochastic Calculus

Discontinuous Dynamics In 1942 Doob [9] proved that the velocity of a Brownian particle is discontinuous for the Ornstein-Uhlenbeck process [38] described by (1.23) and discussed how one can construct a stochastic differential equation to replace the more familiar differential equations of physics up to that time. In particular he replaced (1.23) with

$$d\mathbf{P}(t) = -\lambda \mathbf{P}(t)\, dt + d\mathbf{B}(t), \qquad (1.50)$$

where $d\mathbf{B}(t)$ is an increment of a vector Wiener process. The fluctuating term $\mathbf{f}(t)\,dt$ in (1.23) has been replaced by $d\mathbf{B}(t)$ since, for δ-correlated fluctuations, $\mathbf{B}(t)$ is in fact a nondifferentiable vector Wiener process, that is, a process of unbounded variation. Thus, a more complicated stochastic differential equation, say, of the form

$$d\mathbf{P}(t) = \mathbf{Q}(\mathbf{P})\,dt + G(\mathbf{P})\,d\mathbf{B}(t) \tag{1.51}$$

actually contains the integral

$$\int G(\mathbf{P}(t))\,d\mathbf{B}(t), \tag{1.52}$$

where the momentum $\mathbf{P}(t)$, the solution to (1.51), is not uniquely defined. Consider the solution to (1.51) given in a small time interval $(t + \Delta t, t)$,

$$\mathbf{P}(t + \Delta t) - \mathbf{P}(t) = \mathbf{Q}(\mathbf{P})\,\Delta t + \int_t^{t+\Delta t} G(\mathbf{P}(t'))\,d\mathbf{B}(t'). \tag{1.53}$$

The question arises as to when the momentum experiences the influence of the delta function random force, and therefore what value of the multiplicative function G should be used under the integral. Is it the value just before the impulse, the value just after the impulse, or perhaps it is some average of the two that is coincident with the time of the impulse. This question gave rise to much confusion in the physics literature of the 1960s through the 1980s. In particular, two interpretations are most frequently used for this integral: the Itô interpretation and the Stratonovich interpretation.

Itô Calculus We present here a synopsis of these two approaches, taken from Lindenberg *et al.* [27], that we hope illuminates their differences. Itô defines an increment $\Delta_I \mathbf{B}(t)$ of a Wiener process $\mathbf{B}(t)$ by

$$\Delta_I \mathbf{B}(t) = \mathbf{B}(t + \Delta t) - \mathbf{B}(t), \tag{1.54}$$

where Δt is a small increment of time, that is, $\Delta t \ll t$. It can readily be seen that the distribution of increments $\Delta_I \mathbf{B}(t)$ is independent of $\mathbf{B}(t)$, the state of the process at time t. The use of this *Ansatz* leads to the following expression for the stochastic integral,

$$\int G(\mathbf{P}(t))\,d\mathbf{B}(t) = \lim_{\Delta t \to 0} \sum_{j=0}^{N} G[\mathbf{P}(t_j)]\,[\mathbf{B}(t_j + \Delta t) - \mathbf{B}(t_j)], \tag{1.55}$$

with the initial time being zero. The function G thus enters only as a function of the earliest time index in each interval $(t_j, t_j + \Delta t)$. One consequence of the definition (1.55) is that the average of the integral over the fluctuations vanishes since the system variables, by causality, cannot anticipate the presence of δ-correlated fluctuations. Thus, we can write

$$\mathbf{P}(t + \Delta t) - \mathbf{P}(t) = \mathbf{Q}(\mathbf{P}) \Delta t + G(\mathbf{P}(t)) \int_t^{t+\Delta t} d\mathbf{B}(t') . \qquad (1.56)$$

The detailed properties of (1.56) have been explored by a number of investigators; see, for example, Lindenberg and West [26] for a list of references.

Stratonovich Calculus On the other hand, rather than (1.54), the Stratonovich definition of an increment of a vector Weiner process $\Delta_S \mathbf{B}(t)$ would be

$$\Delta_S \mathbf{B}(t) = \mathbf{B}(t + \Delta t/2) - \mathbf{B}(t - \Delta t/2) . \qquad (1.57)$$

The distribution of increments now does depend on $\mathbf{B}(t)$. The Stratonovich *Ansatz* leads to the following representation of the stochastic integral,

$$\int G(\mathbf{P}(t)) \, d\mathbf{B}(t) = \lim_{\Delta t \to 0} \sum_{j=0}^{N} G\left[\overline{\mathbf{P}}(t_j)\right] \left[\mathbf{B}(t_j + \Delta t) - \mathbf{B}(t_j)\right], \qquad (1.58)$$

where

$$\overline{\mathbf{P}}(t_j) \equiv \{\mathbf{P}(t_j + \Delta t) + \mathbf{P}(t_j)\} / 2.$$

The elements of the function G here depend on the average value of the momentum in the time interval $(t_j, t_j + \Delta t)$. One consequence of (1.58) is that the average of the integral over the fluctuations, in general, does not vanish unless $G(\mathbf{P})$ is a constant independent of the momentum.

Finite Correlation Time The Stratonovich integration rule (1.58) is consistent with fluctuating processes that have a finite correlation time τ_c, a fact that is built, through the *Ansatz*, into the dependence of $\Delta_S \mathbf{B}(t)$ on $\mathbf{B}(t)$ even in the limit of vanishing correlation time. The Stratonovich interpretation corresponds to the use of the ordinary rules of calculus and allows one to perform nonlinear variable transformations on the dependent variable in the usual way. Therefore we obtain

$$\mathbf{P}(t + \Delta t) - \mathbf{P}(t) = \mathbf{Q}(\mathbf{P}) \Delta t + G\left(\overline{\mathbf{P}}(t)\right) \int_t^{t+\Delta t} d\mathbf{B}(t') .$$

This new mathematics, particularly in the Stratonovich form in the physical sciences, has become so accepted that we now write the Langevin equation (1.23), but interpret it as the Doob equation (1.50) with the appropriate definition for the stochastic integral.

1.3   Comments on Physics of Fractional Calculus

Separation of Time Scales   A process appears random on the macroscopic level because of the separation of time scales between the microscopic and macroscopic dynamics. When such a separation exists, the Langevin equation, using the interpretation of Doob, adequately describes the dynamics of the physical phenomenon. On the other hand, when this separation of time scales does not exist, ordinary statistical physics is no longer adequate to describe the phenomenon, as discussed, for example, by Grigolini et al. [18]. In particular, a lack of time-scale separation may induce a fractional, stochastic, differential equation on the macroscopic level [44].

Fractal Statistics   What the Weierstrass and other nonanalytic functions have in common with the statistics of complex stochastic processes, such as those described by the Lévy distribution, is the lack of a differential description of the time evolution of the phenomena they are intended to model. It was only a little over a decade ago that it was recognized by Shlesinger et al. [37] that Lévy flights can be used to model the randomly varying velocity field of turbulent flow. The key to this understanding is that the equation of evolution of the probability density is a fractional diffusion equation whose solution is a truncated Lévy distribution [2, 42].

Fractional Stochastic Dynamics   Fractional diffusion equations have been used to model the evolution of stochastic phenomena with long-time memory, that is, phenomena with correlations that decay as inverse power laws rather than exponentially in time [8, 7, 43, 44]. It has been known for quite some time in the economics literature that such statistical properties can be successfully described using finite difference equations in which the finite difference is of fractional order, see Hosking [3, 19]. The continuum limit of such fractional difference stochastic equations is fractional differential stochastic equations, of the possible form

$$\frac{d^\alpha \mathbf{P}(t)}{dt^\alpha} = \mathbf{f}(t). \tag{1.59}$$

In (1.59) α is not an integer, and this fractional differential form of the dynamics is used to accommodate the nondifferentiability of the underlying phenomena. The evolution of the probability density associated with the momentum in (1.59) is a fractional partial differential equation in the phase space for the phenomenon that describes the evolution of the probability density; see, for example, West and Grigolini [43] and also Zaslavsky et al. [46]. We have a great deal more to say about these things later.

1.3.1   Markov Approximation

Van Hove's Argument   In a previous lecture we demonstrated how to start from a coupled Hamiltonian description of a Brownian particle and replace the

microscopic part with its evolution from a microscopic distribution of initial
states (harmonic oscillator heat bath) to obtain a macroscopic stochastic force.
This approach is one way to obtain statistical mechanics from classical mechanics. Another approach was introduced by Van Hove [39] and concerns the taking
of certain limits of the microscopic equations. The Van Hove limit of a microscopic process has been used as the essential step for the derivation of statistical
mechanics from microscopic dynamics in a number of places; see, for example,
Zwanzig [47] who discusses the uses and abuses of this approach. We review
the Van Hove approach and how it works in the case of a specific model rather
than in general. Let us start by recalling the meaning of the Van Hove limit.
Consider the integro-differential equation

$$\frac{dg(t)}{dt} = -\lambda^2 \int_0^t K(t - t', \lambda)g(t')dt', \qquad (1.60)$$

where g is some quantity of interest, K is a memory kernel, and λ is a parameter. Equation(1.60) is a typical non-Markovian equation obtained in studying
physical systems coupled to an environment, and whose environmental degrees
of freedom have been averaged over. In this case the parameter λ can be re-
garded as the strength of the perturbation induced by the environment on the
system of interest. This is what we did earlier in the derivation of the generalized
Langevin equation from the Hamiltonian model of the coupling of a Brownian
particle to a heat bath of harmonic oscillators. Thus, what we deduce about
the general properties of (1.60) may also be applied to the generalized Langevin
equation.

The Way Markov Did It Wide use is made of the Markov approximation
in the literature which replaces the integro-differential equation (1.60) with the
rate equation

$$\frac{dg(t)}{dt} = -\left(\lambda^2 \int_0^\infty K(t', \lambda)dt'\right)g(t) \qquad (1.61)$$

and this is essentially what was done when the memory kernel in the generalized
Langevin equation was replaced by a delta function in time. Let us now consider
an apparently different technique that is due to Van Hove. Notice that the
coefficient in braces is independent of the time t and therefore defines an overall
rate coefficients.

The way Van Hove did it The Van Hove approach [47] to solving (1.60)
consists of taking the limit $\lambda \to 0$ and $t \to \infty$ in such a way that the product of
the interaction coefficient and time remains constant

$$\lim_{\substack{\lambda \to 0 \\ t \to \infty}} \lambda^2 t = \text{constant.}$$

In this way we can introduce the new variable

$$x = \lambda^2 t$$

and the new function

$$G(x) = g(t),$$

to rewrite (1.60) as

$$\frac{dG(x)}{dx} = -\int_0^{x/\lambda^2} K(t', \lambda) G(x - \lambda^2 t') dt'. \tag{1.62}$$

Now, the adoption of the limit

$$x = \lim_{\substack{\lambda \to 0 \\ t \to \infty}} \lambda^2 t = \text{constant} \tag{1.63}$$

makes it possible for us to expand the function under the integral in a Taylor series and replace the time convolution form of (1.62) with

$$\frac{dg(t)}{dt} = -\tau_c \lambda^2 g(t). \tag{1.64}$$

The solution to (1.64) is, of course,

$$g(t) = g(0) e^{-\tau_c \lambda^2 t}, \tag{1.65}$$

where the time scale τ_c is determined by the memory kernel

$$\tau_c = \int_0^\infty dt K(t). \tag{1.66}$$

Thus, we obtain the same exponential solution using the Van Hove approach as we obtain by means of the Markov approximation (1.61).

An Inconsistency   Following the argument of Zwanzig [47] we insert the solution (1.65) into the dynamical equation (1.60) to obtain in the asymptotic, $t \to \infty$, limit

$$-\lambda^2 \tau_c e^{-\tau_c \lambda^2 t} = -\lambda^2 \int_0^\infty K(t', \lambda) e^{-\tau_c \lambda^2 (t' - t)} dt'. \tag{1.67}$$

We factor the exponential out of (1.67) to obtain

$$\tau_c = \int_0^\infty K(t', \lambda) e^{\tau_c \lambda^2 t'} dt'$$

so that if τ_c is to remain finite, then the memory kernel must asymptotically behave as the decaying exponential

$$\lim_{t \to \infty} K(t, \lambda) \approx e^{-\tau_c \lambda^2 t}. \tag{1.68}$$

The question is whether this exponential relaxation can be achieved by a Hamiltonian system. Zwanzig demonstrates that the Laplace transform of the memory kernel for a Hamiltonian system has poles on the imaginary axis so that the memory kernel always has an oscillatory component and cannot in general decay exponentially. Thus, the Van Hove approximation, and its mathematical equivalent, the Markov approximation, are inconsistent with a Hamiltonian description of the system dynamics. In other words, the Langevin equation is not derivable from a Hamiltonian model of a Brownian particle coupled to a heat bath, unless there is some way to consistently reduce the generalized Langevin equation to the zero memory Langevin equation without making the invalid Markov assumption [17]. We do that now.

Various Limits   Taking the limit $\lambda \to 0$ corresponds to assuming the coupling of the system to the environment is weak, whereas the limit $t \to \infty$ means that the observation time is much larger than other temporal scales present in the system. Specifically this time must be larger than the microscopic time τ_c. This observation allows us to reformulate the Van Hove limit in a slightly different way, more suitable for our purposes. First, instead of taking the asymptotic $t \to \infty$ limit, we take the vanishing correlation time limit $\tau_c \to 0$. Also, we replace the limit $\lambda \to 0$ with the equivalent limit $\Delta \to \infty$, where Δ is a new coupling constant, to be specified subsequently. The quantity to be kept constant in carrying out the limit is just the product $\Delta^2 \tau_c$. Notice that connecting Δ, τ_c, and λ as $\lambda = \Delta \tau_c$ makes it possible to keep $\Delta \ll \tau_c^{-1}$ (in such a way that $\lambda \ll 1$) and at the same time to perform the limit $\Delta^2 \to \infty$ so as to make $\Delta^2 \tau_c \to$ constant.

1.3.2   Inverse Power-Law Memory

Time Scale Separation   We now want to examine the situation when the approximations considered in the previous lecture no longer apply. This occurs when the microscopic time scale no longer separates from the macroscopic time scale and the phenomenon exhibits long-term memory. We investigate this situation by going back to the generalized Langevin equation. The generalized Langevin equation in the potential-free case $V_m = 0$ has the form

$$\dot{P} = -\Delta^2 \int_0^t \Phi_1 (t - \tau) P(\tau) \, d\tau + f(t), \qquad (1.69)$$

where the physics of the underlying process is contained in the relaxation kernel, the fluctuations, and the relationship between the two. The relaxation kernel is actually the correlation function for the fluctuations, and is related to the memory kernel by

$$K(t) = \langle f(t') f(t' + t) \rangle = \Delta^2 \Phi_1 (t). \qquad (1.70)$$

The autocorrelation function for the momentum of the Brownian particle,

$$\Phi_0 (t) \equiv \frac{\langle P(t') P(t' + t) \rangle}{\left\langle P(t)^2 \right\rangle}, \qquad (1.71)$$

satisfies a similar equation [17]

$$\dot{\Phi}_0 = -\Delta^2 \int_0^t \Phi_1 (t - \tau) \Phi_0 (\tau) d\tau. \tag{1.72}$$

In order to derive both normal and anomalous relaxation properties, we are interested in making a non-trivial choice of the correlation function of the fluctuations in the random force $\Phi_1 (t)$. However, we also want our choice to be compatible with a completely dynamical approach; see, for example, Grigolini et al. [17], where we find that $\Phi_1 (t)$ must be an infinitely differentiable function of time.

Correlation Function We decided to focus our attention on an autocorrelation function that has received a great deal of attention in the literature recently, that being, the inverse power law:

$$\Phi_1 (t) = \frac{\overline{T}^\beta}{(\overline{T}^2 + t^2)^{\beta/2}}. \tag{1.73}$$

Thus, it is straightforward to prove via successive time differentiation of (1.73) that the odd moments vanish. This is another important mathematical property necessary to make the relaxation process compatible with Hamiltonian dynamics [17]. As pointed out above, one familiar example of a relaxation function that is incompatible with Hamiltonian dynamics is the exponential: a case exhaustively discussed by Lee [24] and much earlier by Zwanzig [47]. It is easy to prove that the choice for the autocorrelation function given by the inverse power law (1.73) is compatible with the conditions established by Grigolini et al. [17] and consequently does not conflict with the derivation of the generalized Langevin equation from a Hamiltonian picture.

Ordinary Statistical Mechanics There are two different situations that may arise depending on the choice of the power-law index, ordinary statistical mechanics and anomalous statistical mechanics, depending on whether the microscopic time scale is finite. Ordinary statistical mechanics can be recovered from the generalized Langevin equation (1.69), using the inverse power-law autocorrelation function (1.73), along with the integrability condition on the power-law index,

$$\beta > 1. \tag{1.74}$$

The microscopic time scale τ_c is given in this case in terms of the parameter $\overline{T}$,

$$\tau_c = \int_0^\infty dt \frac{\overline{T}^\beta}{(\overline{T}^2 + t^2)^{\beta/2}} = \frac{\sqrt{\pi}}{2} \frac{\Gamma\left(\frac{\beta-1}{2}\right)}{\Gamma\left(\frac{\beta}{2}\right)} \overline{T}. \tag{1.75}$$

Therefore, the Van Hove limit in the form

$$\text{Lim}_{\text{VH}}\Delta^2 \tau_c \equiv \lim_{\tau \to 0, \Delta^2 \to \infty} \Delta^2 \tau_c , \qquad (1.76)$$

is achieved in the limit $\overline{T} \to 0$,

$$\vartheta = \frac{\sqrt{\pi}}{2} \frac{\Gamma\left(\frac{\beta-1}{2}\right)}{\Gamma\left(\frac{\beta}{2}\right)} \text{Lim}_{\text{VH}}\Delta^2 \overline{T}. \qquad (1.77)$$

This limiting procedure results in exponential relaxation for the correlation function $\Phi_0(t)$ and allows us to safely interpret (1.69) as identical to the ordinary Langevin equation,

$$\dot{P} = -\vartheta P + f(t). \qquad (1.78)$$

This is the traditional result for the Langevin equation obtained using the Van Hove method along with the inverse power-law relaxation function, something that could not be done with an exponential relaxation function.

Nonordinary Statistical Mechanics Anomalous statistical mechanics is a consequence of the microscopic time scale diverging and thereby washing out the separation between the dynamics in the microscopic and macroscopic domains. In the anomalous case, where the power-law index of the correlation function is in the interval

$$0 < \beta < 1, \qquad (1.79)$$

the nonintegrability of the correlation function (1.75) prevents us from adopting the above approach, which is to say that the correlation time diverges

$$\tau_c = \int_0^\infty dt \frac{\overline{T}^\beta}{(\overline{T}^2 + t^2)^{\beta/2}} = \infty.$$

In this case we are forced to look for a different procedure to go from a microscopic to a macroscopic description of the system. This procedure can be derived in a natural way from the original Van Hove approach. Let us consider the following limit,

$$Q = \lim_{\overline{T} \to 0, \Delta^2 \to \infty} \Delta^2 \overline{T}^\beta \equiv \text{Lim}_{\text{GVH}}\Delta^2 \overline{T}^\beta , \qquad (1.80)$$

where Q is a finite constant. We refer to this limit as the generalized Van Hove (GVH) limit.

Generalized Van Hove Limit Adopting the *Ansatz* (1.80), and inserting the inverse power-law autocorrelation function (1.73) into the evolution equation for the momentum autocorrelation function (1.72), we obtain in the generalized Van Hove limit

$$\dot{\Phi}_0 = -\text{Lim}_{\text{GVH}}\Delta^2 \overline{T}^\beta \int_0^t \frac{\Phi_0(t')dt'}{(\overline{T}^2 + (t-t')^2)^{\beta/2}} = -Q \int_0^t \frac{1}{(t-t')^\beta}\Phi_0(t')dt'. \qquad (1.81)$$

For dimensional reasons, it is convenient to write

$$Q \equiv \Delta^2 \tau_c^\beta, \tag{1.82}$$

where we have introduced a finite parameter with the dimension of time. Now, we want to relate the momentum autocorrelation function given by (1.81) to the fractional relaxation equation obtained by Glöckle and Nonnenmacher [13, 14] for viscoelastic materials. The fractional relaxation equation for a function $\mathcal{E}_\alpha(t)$ is given by

$$\mathcal{E}_\alpha(t) - \mathcal{E}_\alpha(0) = \frac{1}{\tau_c^\alpha} \frac{d^{-\alpha}}{dt^{-\alpha}} \mathcal{E}_\alpha(t), \tag{1.83}$$

where the strange symbol $d^{-\alpha}/dt^{-\alpha}$ in (1.83) denotes the Riemann-Liouville fractional integral,

$$\frac{d^{-\alpha}}{dt^{-\alpha}} f(t) = \frac{1}{\Gamma(\alpha)} \int_0^t \frac{f(t')dt'}{(t-t')^{1-\alpha}}. \tag{1.84}$$

This is not the only definition for a fractional integral and it, as well as other choices, is discussed at length in subsequent lectures. For the moment we merely remark that with this choice of fractional integral (1.83) may be solved using Laplace and Mellin transforms [15]. The solution to (1.83), as we discuss, gives the explicit form in terms of the so-called Mittag-Leffler function,

$$\mathcal{E}_\alpha(t) \equiv E_\alpha \left(-\left(\frac{t}{\tau_c} \right)^\alpha \right) = \mathcal{E}_\alpha(0) \sum_{k=0}^\infty \frac{(-1)^k}{\Gamma(\alpha k + 1)} \left(\frac{t}{\tau_c} \right)^{\alpha k}, \tag{1.85}$$

which exhibits interesting behaviors at both long and short times compared with the time constant

$$\tau_c \equiv \left(\frac{Q}{\Delta^2} \right)^{1/\beta}.$$

Note that the Mittag-Leffler function is defined by

$$E_\alpha(z) \equiv \sum_{k=0}^\infty \frac{z^{\alpha k}}{\Gamma(\alpha k + 1)} \tag{1.86}$$

which we review in full detail at the appropriate place.

Asymptotic Limits Figure 1.2 depicts a log-log plot of the Mittag-Leffler function for $\alpha = 0.6$ (solid line). The dashed line indicates the stretched exponential function, which is the short-time representation of (1.85):

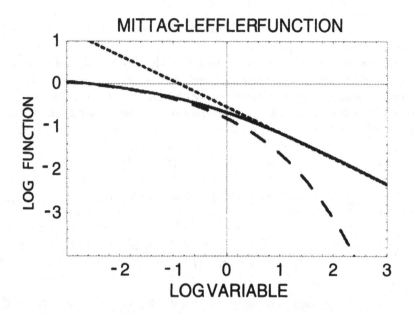

Figure 1.2: *The Mittag-Leffler function is graphed (solid line) and compared with the stretched exponential (large dashed line) at early times and the inverse power law (small dashed line) at late times. We see that the Mittag-Leffler function extrapolates smoothly between the two asymptotic forms.*

$$\lim_{t \to 0} \mathcal{E}_\alpha(t) \approx E_0 e^{-(t/\tau_c)^\alpha}. \tag{1.87}$$

The straight line in Figure 1.2 marked by short dashes represents the inverse power law, which is the long-time representation of (1.85):

$$\lim_{t \to \infty} \mathcal{E}_\alpha(t) \approx E_\infty \left(\frac{t}{\tau_c}\right)^{-\alpha}. \tag{1.88}$$

The inverse power law indicates that the fractional differential relaxation process denoted by (1.83) has self-similar behavior asymptotically in time, that is, for $t \gg \tau_c$.

Fit to data   It is important to stress that the Mittag-Leffler function has been used by Glöckle and Nonnenmacher [16] to fit with remarkable accuracy the relaxation curves of stress experiments on glassy materials. This suggests that the dynamical randomness without time-scale separation takes the shape of time fractional derivatives and becomes experimentally detectable at the macroscopic level. We discuss the physical implications of these and other related ideas subsequently.

1.4   Commentary

Fractional Differential Equation   In these first few lectures we have suggested, but not shown, that the dynamics of complex phenomena, described by fractal functions, can be expressed in terms of fractional differential equations of motion. We show how to do this using the Weierstrass function in the next few lectures. Another approach for understanding the dynamics of nonanalytic phenomena is that of Nonnenmacher and colleagues [13] through [16], who generalize traditional rheology to fractal models of viscoelasticity in terms of the fractional calculus and subsequently solve fractional initial value problems. The solutions to their fractional-differential equations of motion yield physical observables, such as stress relaxation, that are in excellent agreement with experiment, as we review. Other applications of these ideas, not discussed as yet, have generalized the analysis of wave phenomena, starting with the diffractals of Berry [4], up to and including the fractional wave equation of Schneider and Wyss [36]. We return to these representations of the dynamics of complex phenomena in due course.

Modest Goals   Our goal in these lectures is relatively modest in that we do not attempt to describe any particular phenomenon in complete detail. On the other hand it is ambitious in that we argue for the generic result that fractal functions have fractional derivatives and therefore complex phenomena having a fractal dimension are more reasonably modeled using fractional equations of motion than they are using ordinary differential equations of motion such as given by Newton's laws. Part of that argument has to do with the need for the separation of time scales in order for the macroscopic properties of a process to be separated from the microscopic properties. The details of the microscopic irregularities become irrelevant when the time scales separate, in which case the large scale dynamics may be smooth. This is the basis of the traditional stochastic calculus, where the microscopic irregularities lead to macroscopic random fluctuations, but where we can average over the stochastic equations to obtain well-behaved dynamics for averages of the physical observables. On the other hand, when no time-scale separation exists, the irregular microscopic motion may in fact be amplified to the scale of macroscopic observation and no smoothing of the dynamics can be made.

A Calculus of Complexity   In these lectures we propose a calculus, "that is a process of reasoning by computation of symbols" [40], for the understanding of complexity, "..having many varied interrelated parts, patterns, or elements and consequently hard to understand fully" [40], that consists of fractional differentials and fractional integrals in addition to statistical fluctuations. Of course, the fractional calculus has existed in many forms for hundreds of years, but as a mathematical, not a physical, discipline. Herein we try to make that connection between the mathematics and the physics. Leo Kadanoff, in a *Physics Today* article [21], asked, "Fractals: Where's the physics? ", alluding to the, then popular, commercial on "Where's the beef? ", regarding the lack of meat in fast-food hamburgers. His concern was over the flood of papers on fractals in physics journals in the middle 1980s that had little or no regard for the physical mechanisms generating the fractals. Kadanoff's dissatisfaction was apparently directed towards our lack of understanding of how the fractal nature of a given phenomenon could be tied to some more fundamental generating process such as Wilson's renormalization approach [45]. From his view, until such a connection can be made, the physics of fractals is a subject waiting to be born. It is our hope that these lectures may contribute to the birth of that subject, and failing such a birth, then we hope these lectures at least contribute to its conception.

Real Systems Have Long Memories The lesson to draw from these first few lectures is that all derivations of thermodynamics from mechanics (Hamilton's equations), to date, require the Markov approximation. Furthermore, the Markov assumption, in the form of an exponential relaxation, is fundamentally incompatible with Hamiltonian dynamics. Thus, real physical systems are not Markovian, but have nonexponential memory, and one way to capture the influence of that memory is through fractional operators, since such operators are nonlocal in time. This idea of fractional operators is also closely tied to the more recent developments of fractal processes and fractal functions, as we show.

Bibliography

[1] R. J. Abraham and C. D. Shaw, *Dynamics - The Geometry of Behavior, Part 1* (1982); *Part 2* (1983); *Part 3* (1985) and *Part 4* (1988), Aerial Press, Santa Cruz, CA

[2] P. Allegrini, P. Grigolini and B. J. West, Dynamical approach to Lévy processes, *Phys. Rev. E* **54**, 4760-67 (1996).

[3] J. Beran, *Statistics of Long-Memory Processes*, Monographs on Statistics and Applied Probability 61, Chapman & Hall, New York (1994).

[4] M. Berry, Diffractals, *J. Phys. A: Math. Gen.* **12**, 781-797 (1979).

[5] M. Bologna, P. Grigolini and B. J. West, *J. Chem. Phys.* (in press, 2002).

[6] L. Boltzmann, *Lectures on the Principles of Mechanics, Vol.1*, 66, Leipzig: Barth, (1987,1904).

[7] A. S. Chaves, Fractional diffusion equation to describe Lévy flights, *Phys. Lett. A* **239**, 13 (1998).

[8] A. Compte, Stochastic foundations of fractional dynamics, *Phys. Rev. E* **53**, 4191 (1996).

[9] J. L. Doob, The Brownian Movement and Stochastic Equations, *Ann. Math.* **43**, 351 (1942).

[10] G. W. Ford, M. Kac and P. Mazur, *J. Math. Phys.* 6, 504 (1965).

[11] U. Frisch and G. Parisi, in *Turbulence and Predictability in Geophysical Fluid Dynamics and Climate Dynamics*, edited by M. Ghil, R. Benzi and G. Parisi, North-Holland, Amsterdam (1985).

[12] D. V. Giri, *Dirac Delta Functions*, unpublished report (1979).

[13] W. G. Glöckle and T. F. Nonnenmacher, *Macromolecules* **24**, 6426 (1991).

[14] W. G. Glöckle and T. F. Nonnenmacher, A fractional Calculus approach to self-similar protein dynamics, *Biophys. J.* **68**, 46-53 (1995).

[15] W. G. Glöckle and T. F. Nonnenmacher, Fox function representation of non-Debye relaxation processes, *J. Stat. Phys.* **71**, 741 (1993) .

[16] W. G. Glöckle and T. F. Nonnenmacher, Fractional relaxation and the time-temperature superposition principle, *Rheol. Acta* **33**, 337 (1994).

[17] P. Grigolini, G. Grosso, G. Pastori-Parravicini, and M. Sparpaglione, *Phys. Rev.* **B27**, 7342 (1983).

[18] P. Grigolini, A. Rocco and B. J. West, Fractional calculus as a macroscopic manifestation of randomness, *Phys. Rev. E* **59**, 2303-2306 (1999).

[19] J. T. M. Hosking, Fractional Differencing, *Biometrika* **68**, 165-178 (1981).

[20] B. D. Hughes, *Random Walks and Random Environments, Vol. 1: Random Walks*, Oxford Science Publications, Clarendon Press, Oxford (1995).

[21] L. P. Kadanoff, Fractals: Where's the beef?, *Physics Today*/Feb., 6 (1986).

[22] C. Lanczos, *The Variational Principles of Mechanics*, 4th edition, Dover, New York (1970).

[23] P. Langevin, C.R. *Acad. Sci.* **530**, Paris (1908).

[24] M. H. Lee, *Phys. Rev. Lett.* **51** (1983) 1227.

[25] P. Lévy, *Calcul des probabilities*, Guthier-Villars, Paris (1925); *Théorie de l'addition des variables aléatoires*, Guthier-Paris (1937).

[26] K. Lindenberg and B. J. West, *The Nonequilibrium Statistical Mechanics of Open and Closed Systems*, VCH, Berlin (1990).

[27] K. Lindenberg, K. E. Shuler, V. Seshadre and B. J. West, in *Probabilistic Analysis and Related Topics*, Vol. 3, A.T. Bharucha-Reid, ed., Academic Press, New York (1983).

[28] B. B. Mandelbrot, *The Fractal Geometry of Nature*, W.H. Freeman, San Francisco (1982).

[29] B. B. Mandelbrot, *Fractals, form, chance and dimension*, W.H. Freeman, San Francisco (1977).

[30] P. Meakin, *Fractals, scaling and growth far from equilibrium*, Cambridge Nonlinear Science Series 5, Cambridge University Press, Cambridge, MA (1998).

[31] E. W. Montroll and B. J. West, On an enriched collection of stochastic processes, in *Fluctuation Phenomena*, pp.61-206, E.W. Montroll and J.L. Lebowitz, eds., second edition, North-Holland Personal Library, North-Holland, Amsterdam, 61-206 (1987); first edition (1979).

[32] E. W. Montroll and M. F. Shlesinger, On the wonderful world of random walks, in *Nonequilibrium Phenomena II: From Stochastics to Hydrodynamics*, pp. 1-121, E.W. Montroll and J.L. Lebowitz, eds., North-Holland, Amsterdam (1983).

[33] J. Perrin, Mouvement brownien et réalité moléculaire, *Annales de chimie et de physique* VIII 18, 5-114: Translated by F. Soddy as *Brownian Movement and Molecular Reality*, Taylor and Francis, London (1925).

[34] L. F. Richardson, Atmospheric diffusion shown on a distance-neighbour graph, *Proc. Roy. Soc. London* A **110**, 709-737 (1926).

[35] M. Schroeder, *Fractals, Chaos, Power Laws*, W.H. Freeman, New York (1991).

[36] W. R. Schneider and W. Wyss, *J. Math. Phys.* **30**, 134 (1989).

[37] M. F. Shlesinger, B. J. West and J. Klafter, Lévy dynamics for enhanced diffusion: an application to turbulence, *Phys. Rev. Lett.* **58**, 1100-03 (1987).

[38] G. E. Uhlenbeck and L. S. Ornstein, On the theory of the Brownian motion, *Phys. Rev.* **36**, 823 (1930).

[39] L. Van Hove, *Physica* **21** (1955) 517.

[40] P.B. Gove, Ed., *Webster's Third New International Dictionary*, G. & C. Merriam, Springfield, Mass. (1981).

[41] B. J. West and W. Deering, Fractal Physiology for Physicists : Lévy Statistics, *Phys. Repts.* **246**, 1-100 (1994).

[42] B. J. West, P. Grigolini, R. Metzler and T. F. Nonnenmacher, Fractal diffusion and Lévy stable processes, *Phys. Rev. E* **55**, 99 (1997).

[43] B. J. West and P. Grigolini, Fractional differences, derivatives and fractal time series, in *Applications of Fractional Calculus in Physics*, Ed. R. Hilfer, World Scientific, Singapore (1998).

[44] B. J. West, *Physiology, Promiscuity and Prophecy at the Millennium : A Tale of Tails*, Studies of Nonlinear Phenomena in the Life Sciences Vol. 7, World Scientific, Singapore (1999).

[45] K. Wilson, The renormalization group and critical phenomena, in *Nobel Lectures Physics 1981-1990*, World Scientific, New Jersey (1993).

[46] G. M. Zaslavsky, M. Edelman and B. A. Niyazov, Self-similarity, renormalization, and phase space nonuniformity of Hamiltonian chaotic dynamics, *Chaos* **7**, 159 (1997).

[47] R. W. Zwanzig, in *Lectures in Theoretical Physics*, Boulder, **III** (1960) 106, Interscience, New York, (1961).

Chapter 2

Failure of Traditional Models

Enter Fractals The fractal concept was formally introduced into the physical sciences by Beniot Mandelbrot over 20 years ago and has since then captured the imagination of a generation of scientists. Mandelbrot had, of course, been working on the development of the idea for over a decade before he was finally willing to expose his brainchild to the scrutiny of the scientific community at large. His first monograph on fractals [16] brings together the experimental and physical arguments that undermine the traditional picture of the physical world. Since the time of Lagrange (1759) it has been accepted that celestial mechanics and physical phenomena are, by and large, described by smooth, continuous, and unique functions. This belief is part of the conceptual infrastructure of the physical sciences. The evolution of physical processes is modeled by systems of dynamical equations and the solutions to such equations are continuous and differentiable at all but a finite number of points. Therefore the phenomena being described by these equations were thought to have these properties of continuity and differentiability as well. Thus, the solutions to the equations of motion such as the Euler-Lagrange equations, or Hamilton's equations, are analytic functions and such functions were thought to represent physical phenomena in general.

Data Are Not Smooth From the phenomenological side, Mandelbrot called the accuracy of the traditional perspective into question, by pointing to the failure of the equations of physics to explain such familiar phenomena as turbulence and phase transitions. In his books [15, 16] he catalogued and described dozens of physical, social, and biological phenomena that cannot be properly described using the familiar tenets of dynamics in physics. The functions required to explain these complex phenomena have properties that for a hundred years had been thought to be mathematically pathological. Mandelbrot argued that, rather than being pathological, these functions capture essential properties of reality and are therefore better descriptors of the physical world than the

37

traditional analytical functions of theoretical physics.

Trajectories Are Not Smooth   On the theoretical physics side, the Kolmogorov-Arnold-Moser (KAM) theory for conservative dynamical systems describes how the self-similar continuous trajectories of a particle break up into a chaotic sea of random disconnected points. Furthermore, the strange attractors for dissipative dynamical systems have a fractal dimension in phase space. Both these developments in classical dynamics, KAM theory and strange attractors, emphasize the importance of nonanalytic functions in the description of the evolution of deterministic nonlinear dynamical systems. We do not discuss the details of such dynamical systems in these lectures, but refer the reader to a number of excellent books on the subject ranging from the mathematicaly rigorous, but readable [19], to provocative picture books [1], and to extensive applications [28].

Smooth Does Not Mean Differentiable   In the late 1800s, most mathematicians felt that a continuous function must have a derivative almost everywhere, which means the derivative of a function is singular only on a set of points whose total length (measure) vanishes. However, some mathematicians wondered if functions existed that were continuous, but did not have a derivative at any point (continuous everywhere but differentiable nowhere). It is interesting to note that mathematicians at the time were reluctant to consider such unusual functions as being worthy of serious research attention. Similarly, there is a similar resistance today to the shifting emphasis that has occurred in research, and that emphasis is just beginning in the classroom, with regard to placing general nonlinear analysis and nonanalytic functions in the forefront of the physics curriculum.

Mathematical Pathologies   The motivation for considering pathological functions, such as those that are continuous, but not differentiable, was initiated within mathematics and not in the physical or biological sciences, the insights of Boltzmann and Perrin notwithstanding. After all, what possible use can there be for a function that is so jagged that it has no tangents at all? In 1872, Karl Weierstrass (1815-1897) gave a lecture to the Berlin Academy in which he presented functions that had the aforementioned continuity and non-differentiability properties. Thus, these functions had the symmetry of self-similarity, as we show. Twenty-six years later, Ludwig Boltzmann, who elucidated the microscopic basis of entropy, said that physicists could have invented such functions in order to treat collisions among molecules in gases and fluids. Boltzmann had a great deal of experience thinking about such things as discontinuous changes of particle velocities that occur in kinetic theory and to wonder about their proper mathematical representation. He had spent many years trying to develop a microscopic theory of gases and he was successful in developing such a theory, only to have his colleagues reject his contributions. Although kinetic theory led to acceptable results (and provided a suitable microscopic definition of entropy), it was based on time-reversible equations; that is, entropy distinguishes the past from the future, whereas the equations of classical mechanics do not [5]. This basic inconsistency between analytic dynamics and thermodynamics remains unresolved today, although there are indications that the resolution of this old chestnut lies in microscopic chaos.

Brownian Motion It was assumed in the kinetic theory of gases that molecules are materially unchanged as a result of interactions with other molecules, and collisions are instantaneous events as would occur if the molecules were impenetrable and perfectly elastic. As a result, it seemed quite natural that the trajectories of molecules would sometimes undergo discontinuous changes. Robert Brown, in 1827, observed the random motion of a speck of pollen immersed in a water droplet [6]. Discontinuous changes in the speed and direction of the motion of the pollen were indicated, but the mechanism causing these changes was not understood.

Not Analytic Albert Einstein published a paper in 1905 that, although concerned with diffusion in physical systems, ultimately explained the source of Brownian motion as being due to the net imbalance of the random collisions of the lighter particles of the medium with the surface of the pollen mote. Jean Baptiste Perrin, of the University of Paris, experimentally verified Einstein's predictions and received the Nobel Prize for his work in 1926. Perrin [21], giving a physicist's view of mathematics in 1913, stated that curves without tangents are more common than those special, but interesting ones, like the circle, that have tangents. In fact he was quite adamant in his arguments emphasizing the importance of nonanalytic functions for describing complex physical phenomena, such as Brownian motion. Thus, there are valid physical reasons for looking for these types of functions, but the scientific reasons became evident to the general scientific community only long after the mathematical discovery was made by Weierstrass.

Fractals as Models In previous lectures we discussed some of the difficulties associated with describing the macroscopic motion of a particle coupled to a complex environment. In particular we observed how Hamiltonian systems are used, and sometimes misused, in modeling irregular phenomena. We also discussed the ambiguity associated with the interpretation of differentials in stochastic differential equations. Now we step back and take a different approach to understanding complex phenomena such as the above Brownian motion, using the concept of fractals.

Fractal Functions and Differentiability In the following lectures we describe some of the essential features of fractal functions starting from the simple generators of fractal geometrical objects to processes that change over time. The former objects are described by Cantor sets of various kinds, that we characterize by means of fractional dimensions. The dynamical processes are described by functions that are fractal, such as the Weierstrass function, that are continuous everywhere but are nowhere differentiable. This idea of nondifferentiability leads to the introduction of the elementary definitions of fractional integrals and fractional derivatives starting from the limits of appropriately defined sums. We find that the relation between fractal functions and the fractional calculus is a deep one. For example, the fractional derivative of a regular function yields a fractal function of dimension determined by the order of the fractional derivative. Thus, the evolution of phenomena that are best described by a fractal function are probably best described by fractional equations of motion, as well. In any event, this latter perspective is the one we develop in these lectures. Others, such as Ford et al. have also made inquires along this line [11]:

> It is interesting to investigate whether fractional calculus, which generates the operation of derivation and integration to fractional order, can provide a possible calculus to deal with fractals. In fact there has been a surge of activity in recent times which supports this point of view. This possible connection between fractals and fractional calculus gives rise to various interesting questions.......This leads to a more specific problem of studying fractional differentiability properties of nowhere differentiable functions, and, the possible relation of the order of differentiability with the dimension of the graph of the function.

Fractional Derivatives and Fractal Dimensions   We have made a number of allusions to the properties of nonanalytical functions. So let us devote some space to discussing the notions of Cantor sets and fractal dimensions, and show that the Weierstrass function is fractal. We then turn our attention to the construction of fractional derivatives and fractional integrals. We begin this latter discussion with the series representation of these fractional operators and apply these definitions to a few simple functions. Finally, we apply the fractional integral (derivative) to the Weierstrass function and show that it yields a new fractional function with a new fractal dimension that depends on the order of the fractional integral (derivative). Thus, although the Weierstrass function does not have an ordinary derivative it does have a finite fractional derivative. Here we do not actually use the original form of the Weierstrass function, but for technical reasons we use a generalization of that definition, that we called the generalized Weierstrass function in Chapter 1.

2.1   Fractals; Geometric and Otherwise

Weierstrass' Student, Cantor   Soon after the Weierstrass function appeared, Georg Cantor (1845-1918), a student and later a colleague of Weierstrass, who had in fact challenged Weierstrass to construct such a function, himself constructed a set of points that provided another surprising result for mathematicians [25]. Cantor provided the first example of a proper subset of the unit interval {0,1}, which contained an uncountable number of points (as did the entire interval), and which also had zero measure (the entire interval had a length of unity). Let us briefly discuss the construction of one of these Cantor sets, depicted in Figure 2.1.

Cantor Sets   The initial stage ($z = 0$) of the middle-third Cantor set is a line segment of unit length; see Figure 2.1. The next stage ($z = 1$) is obtained by discarding the middle third of the line, leaving the two segments {0,1/3} and {2/3,1}. The $z = 2$ stage is obtained by removing the middle third of each of the two intervals from the $z = 1$ stage, leaving the four intervals {0,1/9}, {2/9,1/3}, {2/3,7/9}, {8/9,1}. Repeating this procedure over and over again ($z \to \infty$) eventually produces the Cantor set, which Mandelbrot poetically referred to as Cantor dust, consisting of an uncountably infinite number of points that are separated from one another. That is, in every neighborhood of one of the Cantor points, no matter how small, there are points that do not belong to the Cantor set. So, the set is not continuous, even though it has the same number of points as the entire continuous interval {0,1}. In fact, the Cantor set is so full of holes that its measure is zero and the deleted set of points has a measure unity, just as does the original interval. The elements of the Cantor set are effectively self-similar. Note that any stage in the generation procedure can be obtained from taking only the left half of the next stage and multiplying by three; see Figure 2.1. The entire set can be obtained by appropriately rescaling any portion of the set.

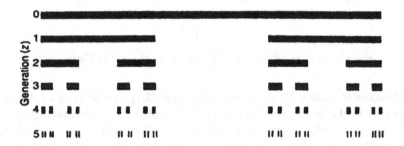

Figure 2.1: *Here we depict the procedure for generating a Cantor set resulting from discarding the middle third of each line segment in going from generation to generation. In the limit the resulting set of points is a Cantor set of dimension $D = \log 2/\log 3 \approx 0.6309$.*

Similarity Dimensions   If the mathematicians were surprised at the existence of a Cantor set, imagine the excitement experienced by physical scientists during the last two decades as they discovered phenomenon after phenomenon described by these abstract mathematical objects, particularly the relation between such sets and dynamical systems. The similarity dimension of a set of points may be defined for self-similar sets. Such sets can be covered by translating a generating element throughout the set, which is the basic idea in making a measurement of the length of a continuous object with a ruler. We introduce the idea of self-similarity using the notion of a reference structure repeating itself over many scales, telescoping both upward and downward in scale. Here we use this scaling argument to develop one definition of dimension that although apparently consistent with our intuition, leads to conclusions that are in apparent conflict with our expectations. Consequently we find that it is our intuition that needs some readjustments and not our definitions. Let us start with some simple geometrical forms and develop our definition from their analysis.

2.1.1   Fractal Dimension

Self-Similar Dimension   Consider a straight line segment such as shown in Figure 2.2a. Dividing the segment into N self-similar pieces by applying a ruler of length η, the length of the interval is then $L(\eta) = N\eta$. If $L = 1$, then the ruler must have length $\eta = 1/N$ to exactly cover the line. Similarly an area L^2 can be covered by N elements, each of area η^2, so that $L^2 = N\eta^2$, and $\eta = 1/N^{1/2}$ for $L^2 = 1$; cf. Figure 2.2b. In three dimensions, a unit cube is covered by N elementary cubes of side $\eta = 1/N^{1/3}$ as shown in Figure 2.2c. Note that in each of these examples we constructed smaller objects of the same geometrical shape as the larger object in order to cover it. This geometrical equivalence is the basis of our notion of self-similarity. In particular, the number of self-similar objects required to cover an object of dimension D is given by

$$N = \eta^{-D}, \tag{2.1}$$

where in the above examples $D = 1, 2$, and 3, as shown in Figure 2.2.

A Mathematician's Dimension   Equation (2.1) can be used to define the dimension D of a set in terms of the number N of elementary covering elements (of length, area, volume, etc.) that are constructed from basic intervals of length η. In the figure the length of the ruler is taken to be unity. Taking the logarithm of both sides of (2.1) and rearranging yields

$$D = \frac{\ln N}{\ln(1/\eta)}. \tag{2.2}$$

The relation for the dimension in terms of the number of rulers of a given length (2.2) is mathematically rigorous only in the limit of the ruler size vanishing, $\eta \to 0$. Now here is a remarkable implication of this equation. Although (2.2) defines the dimension of a self-similar object, there is nothing in the definition that guarantees D has an integer value. In fact, in general, D is noninteger. Opening up our minds to the possible noninteger values of D is the generalization of the above definition of dimension. As the ruler size goes to zero the number of rulers necessary to cover the object diverges to infinity in such a way that the ratio in (2.2) remains finite and gives a noninteger value of D for a self-similar object. Let us apply this definition to a geometrical fractal.

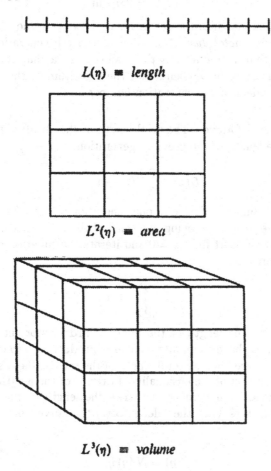

$L(\eta) \equiv length$

$L^2(\eta) \equiv area$

$L^3(\eta) \equiv volume$

Figure 2.2: *Here we depict the number of hypercubes needed to cover one object in one, two, and three dimensions with the volume of the hypercube being unity in each case. The size of one edge of the hypercube is η and its volume is η^D for both integer and noninteger dimension D.*

Calculation of dimension   Let us again consider the middle-third Cantor set depicted in Figure 2.1. This set is constructed by starting with a line segment of unit length and removing the middle third. This leaves two line segments, each of length $\eta(1) = 1/3$ at the first generation, $z = 1$. We then remove the middle third from each of these two line segments, leaving four line segments, each of length $\eta(2) = 1/9$, at the second generation, $z = 2$. Continuing this process, at the zth generation there are a total of $N(z) = 2^z$ line segments, each of length $\eta(z) = 3^{-z}$. Using the values of N and η at the zth generation and then taking the limit as the number of generations goes to infinity, $z \to \infty$, we obtain for the self-similar dimension

$$D = \lim_{z \to \infty} \frac{\ln N(z)}{\ln(1/\eta(z))} = \lim_{z \to \infty} \frac{z \ln 2}{z \ln 3} \approx 0.6309. \tag{2.3}$$

Thus, this dimension classifies the set as being between a line ($D = 1$) and a point ($D = 0$); it is *fractal dust*. Note that at the zth generation we have a prefractal, rather than a fractal, but the scaling is such that the ratio of the logarithm of the number of segments and the logarithm of the length of the segments is independent of the generation number z.

Length of a Fractal Curve   The total length of the set after z generations, $L(z)$, is 2/3 of the length of the previous generation, $L(z-1)$, so that

$$L(z) = \frac{2}{3} L(z-1) \tag{2.4}$$

which is clearly a scaling relation from one generation to the next. The relation (2.4) arises because we are removing one third of the total length of the set at each generation. It we start from $z = 0$ and iterate the equation we obtain the algebraic expression

$$L(z) = \left(\frac{2}{3}\right)^z L(0). \tag{2.5}$$

In the limit $z \to \infty$ the length of the middle third Cantor set goes to zero exponentially, since we have an infinite number of products of a fraction less than one. Another way to see how the total length of the set vanishes asymptotically is by using the definition of dimensionality. In terms of the length of the ruler, rather than the generation number, we write the length of the curve as the number of times one sets down the ruler to cover the curve times the length of the ruler

$$L(\eta) = N(\eta)\,\eta. \tag{2.6}$$

With the number in (2.6) being given by $N(\eta) = \eta^{-D}$ from (2.1) we obtain

$$L(\eta) \sim \eta^{1-D}, \tag{2.7}$$

which vanishes as $\eta \to 0$ because $(1 - D) > 0$. This expression for the length shows the explicit dependence of the length of the curve on the size of the measuring ruler (η) that is used. The dependence of the length of the curve on the size of the ruler used to measure the curve is a property of self-similar objects. In this case, the length vanishes as the ruler length vanishes, but as the next example shows, this is not the general situation.

Geometrical Fractals   Another example of mathematical pathology is the snowflake curve first constructed in 1906 by Helge Van Koch. This closed plane curve has an infinite length, but encloses a finite area. Starting with an equilateral triangle (the generator), the second stage is generated by replacing the middle third of each line in the generator by a scaled down version of the generator. In Figure 2.3 the scaled-down version of the triangle is 1/3 of the size of the generator in the preceding generation. Continuing this procedure results in a curve that, in the limit of an infinite number of generations, is infinitely long and continuous, but is without tangents anywhere. The curve is self-similar and has a topological dimension of unity, since it is topologically equivalent to a circle. In fact the snowflake can be inscribed within a circle of finite radius and therefore confines a finite area, even though the perimeter of the snowflake becomes infinitely long. Unlike the case of the middle third Cantor set, where each line segment generates two new line segments for each line segment at the preceding stage, the Koch snowflake generates four new line segments for each line segment at the preceding stage. Thus, in the Koch snowflake, the length of a line segment at the zth stage is $\eta(z) = 3^{-z}$ just as before, however the number of line segments is $N(z) = 4^z$. The dimensionality of the limiting set is therefore given by

$$D = \lim_{z \to \infty} \frac{\ln N(z)}{\ln(1/\eta(z))} = \lim_{z \to \infty} \frac{z \ln 4}{z \ln 3} \approx 1.2618 \qquad (2.8)$$

so that the self-similar dimension of the Koch snowflake is twice that of the middle-third Cantor set. In addition, since the dimension is greater than unity the length of the perimeter of the snowflake as given by (2.7) diverges; that is,

Figure 2.3: *The figure depicts three generations in the construction of the Koch snowflake using a triangle as the generator. The length of each leg of the original triangle is three and this length becomes four when the new triangle is inserted. The perimeter of the first triangle is three units and that of the second triangle is nine units and so on.*

$$\lim_{\eta \to 0} L(\eta) \sim \lim_{\eta \to 0} \eta^{1-D} \to \infty, \text{ for } D > 1. \qquad (2.9)$$

The length of the perimeter of the snowflake becomes infinite as the size of the ruler goes to zero because $D > 1$. Note that here two points separated by a finite distance can be connected by a curve that is infinitely long. Or, as in the case of the Koch snowflake, an infinitely long, closed curve can enclose a finite area.

Temporal Fractal   A fractal property can be spatial, as in the fixed geometry of the mathematical examples above; it can be temporal, as in a sequence of data taken from a system over an interval of time; and it can be exact or statistical. Furthermore, as is usual in the application of mathematical models to nature, natural fractals are less restricted than mathematical ones. The ideal elements (infinite lines, smooth planes, etc.) of Euclidean geometry are never realized in nature, and neither are the ideal elements of fractal geometry, although the latter are closer to nature than the former. The definitions of the dimension of the mathematical fractals given previously require that the sizes of the elementary covering elements vanish with the size of the ruler. These definitions of similarity dimensions are examples of a general dimension defined by Hausdorff, which is now encompassed by the term fractal dimension [8], applicable even to sets that may not be self-similar over all ranges of space or time. Independence of scale for properties of systems containing mass is limited to finite spatial and/or temporal intervals. For example, molecular diameters and typical periods of molecular motions may set lower limits to self-similarity, although those may be extreme restrictions in a particular situation. However, it is still possible and useful to apply the general idea of a fractal to a natural system and define its fractal dimension. Knowing that the fractal dimension of an object or process is closer to unity than it is to two, for example, indicates a system or process is closer to being a smooth curve, with respect to some set of variables, than it is to being a smooth plane.

Fractals Were Shunned   Many prominent mathematicians sought to avoid these fractal functions, while recognizing their existence. Poincaré (1854-1912), who contributed to all areas of mathematics and laid the groundwork for much of what we today call chaos theory and nonlinear dynamics, referred to these functions as a gallery of monsters. Hermite (1822-1901) was of a similar mind and tried to convince Lebesgue (1875-1941), inventor of the modern theory of integration, not to publish in this area.

Fractals Are Everywhere   Fractals are ubiquitous in complex natural phenomena [2]. They are observed in the architecture of the mammalian lung, they determine the interbeat interval in human heartbeats and the variation in human strides, they influence the information content of DNA sequences, and they describe the branching of trees and the root systems in plants, as well as the growth of bacterial colonies and many other biological systems [29, 30] . In physical phenomena they are also seen everywhere, in viscous fingering, dielectric breakdown, snowflake growth, and on and on [18].

Measure of Quality   The dimension of a naturally occurring fractal is a quantitative measure of a qualitative property of a structure that is self-similar over some region of space or interval of time. There may be more than one region of self-similarity, depending on the parameters of the system. A finger of frost on the windowpane is a crystalline solid that has a fractal dimension less than two, but greater than one. This dimension is a consequence of the process being constrained to unfold within a two-dimensional space. If the phase transition had unfolded in free space, the result would be falling snow, with the snowflake having a fractal dimension greater than two.

2.2 Generalized Weierstrass Function

A Physical Fractal Let us now turn our attention to functions that have the scaling properties discussed in the preceding lectures. In his 1926 investigation of turbulence, Richardson observed that the velocity field of the atmospheric wind is so erratic that it probably cannot be described by an analytic function. In this paper Richardson asks: *Does the wind posses a velocity?* [22]. He suggested a Weierstrass function as a candidate to represent the velocity field, since it is continuous everywhere, but nowhere differentiable, properties he observed in the wind data. We have, since that time, come to realize that Richardson's intuition was superior to a half century of analysis regarding the nature of turbulence. The Weierstrass function was introduced during our discussion of fractals as the first example of a function that is continuous everywhere but is nowhere differentiable, a truly nonanalytic function. We can use this function, as well as suitable generalizations of it, to model phenomena having long-time correlations, such as the turbulent fluctuations in the velocity of the wind.

Explicit Form Here we investigate a generalization of the Weierstrass function because it makes some of the technical discussion simpler. One form of a generalized Weierstrass function (GWF) is

$$W(t) = \sum_{n=-\infty}^{\infty} \frac{1}{a^n} \left(1 - e^{ib^n t}\right) e^{i\phi_n}, \qquad (2.10)$$

where a and b are real parameters and ϕ_n is a phase in the interval $[0, 2\pi]$. The set of phases $\{\phi_n\}$ in the series (2.10) may be selected either deterministically or randomly depending on whether one wants a deterministic or random fractal function, respectively. If the phases are chosen such that $\phi_n = n\mu$, with μ a constant, then (2.10) obeys the self-affine scaling law

$$W(bt) = ae^{-i\mu}W(t). \qquad (2.11)$$

The scaling in (2.11) implies that the entire generalized function can be reconstructed from its values in an interval, say, $t_0 \le t \le bt_0$, as we show below. Note that (2.11) can be expressed in terms of a complex exponent

$$H = \frac{\ln a}{\ln b} - i\frac{\mu}{\ln b}, \qquad (2.12)$$

such that the right-hand side of (2.11) is $b^H W(t)$.

Scaling Let us examine the properties of the real part of the GWF when all the phases are chosen to be zero,

$$F(t) \equiv \operatorname{Re} W(t)_{\phi=0}$$

$$= \sum_{n=-\infty}^{\infty} \frac{1}{a^n} [1 - \cos(b^n t)]. \tag{2.13}$$

In Figure 2.4 we depict $F(t)$ for various values of the parameters a and b for $0 \le t \le 0.5$. If we scale the time in (2.13) by the parameter b we obtain

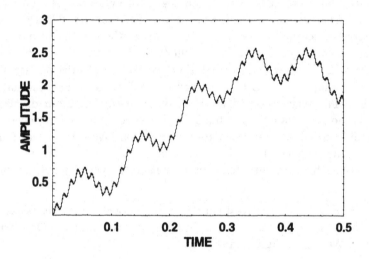

Figure 2.4: *The real part of the generalized Weierstrass function is graphed over the unit time interval of the scaling parameters given by $a = 4$ and $b = 8$.*

$$F(bt) = \sum_{n=-\infty}^{\infty} \frac{1}{a^n} [1 - \cos(b^{n+1}t)]$$

$$= a \sum_{n=-\infty}^{\infty} \frac{1}{a^{n+1}} [1 - \cos(b^{n+1}t)] \tag{2.14}$$

so that shifting the index in the summation by one we can rewrite (2.14) as

$$F(bt) = aF(t). \tag{2.15}$$

Equation (2.15) has the form of a renormalization group scaling relation [18]. Here we can treat (2.15) as an empirical relation, more general than, but consistent with, the GWF. The solution to the scaling equation, (2.15), is given by the functional form [18, 29]

$$F(bt) = A(t) t^H \tag{2.16}$$

which when inserted into (2.15) yields the equation for the amplitude and the scaling index

$$A(bt) = A(t), \quad a = b^H. \tag{2.17}$$

Thus, we see that the power-law index must be related to the two parameters in the series expansion by

$$H = \frac{\log a}{\log b} \tag{2.18}$$

which is the ratio of the influence of the relative scale size to the scale size itself in the time series $F(t)$. The amplitude function $A(t)$ must be equal to $A(bt)$ so that it is periodic in the logarithm of the time interval with period $\log b$, just as $B(t + \overline{T}) = B(t)$ is periodic in t with period $\overline{T}$. The general form of the solution to the modulation function relation (2.17) is given by

$$A(t) = \sum_{n=-\infty}^{\infty} A_n \exp\left[i 2\pi n \frac{\ln t}{\ln b}\right] \tag{2.19}$$

where the set of time-independent coefficients $\{A_n\}$ is determined by the time series. To verify that (2.19) satisfies (2.17), we merely scale the time by the parameter b and verify that this procedure shifts the phase by 2π, thereby leaving the value of the function unchanged.

Complex Exponent Note that we can write the solution to the scaling relation (2.16) in terms of a complex exponent as

$$F(t) = \sum_{n=-\infty}^{\infty} A_n t^{H_n}, \tag{2.20}$$

where the exponent is indexed by the integer n

$$H_n = \frac{\ln a}{\ln b} - i \frac{2\pi n}{\ln b}. \tag{2.21}$$

This exponent has been related to a complex fractal dimension in the architecture of the human lung [29], in many other physiological systems [31], as well as in earthquakes, turbulence, and financial crashes [27].

Weierstrass Is Fractal The real part of the GWF is plotted in Figure 2.4 for parameter values $a = 4.0$ and $b = 8.0$. The structure of the function is built up by the addition of successively faster (higher frequency because $b > 1$) and smaller amplitude terms (smaller amplitudes because $a > 1$) in the series. The Weierstrass function is a superposition of harmonic terms: a fundamental term with a unit frequency and a unit amplitude, a second periodic term of frequency

b and an amplitude of $1/a$, a third periodic term with a frequency b^2 and an amplitude of $1/a^2$, and so on. These parameters can be related to the dimension of a Cantor set D, if we choose $a = b^{D-2}$. Thus, in giving a functional form to Cantor's ideas, Weierstrass became the first scientist to construct a fractal function. Note that for this concept of a fractal function, or fractal set, there is no smallest scale. For $b > 1$ in the limit of an infinite number of terms, the highest frequency contribution to the Weierstrass function diverges.

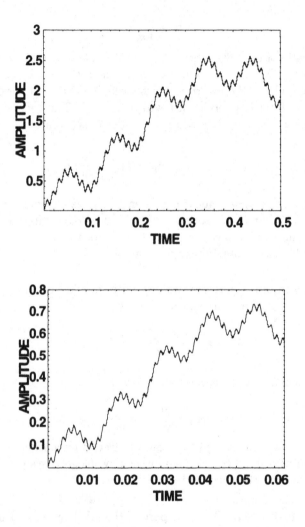

Figure 2.5: *The upper graph is the same as that contained in Figure 2.4. The lower graph is a scaled region of the upper graph. Note the identical behavior in the two graphs in spite of the change in scale.*

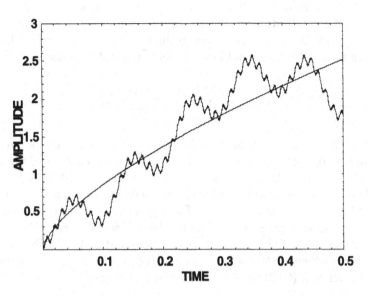

Figure 2.6: *The erratic behavior of the generalized Weierstrass function from Figure 2.4 is depicted. Superimposed on these fluctuations is the dominant power-law behavior determined by (2.16) and found in [3].*

How It Looks   The self-similarity of $F(t)$ defined by (2.15) is shown in Figure 2.5 where the changes in scale between parts (a) and (b) of the figure, $F(t) \rightarrow F(t/8)/a = 0.25F(0.125t)$, clearly demonstrate that the function reproduces itself at ever smaller scale sizes. The fractal dimension of the curve is $D = 2 - \ln a/\ln b = 2 - H$. The algebraic increase in $F(t)$ given by (2.16) is a consequence of the scaling of the GWF with the time increment. The scaling relation (2.15) does not, in and of itself, guarantee that $F(t)$ is a fractal function, but Berry and Lewis [4] have studied it and concluded that the GWF is a fractal function. Mauldin and Williams [17] examined the formal properties of such functions and concluded that for $b \geq 1$ and $1 < a < b$ the dimension is in the interval $[2 - H - C/\ln b, 2 - H]$ where C is a positive constant and b is sufficiently large.

Variability of Function   It is clear from Figure 2.5 that the GWF becomes increasingly variable as the fractal dimension increases. It is equally obvious that the scaling solution (2.16) gives the smoothed dominant behavior of the function depicted in Figure 2.6. In this figure we see the self-similarity of the GWF due to its fractal nature. For example, the apparently erratic variations in the curve repeat themselves at multiple scales, but one feature that dominates the short time behavior of the function is a geometrical increase of the function with time. This is seen in Figure 2.6 where the algebraic increase in the GWF is superimposed on the natural erratic shape of the curve. The exponent of the time is given by the ratio of logarithms (2.18) yielding 2/3 which appears to do a credible job of mimicking the average growth of $F(t)$.

Weierstrass Function and Fractal Dimensions   The Weierstrass function was the first exemplar of a function that is continuous and nondifferentiable. A number of generalizations of this fractal function have been discussed in the literature [7, 15, 30], the most general including random behavior. The above GWF was defined by Berry and Lewis [4] in a different notation as

$$W(t) = \sum_{n=-\infty}^{\infty} \frac{(1 - e^{i\gamma^n t})e^{i\phi_n}}{\gamma^{(2-D)n}}, \tag{2.22}$$

where $1 < D < 2$, $\gamma > 1$,and, as before, ϕ_n is an arbitrary phase. They show using a combination of numerical calculations and analysis of the Fourier series (2.22), that for these values of the parameters, the GWF is continuous everywhere, not differentiable anywhere, has no characteristic scale, and almost certainly has a fractal dimension D.[1] The argument as to the fractal nature of the GWF was given mathematical rigor by Mauldin and Williams [17] somewhat later. Furthermore, if the set of phases $\{\phi_n\}$ is uniformly distributed on the interval $(0, 2\pi)$ then $W(t)$ is a stochastic fractal function. The properties of a modified version of (2.22) were considered by Falconer [7].

Dimension of the Trail   It is well known that (2.22) has a fractal graph, that is, the trails left by the GWF in $(\operatorname{Re} W, t)$-space or in $(\operatorname{Im} W, t)$-space are fractal curves. The fractal nature of such graphs is manifest in the power-law behavior of the correlation between measurements separated by a time interval τ [7]. Before we consider the increments in the GWF, that is, the difference in measurements separated by a time τ, let us examine the scaling properties of (2.22). Multiplying the time by the parameter γ in (2.22) we obtain

$$W(\gamma t) = \gamma^{2-D} W(t) \tag{2.23}$$

which is obtained by relabeling the series index. As we saw above, the most general solution to the scaling equation (2.23) is

$$W(t) = A(t) t^H, \tag{2.24}$$

where $A(t)$ is a function periodic in the logarithm of t, with period $ln\ \gamma$, and the power-law index is given by

$$H = 2 - D. \tag{2.25}$$

Thus, we see that the dominant behavior of the GWF is a power-law growth in time, and the power-law index is determined by the fractional dimension D.

---

[1] In the following we use the term fractional dimension to avoid technical arguments over whether D is the box counting dimension, the Hausdorff-Besicovitch dimension, the similarity dimension, or the fractal dimension. For our immediate purposes it is sufficient that D is not an integer and is in the interval $1 < D < 2$.

Correlation Coefficient It is possible to connect the expression for the fractional dimension (2.25) with the slope of the correlation function of the increments in the GWF. Let us define such a correlation function in the following way

$$C(\tau) = \langle |\Delta W(t, \tau)|^2 \rangle_\phi, \qquad (2.26)$$

where the increments of the GWF are defined by

$$\Delta W(t, \tau) = W(t + \tau) - W(t) \qquad (2.27)$$

and the average, indicated by the brackets with a ϕ subscript in (2.26) is taken over an ensemble of realizations of the phases $\{\phi\}$ uniformly distributed on the interval $(0, 2\pi)$. We refer to the function (2.27) as the incremental GWF or IGWF. Inserting the IGWF

$$\Delta W(t, \tau) = \sum_{n=-\infty}^{\infty} \gamma^{-(2-D)n} (1 - e^{i\gamma^n \tau}) e^{i(\gamma^n t + \phi_n)} \qquad (2.28)$$

into the phase average in (2.26), carrying out the phase average

$$\langle (..) \rangle_\phi = \prod_{n=-\infty}^{\infty} \int_0^{2\pi} \frac{d\phi_n}{2\pi} (..), \qquad (2.29)$$

which for a uniform distribution of phases yields

$$\left\langle e^{i(\phi_n - \phi_{n'})} \right\rangle_\phi = \delta_{n,n'}, \qquad (2.30)$$

we obtain the correlation function

$$C(\tau) = 2 \sum_{n=-\infty}^{\infty} \gamma^{-2(2-D)n} (1 - \cos \gamma^n \tau). \qquad (2.31)$$

Note, since the correlation of the IGWF is independent of the initial time t, these increments are a realization of a stationary stochastic process.

Correlations Scale The dominant behavior of the IGWF correlation function (2.31) is determined by the solution to the exact scaling relation

$$C(\gamma \tau) = \gamma^{2(2-D)} C(\tau) \qquad (2.32)$$

obtained from (2.31) by again relabeling the series. The most general solution to (2.32) is

$$C(\tau) = A(\tau) \tau^{2H}, \qquad (2.33)$$

where, as before, $A(\tau)$ is a function periodic in the logarithm of its argument and the power-law index is given by (2.25) [18, 30]. Note that H plays the same role here as the Hurst exponent does in random walk processes. In fact if $H = 1/2$, then the correlation function increases linearly with time, so that the IGWF, ΔW, would be a normal diffusive process with Gaussian statistics that is stationary in time. We discuss this connection to random walks subsequently.

Nondifferentiability of Phenomena   Recall that the GWF is the kind of nonanalytic function that Richardson wanted to use to describe the turbulent velocity field of the wind. However, in choosing to model the turbulent flow field in this way, must we abandon such fundamental ideas as the equation of evolution for the velocity field? At first glance it seems that we must relinquish the equations of motion since the GWF is nondifferentiable. However, the lack of differentiability has only been established for integer derivatives. It has recently been shown that the fractional derivative (fractional integral) of the GWF is another fractal function with a greater (lesser) fractal dimension, and that the GWF itself is a solution to a fractional differential stochastic equation of motion [24]. So before discussing non-analytic functions further, let us review some of the properties of fractional integrals and fractional derivatives.

2.3   Fractional Operators

There Are Many Definitions   We mentioned the possibility of the fractional calculus being of value in describing the changes in fractal processes over time. However, we should be cognizant of the fact that there is no one fractional calculus, rather there is a collection of fractional differentials and fractional integrals that have been found to reduce to the standard calculus when the appropriate fractional index becomes integer and the functions being acted upon have the specified form. We find it convenient to first review the series representation of fractional operators, since this procedure is the closest to the approach we learn in introductory courses on the traditional calculus. This is a natural starting point for learning about the Riemann-Liouville fractional operators, which is by far the most popular formalism among those that use the fractional calculus to describe complex phenomena.

2.3.1   Series Representations of Fractional Operators

Limits and Derivatives   Here we utilize the standard definitions of derivatives and integrals in terms of sums and differences and establish how these familiar quantities may be generalized to the fractional calculus. We follow in part the strategy of Oldham and Spanier [20] in that we construct a formalism that encompasses both derivatives and integrals within a single methodology. The traditional definition of a derivative of a continuous function is expressed in terms of a limiting procedure and we do the same thing for fractional derivatives. Consider the Taylor expansion of the continuous function $X(t)$ given by

$$X(t - \tau) = X(t) - \tau \mathcal{D} X(t) + \frac{\tau^2}{2!} \mathcal{D}^2 X(t) - \cdots \frac{(-\tau)^n}{n!} \mathcal{D}^n X(t) + \cdots, \quad (2.34)$$

where $\mathcal{D}$ is the derivative operator that can be expressed in terms of the upshift operator E_τ,

$$E_\tau X(t) = X(t - \tau) \tag{2.35}$$

so that (2.34) can be written as

$$E_\tau X(t) = e^{-\tau \mathcal{D}} X(t). \tag{2.36}$$

We define $X(t)$ on the positive real line as would be the case for a continuous time series for a physical observable. We therefore have the symbolic equivalence between the upshift operator and the derivative operator

$$E_\tau = e^{-\tau \mathcal{D}} \tag{2.37}$$

so we can write the left-side difference operator as

$$E_\tau = 1 - \Delta_+ = e^{-\tau \mathcal{D}}. \tag{2.38}$$

In this way we have the definition of the derivative operator

$$\lim_{\tau \to 0} \frac{\Delta_+}{\tau} = \lim_{\tau \to 0} \frac{(1 - e^{-\tau \mathcal{D}})}{\tau} = \mathcal{D}, \tag{2.39}$$

where we have expanded the middle term in (2.39) in powers of the increment τ to obtain the rightmost equality. It is clear from this discussion that if t is the time, then $\mathcal{D}$ is the time derivative operator. When this operator is applied to a continuous function of time we obtain

$$\frac{dX(t)}{dt} = \lim_{\tau \to 0} \frac{\Delta_+}{\tau} X(t) = \mathcal{D} X(t). \tag{2.40}$$

Standard Sries for Derivatives It is a simple matter to generalize this definition of the time derivative to higher order derivatives. Symbolically we can write for integer n

$$\frac{d^n X(t)}{dt^n} = \lim_{\tau \to 0} \frac{\Delta_+^n}{\tau^n} X(t), \tag{2.41}$$

which in terms of the upshift operator can be expressed as

$$\frac{d^n X(t)}{dt^n} = \lim_{\tau \to 0} \frac{(1 - E_\tau)^n}{\tau^n} X(t). \tag{2.42}$$

The difference in (2.42) can be simplified using the binomial coefficient to write

$$\frac{d^n X(t)}{dt^n} = \lim_{\tau \to 0} \frac{1}{\tau^n} \sum_{j=0}^{n} \binom{n}{j} (-E_\tau)^j X(t) \tag{2.43}$$

so that using the definition of the upshift operator we have

$$\frac{d^n X(t)}{dt^n} = \lim_{\tau \to 0} \frac{1}{\tau^n} \sum_{j=0}^{n} \binom{n}{j} (-1)^j X(t - j\tau). \tag{2.44}$$

If the nth derivative of the function exists, Equation (2.44) defines the nth derivative as an unrestricted limit, which is to say, that the limit as $\tau \to 0$ can be taken where τ can assume any value in achieving the limit.

Include the Interval In order to put (2.44) in a form that can be directly related to the integral of the same function, we restrict the values that τ can take on as it goes to zero. This restriction is also part of our eventual discussion of fractional derivatives. Consider the interval (a, t) over which the function is defined. We divide this interval into N parts of equal size

$$\tau_N \equiv \frac{t - a}{N} \tag{2.45}$$

so that now the limit $\tau \to 0$ only assumes discrete values as $N \to \infty$. Using this value of the differential time interval we rewrite the derivative in (2.44) as

$$\frac{d^n X(t)}{dt^n} = \lim_{\tau_N \to 0} \frac{1}{\tau_N^n} \sum_{j=0}^{n} \binom{n}{j} (-1)^j X(t - j\tau_N)$$

and using the property of binomial coefficients that

$$\binom{n}{j} = 0, \text{ if } j > n$$

when n is integer, (2.44) may be written

$$\frac{d^n X(t)}{[d(t - a)]^n} = \lim_{N \to \infty} \left[\frac{t - a}{N}\right]^{-n} \sum_{j=0}^{N-1} \binom{n}{j} (-1)^j X\left(t - j\left[\frac{t - a}{N}\right]\right). \tag{2.46}$$

Here we have introduced the notation of bracketing the differential to denote that the limit exists in the usual unrestricted sense, and that the increments are obtained in the interval (a, t).

Standard Series for Integrals Now let us look at the definition of integrals. We begin with the usual definition of an integral as the limit of a Riemann sum. We use the notation of Oldham and Spanier to write

$$\frac{d^{-1}}{[d(t - a)]^{-1}} [X(t)] \equiv \int_a^t X(t') \, dt'$$

$$= \lim_{\tau_N \to 0} \left\{ \tau_N \sum_{j=0}^{N-1} E_\tau^j X(t) \right\} = \lim_{\tau_N \to 0} \left\{ \tau_N \sum_{j=0}^{N-1} X(t - j\tau_N) \right\} \tag{2.47}$$

so the upper limit of the integral is $t = a + N\tau_N$, and the increment has the same definition as previously. Application of the definition (2.47) to a double integral gives

$$\frac{d^{-2}}{[d\,(t-a)]^{-2}}\,[X\,(t)] \equiv \int_a^t dt_1 \int_a^{t_1} X\,(t')\,dt'$$

$$= \lim_{\tau_N \to 0} \tau_N^2 \sum_{j=0}^{N-1} (j+1)\,E_\tau^j X\,(t) = \lim_{\tau_N \to 0} \tau_N^2 \sum_{j=0}^{N-1} (j+1)\,X\,(t-j\tau_N)\,.$$

$$(2.48)$$

Continuing this procedure for higher-order integrals yields coefficients that build up as

$$\binom{j+n-1}{j},$$

where n is the order of integration, and all the signs are positive. Therefore, we have for the general expression

$$\frac{d^{-n}}{[d\,(t-a)]^{-n}}X\,(t) \equiv \lim_{\tau_N \to 0} \tau_N^n \sum_{j=0}^{N-1} \binom{j+n-1}{j} X\,(t-j\tau_N)\,. \qquad (2.49)$$

We can now express the binomial coefficients in (2.49) in terms of gamma functions, as well as those in (2.46),

$$(-1)^j \binom{n}{j} = \binom{n-j-1}{j} = \frac{\Gamma\,(j-n)}{\Gamma\,(-n)\,\Gamma\,(j+1)} \qquad (2.50)$$

so that both the nth derivative and the nth integration equations are summarized in the single expression

$$\frac{d^q}{[d\,(t-a)]^q}X\,(t) \equiv \lim_{N \to \infty} \frac{\left[\frac{t-a}{N}\right]^{-q}}{\Gamma\,(-q)} \sum_{j=0}^{N-1} \frac{\Gamma\,(j-q)}{\Gamma\,(j+1)}X\,\left(t-j\left[\frac{t-a}{N}\right]\right)\,, \qquad (2.51)$$

where q is an integer of either sign. In this form we refer to (2.51) as the *differintegral*, a word coined by Oldham and Spanier, but which does not seem to have been universally embraced by the scientific community.

Composition Rule We do not prove all the relations necessary to form a calculus, but rather refer the student to the literature where appropriate. We do, however, highlight some analyses that we feel emphasize important relations that might not be readily apparent, but which we think are important in understanding the applications we make of the fractional calculus. For example, the composition rule

$$\frac{d^q}{[d\,(t-a)]^q}\left\{\frac{d^Q}{[d\,(t-a)]^Q}X\,(t)\right\} = \frac{d^{q+Q}}{[d\,(t-a)]^{q+Q}}X\,(t) \qquad (2.52)$$

(where q and Q are integers) holds unless $Q > 0$ and $q < 0$, which is to say, the composition rule holds unless the function $X(t)$ is first differentiated and then integrated. This is quite familiar for integration and differentiation between fixed limits. More explicitly

$$\frac{d^{-n}}{[d(t-a)]^{-n}}\left\{\frac{d^N}{[d(t-a)]^N}X(t)\right\} = \frac{d^{N-n}}{[d(t-a)]^{N-n}}X(t), \qquad (2.53)$$

where N and n are nonnegative integers; (2.53) holds true only if $X(a) = 0$ and if all the derivatives of $X(t)$ through the $(N-1)$ are also zero at $t = a$.

Example of the Importance of Ordering A simple example of the breakdown of the composition rule is given for a term of mixed signs. For example, $X(t) = e^{2t} + 1, N = 1, n = 3$, and $a = 0$. In this case

$$\frac{d}{dt}\left[\frac{d^{-3}}{[dt]^{-3}}\left(e^{2t}+1\right)\right] = \frac{d^{-2}}{[dt]^{-2}}\left(e^{2t}+1\right)$$

$$= \int_0^t dt_1 \int_0^{t_1} dt_2 \left(e^{2t_2}+1\right) = \frac{e^{2t}}{4} + \frac{t^2}{2} - \frac{t}{2} - \frac{1}{4} \qquad (2.54)$$

for one ordering of the operators in the differintegral and

$$\frac{d^{-3}}{[dt]^{-3}}\left\{\frac{d}{dt}\left(e^{2t}+1\right)\right\} = \int_0^t dt_1 \int_0^{t_1} dt_2 \int_0^{t_2} dt_3 2e^{2t_3} = \frac{e^{2t}}{4} - \frac{t^2}{2} - \frac{t}{2} - \frac{1}{4} \qquad (2.55)$$

for the other ordering. We see that (2.54) and (2.55) differ by a factor of t^2, so there is a potential problem. Oldham and Spanier show that in general the composition rule that properly takes into account the lower limit of the integral is

$$\frac{d^N}{[d(t-a)]^N}\left\{\frac{d^{-n}}{[d(t-a)]^{-n}}X(t)\right\} = \frac{d^{N-n}}{[d(t-a)]^{N-n}}X(t)$$

$$= \frac{d^{-n}}{[d(t-a)]^{-n}}\left\{\frac{d^N}{[d(t-a)]^N}X(t)\right\} + \sum_{k=n-N}^{n-1}\frac{(t-a)^k}{k!}X^{(k+N-n)}(a),$$

$$(2.56)$$

where the superscript in $X^{(n)}$ denotes the ordinary nth derivative of the function $X(t)$ with respect to t, and $X^{(n)}(a)$ is the nth derivative of $X(t)$ evaluated at $t = a$. So that for $a = 0$ and $n = 3$ this remainder term in (2.56) yields

$$\sum_{k=2}^{2}\frac{t^k}{k!}X^{(k-2)}(0) = \frac{t^2}{2}X(0) = t^2, \qquad (2.57)$$

the term that distinguishes the two orderings of the operators in the example.

Limits Are Important These comments have been made to sensitize the student to the fact that the lower limit in the differintegral can not be simply ignored, or put in at the end of a calculation. Rather, the lower limit enters into all the relations of the calculus and is therefore physically as well as mathematically important.

2.3.2 Riemann-Liouville Fractional Operators

Continuous Fractional Operators We have looked at some of the fanciful definitions of nondifferentiable functions and traced their behavior to the notion of fractal processes and fractal dimensions. Furthermore, we have seen how to extend our ideas of differentiation and integration using analytic series for the possible treatment of such functions. Let us now turn to the definition of continuous fractional integrals and derivatives. The most frequently encountered definition of an integral of fractional order is through an integral transform called the Riemann-Liouville ($\mathcal{RL}$) formula:

$$_aD_t^{(q)}[X(t)] \equiv \frac{1}{\Gamma(-q)} \int_a^t \frac{X(\xi)\,d\xi}{(t-\xi)^{q+1}}, \qquad (2.58)$$

where it is clear that $q < 0$ since the gamma function would otherwise diverge giving a zero result. We wish to extend this definition to the situation where $q > 0$ for which we require that

$$_aD_t^{(q)}[X(t)] = \frac{d^n}{dt^n}\,_aD_t^{(q-n)}[X(t)] \qquad (2.59)$$

where d^n/dt^n is an ordinary n-fold derivative and n is the smallest integer such that $q - n < 0$ and $q - n + 1 > 0$. These two definitions taken together define the operator

$$_aD_t^{(q)}[\cdot] \qquad (2.60)$$

for all q.

Traditional Integrals Let us choose q equal to the negative integer $-n$ in (2.58) to obtain

$$_aD_t^{(-n)}[X(t)] = \frac{1}{\Gamma(n)} \int_a^t \frac{X(\xi)\,d\xi}{(t-\xi)^{1-n}} \qquad (2.61)$$

which is just the $(n-1)$-fold integration of the function

$$\frac{1}{\Gamma(n)} \int_a^t \frac{X(\xi)\,d\xi}{(t-\xi)^{1-n}} \equiv \int_a^t dt_{n-1} \int_a^{t_{n-1}} dt_{n-2} \cdots \int_a^{t_1} X(t_0)\,dt_0 \qquad (2.62)$$

and produces Cauchy's formula for repeated integration. Moreover, on choosing $n = 1$ and $q = 0$ in (2.59) we obtain

$$_aD_t^{(0)}\left[X\left(t\right)\right] = \frac{d}{dt} \;_aD_t^{(-1)}\left[X\left(t\right)\right] = X\left(t\right) \tag{2.63}$$

so that $_aD_t^{(-1)}$ is the inverse of the d/dt operation. Finally, by selecting $n = q$, we establish that

$$_aD_t^{(n)}\left[X\left(t\right)\right] = \frac{d^n}{dt^n} \;_aD_t^{(0)}\left[X\left(t\right)\right] = \frac{d^n}{dt^n} X\left(t\right) \tag{2.64}$$

is the ordinary nth derivative independent of a when n is a nonnegative integer.

Series for Fractional Integral Now we wish to establish that the $\mathcal{RL}$-fractional integral (2.58) and its extension to $q \geq 0$ is equivalent to the differintegral series introduced earlier. That is, do the operators so defined coincide for all functions $X\left(t\right)$? Following Oldham and Spanier we establish that this is, indeed, the case. Thus, we choose $X\left(t\right)$ to be an arbitrary, but fixed, function on the interval $a \leq \xi \leq t$. As before, let $\tau_N \equiv \left(t - a\right)/N$ so that the difference between the two fractional operators is

$$
\begin{aligned}
\Delta &\equiv \frac{d^q X\left(t\right)}{\left[d\left(t - a\right)\right]^q} -_a D_t^{(q)}\left[X\left(t\right)\right] \\
&= \lim_{N\to\infty} \frac{\tau_N^{-q}}{\Gamma\left(-q\right)} \sum_{j=0}^{N-1} \frac{\Gamma\left(j - q\right)}{\Gamma\left(j + 1\right)} X\left(t - j\tau_N\right) - \frac{1}{\Gamma\left(-q\right)} \int_a^t \frac{X\left(\xi\right) d\xi}{\left(t - \xi\right)^{1+q}},
\end{aligned}
\tag{2.65}
$$

where we now replace the $\mathcal{RL}$-fractional operator with its series equivalent and factor out common terms from Δ to obtain

$$\Delta = \lim_{N\to\infty} \frac{\tau_N^{-q}}{\Gamma\left(-q\right)} \sum_{j=0}^{N-1} X\left(t - j\tau_N\right) \left[\frac{\Gamma\left(j - q\right)}{\Gamma\left(j + 1\right)} - \frac{1}{j^{1+q}}\right]. \tag{2.66}$$

Oldham and Spanner use a straight forward argument to show that if $X\left(\xi\right)$ is bounded on $\left(a, t\right)$ and if $q \leq -2$ then

$$\Delta \equiv 0 \tag{2.67}$$

so that the two definitions, when applied to bounded functions, are identical for $q \leq -2$. This fact, coupled with the property

$$\frac{d^n}{dt^n} \frac{d^q X\left(t\right)}{\left[d\left(t - a\right)\right]^q} = \frac{d^{n+q} X\left(t\right)}{\left[d\left(t - a\right)\right]^{n+q}} \;, \; n, q > 0 \tag{2.68}$$

and the extension of the $\mathcal{RL}$-integral (2.59), shows that the two definitions are identical for any q. Indeed, for arbitrary q we know that for any positive integer n,

$$\frac{d^q X(t)}{[d(t-a)]^q} = \frac{d^n}{dt^n}\left\{\frac{d^{q-n} X(t)}{[d(t-a)]^{q-n}}\right\} \tag{2.69}$$

and

$$_aD_t^{(q)}[X(t)] = \frac{d^n}{dt^n}\,_aD_t^{(q-n)}[X(t)]. \tag{2.70}$$

One need only choose n sufficiently large that $q-n \leq -2$ and make use of (2.67) to complete the proof.

2.3.3 Some Elementary Fractional Derivatives and Integrals

Derivative of a Constant Now that we have established that the differintegral of Oldham and Spanier is identical to the $\mathcal{RL}$-fractional operator for both the fractional derivative and fractional integral, we can use the former to evaluate the differintegral of a number of simple functions. Take, for example, the fractional derivative of a constant; that is, $X(t) = C$ in the time interval $b \leq t \leq a$. A straightforward application of (2.51) yields

$$\frac{d^q[C]}{[d(t-a)]^q} \equiv \lim_{N\to\infty}\left[\frac{N}{t-a}\right]^q \sum_{j=0}^{N-1}\frac{\Gamma(j-q)}{\Gamma(-q)\Gamma(j+1)}C, \tag{2.71}$$

where the constant value is factorable from the series. We make use of the relation among binomial coefficients ([10], page 313),

$$\sum_{j=0}^{n}\binom{j-q-1}{j} = \binom{n-q}{q}, \tag{2.72}$$

and expressing this relation in terms of gamma functions using $N = n+1$ we obtain

$$\sum_{j=0}^{N-1}\frac{\Gamma(j-q)}{\Gamma(-q)\Gamma(j+1)} = \frac{\Gamma(N-q)}{\Gamma(1-q)\Gamma(N)}. \tag{2.73}$$

Inserting (2.73) into (2.71) simplifies the differintegral to

$$\frac{d^q[C]}{[d(t-a)]^q} = \lim_{N\to\infty}\left[\frac{N}{t-a}\right]^q\frac{\Gamma(N-q)}{\Gamma(1-q)\Gamma(N)}C = C\frac{[t-a]^{-q}}{\Gamma(1-q)} \tag{2.74}$$

as the final result. Thus, we see that the fractional time derivative of a constant, $q > 0$, is not zero, but is an algebraic function of the time. Equation (2.74) is zero for all positive qs with integer values because the gamma function is singular; it is negative for $1 < q < 2$, $2 < q < 3$, and so on; the fractional derivative is positive for all other values of q. A formula analogous to (2.74) was constructed by Heaviside in the 1890s using a different formalism and in a different context.

Derivative of a Monomial   The next more complicated function is, of course, the first power of the independent variable $X(t) = t - a$, so that the differintegral is

$$\frac{d^q\,[t-a]}{[d\,(t-a)]^q} = \lim_{N\to\infty}\left[\frac{N}{t-a}\right]^q \sum_{j=0}^{N-1} \frac{\Gamma(j-q)}{\Gamma(-q)\,\Gamma(j+1)}\left[t-a-j\frac{t-a}{N}\right]$$

which can be separated into two parts

$$\frac{d^q\,[t-a]}{[d\,(t-a)]^q} = [t-a]^{1-q} \lim_{N\to\infty} N^q \sum_{j=0}^{N-1} \frac{\Gamma(j-q)}{\Gamma(-q)\,\Gamma(j+1)}$$

$$- \lim_{N\to\infty} N^{q-1} \sum_{j=0}^{N-1} j\frac{\Gamma(j-q)}{\Gamma(-q)\,\Gamma(j+1)}.$$

We now employ the summation formula (2.73) in the first series and the summation

$$\sum_{j=0}^{N-1} \frac{\Gamma(j-q)}{\Gamma(-q)\,\Gamma(j)} = \frac{-q\Gamma(N-q)}{\Gamma(2-q)\,\Gamma(N-1)} \tag{2.75}$$

in the second series, since $j\Gamma(j) = \Gamma(j+1)$, to obtain

$$\frac{d^q\,[t-a]}{[d\,(t-a)]^q} = [t-a]^{1-q} \lim_{N\to\infty} N^q \frac{\Gamma(N-q)}{\Gamma(1-q)\,\Gamma(N)}$$

$$+ \lim_{N\to\infty} N^{q-1}\frac{q\Gamma(N-q)}{\Gamma(2-q)\,\Gamma(N-1)}.$$

Therefore using the asymptotic form of the product of gamma functions

$$\lim_{j\to\infty} \frac{\Gamma(\alpha+j)}{\Gamma(\beta+j)} \sim j^{\alpha-\beta}$$

yields

$$\frac{d^q\,[t-a]}{[d\,(t-a)]^q} = [t-a]^{1-q}\left\{\frac{1}{\Gamma(1-q)} + \frac{q}{\Gamma(2-q)}\right\} \tag{2.76}$$

so that using the recursion relation among gamma functions, $z\Gamma(z) = \Gamma(z+1)$ with $z = 1 - q$, we obtain

$$\frac{d^q\,[t-a]}{[d\,(t-a)]^q} = \frac{[t-a]^{1-q}}{\Gamma(2-q)}. \tag{2.77}$$

From a Different Starting Point On the other hand, we can start from the $\mathcal{RL}$-formula to be sure that we obtain the same result. Therefore using

$$_aD_t^{(q)}\,[t-a] = \frac{1}{\Gamma(-q)}\int_a^t \frac{[\xi-a]\,d\xi}{[t-\xi]^{q+1}}$$

so that shifting the variables $u = t - \xi$

$$
\begin{aligned}
_aD_t^{(q)}\,[t-a] &= \frac{1}{\Gamma(-q)}\int_0^{t-a}\frac{[t-a-u]\,du}{u^{q+1}}\\
&= \frac{1}{\Gamma(-q)}\left[\int_0^{t-a}\frac{[t-a]\,du}{u^{q+1}} - \int_0^{t-a}\frac{du}{u^q}\right]\\
&= \frac{1}{\Gamma(-q)}\left[\frac{[t-a]^{1-q}}{-q} - \frac{[t-a]^{1-q}}{1-q}\right]\\
&= \frac{[t-a]^{1-q}}{-q\,(1-q)\,\Gamma(-q)},\ q<0, \quad (2.78)
\end{aligned}
$$

and using a recurrence formula for gamma functions the denominator of (2.78) is $\Gamma(2-q)$ just as we obtained in (2.77).

Generalize to Arbitrary Index We now need to remove the restriction on q in (2.78). This can be done using the equation

$$_aD_t^{(q)}\,[t-a] = \frac{d^n}{dt^n}\,_aD_t^{(q-n)}\,[t-a] \quad (2.79)$$

which for arbitrary q one may select an integer n sufficiently large that $q-n<0$. Using the result (2.78) for the right-hand side of (2.79) yields

$$_aD_t^{(q-n)}\,[t-a] = \frac{[t-a]^{1-q+n}}{\Gamma(2-q+n)}$$

so that we can rewrite (2.79) as

$$
\begin{aligned}
_aD_t^{(q)}\,[t-a] &= \frac{d^n}{dt^n}\left\{\frac{[t-a]^{1-q+n}}{\Gamma(2-q+n)}\right\}\\
&= \frac{\Gamma(2-q+n)}{\Gamma(2-q)}\frac{[t-a]^{1-q}}{\Gamma(2-q+n)} = \frac{[t-a]^{1-q}}{\Gamma(2-q)} \quad (2.80)
\end{aligned}
$$

yielding the appropriate result for arbitrary q.

Special Cases We note, as expected, that formula (2.80) reduces to zero when $q = 2, 3, 4\cdots$; to unity when $q = 1$; to $t-a$ when $q = 0$; and to $[t-a]^n/(n+1)!$ when $q = -n = -1 = -2 = -3\cdots$. Notice also, on comparison of the two differintegral formulas we have so far obtained, that the qth differintegral of $[t-a]$ equals the $(q-1)$th differintegral of unity as it should.

Derivative of a Multinomial   The final function we consider in this lecture is the multinomial

$$X(t) = (t - a)^p,$$

where p is arbitrary. We show however, that p must exceed -1 for differintegration to have the properties we demand of the operator. For integer q of either sign, use the formulas from the traditional calculus. Our first encounter with noninteger q is restricted to negative q so that we may exploit the $\mathcal{RL}$-definition

$$_aD_t^{(q)}\left[(t-a)^p\right] = \frac{1}{\Gamma(-q)} \int_a^t \frac{(\xi - a)^p\, d\xi}{(t - \xi)^{q+1}}. \tag{2.81}$$

To evaluate the integral we shift the limits by transforming the variable to $v = \xi - a$,

$$_aD_t^{(q)}\left[(t-a)^p\right] = \frac{1}{\Gamma(-q)} \int_0^{t-a} \frac{v^p\, dv}{[t - a - v]^{q+1}}. \tag{2.82}$$

A more familiar form of this differintegral is obtained by introducing the second transformation of variables $v = [t - a]\,u$, to obtain

$$_aD_t^{(q)}\left[(t-a)^p\right] = \frac{[t - a]^{p-q}}{\Gamma(-q)} \int_0^1 [1 - u]^{-1-q}\, u^p du. \tag{2.83}$$

The definite integral in (2.83) is recognized as the Beta function

$$\int_0^1 [1 - u]^{r-1}\, u^{p-1} du = B(p, r), \tag{2.84}$$

so that (2.83) becomes

$$_aD_t^{(q)}\left[(t-a)^p\right] = \frac{[t - a]^{p-q}}{\Gamma(-q)} B(p + 1, -q) \tag{2.85}$$

and using the expression of the Beta function in terms of gamma functions

$$B(p, r) = \frac{\Gamma(p)\,\Gamma(r)}{\Gamma(p+r)},$$

we obtain

$$_aD_t^{(q)}\left[(t-a)^p\right] = \frac{\Gamma(p + 1)}{\Gamma(p - q + 1)} (t - a)^{p-q}, \quad q < 1, p > -1. \tag{2.86}$$

Historically, this formula was important in that it was the basis of the concept of fractional differentiation as developed by Gemant [9] for the study of viscoelastic materials. This formulation was used by him, and later used more extensively by Scott Blair et al. [26] in rheology, as we discuss.

2.4 Generalized Weierstrass Function Intervals

Let Us Apply the Formalism Now that we have definitions for fractional integrals and fractional derivatives and have applied them to some simple functions, let us look and see what we can learn about complex processes using this formalism [2]. Consider both the fractional integral and the fractional derivative of the generalized Weierstrass function (GWF) as done by Rocco and West [24]. In applying the fractional calculus, we take note of the fact that the GWF is not a stationary stochastic function. On the other hand, the intervals in the generalized Weierstrass function (IGWF) are stationary. Therefore we apply the fractional calculus to the stationary IGWF and thereby avoid some technical difficulties.

2.4.1 Fractional Integral of the IGWF

Explicit Integration It has been established in a number of places and by a number of different investigators that the derivative of the Weierstrass function diverges; see [32] and the reference therein. Therefore it is of some interest to consider the fractional derivative of this function and see what effect such a fractional derivative has. Let us introduce the Riemann-Liouville definition of a fractional integral of order β of the GWF:

$$_{-\infty}\mathcal{D}_t^{(-\beta)}\left[W\left(t\right)\right] \equiv \frac{1}{\Gamma(\beta)} \int_{-\infty}^t \frac{W(t')dt'}{(t-t')^{1-\beta}}, \qquad (2.87)$$

where $0 < \beta < 1$. Since integration is a linear operation the integral of the difference between two functions is the difference in the integrals of those two functions and we have for the fractional integral of the IGWF:

$$
\begin{aligned}
\Delta W^{(-\beta)}(t,\tau) &\equiv {}_{-\infty}\mathcal{D}_t^{(-\beta)}\left[W\left(t+\tau\right) - W\left(t\right)\right] \\
&= \frac{1}{\Gamma(\beta)} \int_{-\infty}^t \frac{\Delta W\left(t',\tau\right)dt'}{(t-t')^{1-\beta}},
\end{aligned}
\qquad (2.88)
$$

so that inserting (2.28) into (2.88) yields

$$\Delta W^{(-\beta)}(t,\tau) \equiv \frac{1}{\Gamma(\beta)} \sum_{n=-\infty}^{\infty} \frac{e^{i\phi_n}}{\gamma^{(2-D)n}} (1 - e^{i\gamma^n\tau}) \left\{ \int_{-\infty}^t \frac{e^{i\gamma^nt'}dt'}{(t-t')^{1-\beta}} \right\}. \qquad (2.89)$$

The integral between the curly braces yields [24]

$$\Gamma(\beta) \frac{e^{it\gamma^n}}{\gamma^{\beta n}} e^{-i\pi\beta/2} \qquad (2.90)$$

and therefore (2.89) reduces to

---

[2] This section was taken largely from the paper by Rocco and West [24] and suitably modified for the purposes of these lectures.

$$\Delta W^{(-\beta)}(t,\tau) = e^{-i\pi\beta/2} \sum_{n=-\infty}^{\infty} \frac{(1 - e^{i\gamma^n\tau})e^{i(\phi_n+\gamma^n t)}}{\gamma^{(2-D+\beta)n}}. \tag{2.91}$$

Thus, we see that the fractional integral of the IGWF has the same form as the original IGWF, cf. (2.28). The difference between (2.91) and (2.28), up to an overall phase, is $D \to D - \beta$. What is the proper interpretation of this shifting of the parameter value?

Correlation Function for the Integrated IGWF   To answer this question regarding the fractional dimension, we need to go back to the GWF. We address this issue in a subsequent lecture. For the time being, we limit ourselves to noticing that the correlation function related to (2.91) is given by

$$
\begin{aligned}
C(\tau, -\beta) &= \left\langle \left| \Delta W^{(-\beta)}(t,\tau) \right|^2 \right\rangle_\phi \\
&= \sum_{n=-\infty}^{\infty} \frac{2(1 - \cos\gamma^n\tau)}{\gamma^{2(2-D+\beta)n}}.
\end{aligned} \tag{2.92}
$$

If (2.91) corresponds to some difference of properly defined GWFs, by virtue of the scaling index (2.25), these new GWFs would have a new fractional dimension given by $D' = D - \beta$. In the next lecture we make this argument rigorous.

2.4.2   Fractional Derivative of the IGWF

Explicit Differentiation   The calculations in the case of the fractional derivative of the IGWF are similar to those carried out in the case of the fractional integral acting on the IGWF. Let us consider the Riemann-Liouville definition of the β fractional derivative of the GWF:

$$_{-\infty}\mathcal{D}_t^{(\beta)} W(t) \equiv \frac{1}{\Gamma(1-\beta)} \frac{d}{dt} \int_{-\infty}^{t} \frac{W(t')dt'}{(t-t')^\beta}, \tag{2.93}$$

where again $0 < \beta < 1$. The fractional derivative of the IGWF results in

$$
\begin{aligned}
\Delta W^{(\beta)}(t,\tau) &\equiv {}_{-\infty}\mathcal{D}_t^{(\beta)} \left[W(t+\tau) - W(t) \right] \\
&= \frac{1}{\Gamma(1-\beta)} \frac{d}{dt} \int_{-\infty}^{t} \frac{\Delta W^{(\beta)}(t',\tau)dt'}{(t-t')^\beta}.
\end{aligned} \tag{2.94}
$$

The expression (2.94) is now replaced with

$$\Delta W^{(\beta)}(t,\tau) = \frac{1}{\Gamma(\beta)} \sum_{n=-\infty}^{\infty} \frac{e^{i\phi_n}}{\gamma^{(2-D)n}} (1 - e^{i\gamma^n\tau}) \frac{d}{dt} \left\{ \int_{-\infty}^{t} \frac{dt'}{(t-t')^{1-\beta}} e^{i\gamma^n t'} \right\}, \tag{2.95}$$

where the time derivative of the integral in curly braces yields

$$i\Gamma\left(1-\beta\right)\gamma^{\beta n}e^{it\gamma^{n}}e^{-i\pi(\beta-1)/2},\tag{2.96}$$

and therefore the fractional derivative of the IGWF is

$$\Delta W^{(\beta)}(t,\tau) = e^{i\pi\beta/2}\sum_{n=-\infty}^{\infty}\frac{e^{i(\phi_{n}+t\gamma^{n})}}{\gamma^{(2-D-\beta)n}}\left(1-e^{i\gamma^{n}\tau}\right).\tag{2.97}$$

Thus, we see that the fractional derivative of the IGWF has the same form as the original IGWF. The difference between (2.97) and (2.28), up to an overall phase, is that $D \to D + \beta$.

Correlation Function for the Differentiated IGWF   Again after observing that the correlation function related to (2.97) is

$$C\left(\tau,\beta\right) = \left\langle\left|\Delta W^{(\beta)}\left(t,\tau\right)\right|^{2}\right\rangle_{\phi} = \sum_{n=-\infty}^{\infty}\frac{2\left(1-\cos\gamma^{n}\tau\right)}{\gamma^{2(2-D-\beta)n}}\tag{2.98}$$

we make the plausible conjecture that $D + \beta$ corresponds to the new fractional dimension D'' of some properly defined GWF. Again this is made rigorous in the next lecture.

2.4.3   What About the GWF?

Proof for the GWF   We now want to use the fractional integral and fractional derivative of the IGWF to determine these same operations on the GWF. To accomplish this we assume that the IGWF is given, as are its fractional integral and fractional derivative, but the GWF remains to be determined. Using the right-hand side of (2.27) and (2.28) we obtain

$$W\left(t+\tau\right) - W\left(t\right) = \sum_{n=-\infty}^{\infty}\frac{e^{i(\gamma^{n}t+\phi_{n})}}{\gamma^{(2-D)n}} - \sum_{n=-\infty}^{\infty}\frac{e^{i(\gamma^{n}(t+\tau)+\phi_{n})}}{\gamma^{(2-D)n}}\tag{2.99}$$

from which we can define the function

$$f\left(t\right) = \hat{W}\left(t\right) + \sum_{n=-\infty}^{\infty}\frac{e^{i(\gamma^{n}t+\phi_{n})}}{\gamma^{(2-D)n}},\tag{2.100}$$

where $\hat{W}\left(t\right)$ is assumed to be unknown, and (2.99) is replaced with the condition

$$f\left(t+\tau\right) = f\left(t\right).\tag{2.101}$$

The series in (2.100) is divergent, so that in order to regularize the function we write

$$f_N(t) = W_N(t) + \sum_{n=-N}^{N} \frac{e^{i(\gamma^n t + \phi_n)}}{\gamma^{(2-D)n}}, \qquad (2.102)$$

where

$$\begin{aligned}
f(t) &= \lim_{N \to \infty} f_N(t), \\
\hat{W}(t) &= \lim_{N \to \infty} W_N(t). \qquad (2.103)
\end{aligned}$$

The constraint (2.101) is now replaced with

$$\lim_{N \to \infty} f_N(t + \tau) = \lim_{N \to \infty} f_N(t) \qquad (2.104)$$

indicating that the regularized function is either periodic with period τ or is a constant in the limit $N \to \infty$.

Scaling Determines the Function We now wish to establish that knowing the IGWF uniquely determines the function $\hat{W}(t)$, and this is the GWF. To accomplish this we use the constraint given by (2.104). We present the analysis for the case $f_N(t) = A_N$, where A_N is a constant, so that (2.102) can be written

$$W_N(t) = A_N - \sum_{n=-N}^{N} \frac{e^{i(\gamma^n t + \phi_n)}}{\gamma^{(2-D)n}}. \qquad (2.105)$$

In order to determine the constant in (2.105) we impose an additional constraint on the function. Since we want the function $\hat{W}(t)$ to be a fractal we require that $W_N(t)$ satisfy the scaling condition

$$\lim_{N \to \infty} W_N(\gamma t) = \lim_{N \to \infty} b W_N(t), \qquad (2.106)$$

where b is a constant. Note that we exclude the possibility that the function $f_N(t)$ is periodic, since this will not satisfy this additional requirement that the function also scales. Imposing this scaling constraint on (2.105) we have

$$W_N(\gamma t) = A_N - \gamma^{(2-D)} \sum_{n=-N+1}^{N-1} \frac{e^{i(\gamma^n t + \phi_n)}}{\gamma^{(2-D)n}} \qquad (2.107)$$

so if we choose $b = \gamma^{2-D}$ we can satisfy (2.106) in the $N \to \infty$ limit if we also choose

$$A_N = \sum_{n=-N}^{N} \frac{e^{i\theta_n}}{\gamma^{(2-D)n}}, \qquad (2.108)$$

where $\{\theta_n\}$ is a set of arbitrary phases, because

$$\lim_{N\to\infty} A_N = \lim_{N\to\infty} \gamma^{2-D} A_N. \tag{2.109}$$

Thus, the divergences that required the regularization exactly cancel in (2.107) with the choice of constraint (2.108) and we obtain the function

$$\hat{W}(t) = \lim_{N\to\infty} W_N(t) = \sum_{n=-\infty}^{\infty} \frac{\left(1 - e^{i\gamma^n t}\right) e^{i\phi_n}}{\gamma^{(2-D)n}} \tag{2.110}$$

which is the GWF, where we have associated the set of phases with those used earlier in the GWF.

Extend Argument to IGWF This argument can also be applied to the fractional integral of the IGWF given by (2.91) so that we also have for the fractional integral of the GWF

$$W^{(-\beta)}(t,\tau) \equiv \sum_{n=-\infty}^{\infty} \frac{e^{i\phi'_n}}{\gamma^{(2-D+\beta)n}} (1 - e^{i\gamma^n t}), \tag{2.111}$$

where $\varphi'_n = \varphi_n - i\pi\beta/2$. In the same way the fractional derivative of the IGWF given by (2.97) implies that the fractional derivative of the GWF is

$$W^{(\beta)}(t,\tau) \equiv \sum_{n=-\infty}^{\infty} \frac{e^{i\phi''_n}}{\gamma^{(2-D-\beta)n}} (1 - e^{i\gamma^n t}), \tag{2.112}$$

where $\varphi''_n = \varphi_n + i\pi\beta/2$. Thus, our earlier remarks regarding the fractional calculus applied to the IGWF apply equally well to the GWF.

2.4.4 Conclusions About Fractal Functions

New Fractal Dimensions We have determined the fractional dimension of both fractional integrals and fractional derivatives of the GWF. Our result reads in these two cases:

$$\mathrm{Dim}[\mathcal{D}^{(-\beta)}W] = \mathrm{Dim}[W] - \beta, \tag{2.113}$$

$$\mathrm{Dim}[\mathcal{D}^{(\beta)}W] = \mathrm{Dim}[W] + \beta, \tag{2.114}$$

where $\mathrm{Dim}[\cdot]$ denotes the dimension of the function in brackets. This result can be easily interpreted noticing that the fractional dimension gives information about the degree of irregularity of the function under analysis. We have demonstrated that carrying out a fractional integral of the GWF means decreasing its fractional dimension and therefore smoothing the process, whereas carrying out a fractional derivative means increasing the fractional dimension and therefore making the process and its increments more irregular.

Others Have Found These Results Differently A related result was obtained by Kolwankar and Gangal [12, 13], but required the introduction of a local fractional derivative (LFD), that is, a fractional derivative defined such that its nonlocal character is removed. They found that the LFD of a Weierstrass function, that is, the imaginary part of (2.22) with all $\phi_n = 0$, exists up to critical order $2 - D$ and not so for orders between $2 - D$ and 1, where D $(1 < D < 2)$ is the box counting dimension of the graph of the function. It is possible to show that our results are in complete agreement with those of Kolwankar and Gangal [12, 13]. To this aim, let us consider the inequality

$$D_T < D < D_E, \tag{2.115}$$

where D_T is the topological dimension and D_E is the embedding dimension [15]. The condition (2.115) needs to be fulfilled by any function in order to be a fractal. In the specific case of the graph of a fractal function, $D_T = 1$ and $D_E = 2$. The same condition (2.115) must also hold true for the fractional derivative (fractional integral) of the function, in this case the GWF, in order to preserve its fractal properties. Therefore, for the fractional derivative of GWF, we have:

$$D'' < 2 \Rightarrow D + \beta < 2 \Rightarrow \beta < 2 - D. \tag{2.116}$$

This means that the generalized Weierstrass function is fractionally differentiable for all orders less than $2 - D$, which is the result of Kolwankar and Gangal [12, 13].

Moreover, the same kind of reasoning can be applied to the fractional integral of the GWF. In this case, the meaningful condition reads:

$$D' > 1 \Rightarrow D - \beta > 1 \Rightarrow \beta < D - 1. \tag{2.117}$$

This means that the generalized Weierstrass function is fractionally integrable for all orders less than $D - 1$.

2.5   Commentary

Phenomena Are Nonlinear and Random Let us stop and consider what we have accomplished in the last few lectures and consequently what we can deduce from these accomplishments. We have argued, convincingly we hope, that natural phenomena are far too complex in general to be described by Hamilton's equations of motion, and even when the phenomena are sufficiently simple that such equations are applicable, the equations are generically nonlinear and therefore their solutions are chaotic. The existence of chaotic solutions in both nonintegrable Hamiltonian and dissipative systems makes stochastic descriptions imperative. We postponed discussion of the statistics, but took account of the complex character of such phenomena using the concepts of fractals and scaling. Although over 100 years old, the utility of scaling in the physical sciences, in the sense developed herein, has only become apparent to the general scientific community in the last decade or two. In this regard fractals have been found to be ubiquitous in the physical and biological sciences, and we can now see, in part, the reason for that.

Fractal Functions Have Fractional Derivatives   Consider the function

$$f(t) = \sum_{n=-\infty}^{\infty} A_n \left(1 - e^{i\omega_n t}\right) e^{i\phi_n}, \tag{2.118}$$

where $\{\omega_n\}$ is a set of frequencies, $\{\phi_n\}$ is a set of random phases confined to the interval $\{0, 2\pi\}$, and $\{A_n\}$ is a set of real amplitudes. Thus, $f(t)$ is a stochastic function of time with the definite initial condition $f(0) = 0$. If we now use (2.118) as a driving force for a fractional stochastic equation, we can formally write

$$D_t^\alpha \left[F(t)\right] = f(t) \tag{2.119}$$

as our equation of motion. The solution to (2.119) is formally given by the inverse equation

$$F(t) = D_t^{-\alpha} \left[f(t)\right] \tag{2.120}$$

so that using the definition of the fractional integral we obtain the explicit form of the solution [24],

$$F(t) = \sum_{n=-\infty}^{\infty} \frac{A_n}{\omega_n^\alpha} \left(1 - e^{i\omega_n t}\right) e^{i\phi_n'}, \tag{2.121}$$

where we have absorbed an overall phase into the new set of random phases $\{\phi_n'\}$. Therefore, if we choose the parameter values $\omega_n = \gamma^n$, $A_n = 1$ for all n, and $\alpha = 2 - D$, the solution to the fractional stochastic equation (2.121) is a generalized Weierstrass function. Thus, the GWF is the solution to a stochastic fractional differential equation.

Fractional Brownian Motion   In addition if we choose the coefficients in the series representation of the random driving force by $A_n \propto 1/\omega_n$ then (2.118) is a realization of a complex Brownian motion process, which is to say that the statistics are two-dimensional Gaussian and the spectrum is an inverse-square frequency. In this case if we also choose for the parameters in the solution to the fractional stochastic differential equation $\omega_n = \gamma^n$ and $\alpha = 1 - D$, then the solution is again a GWF. In addition, as pointed out by Mandelbrot ([15], 390), the GWF with random coefficients is a good approximation to a fractional Brownian function.

Fractal Stochastic Processes   From this example we see that the fractional derivative of a regular analytic function yields a fractal function. If the original function is random in time, then its fractional derivative is a random stochastic function. Thus, we see how the application of fractional operators on regular functions generates fractal functions, and how the fractional index of the operator is related to the fractal dimension of the generated function.

Bibliography

[1] R. J. Abraham and C. D. Shaw, *Dynamics-The Geometry of Behavior, Part 1* (1982); *Part 2* (1983); *Part 3* (1985) and *Part 4* (1988), Aerial Press, Santa Cruz, CA.

[2] M. F. Barnsley, *Fractals Are Everywhere*, Academic Press, Boston (1988).

[3] M. Berry, Diffractals, *J. Phys. A: Math. Gen.* **12**, 781-797 (1979).

[4] M. Berry and Z. V. Lewis, On the Weierstrass-Mandelbrot fractal function, *Proc. Roy. Soc. Lond.* A **370**, 459-484 (1980).

[5] E. Broda, *Ludwig Boltzmann, Man-Physicist-Philosopher*, Ox Bow Press, Woodbridge (1983).

[6] R. Brown, *Phil. Mag.* **6**, 161 (1829); *Edinburgh J. Sci.* **1**, 314 (1829).

[7] K. Falconer, *Fractal Geometry*, John Wiley, New York (1990).

[8] J. Feder, *Fractals*, Plenum, New York (1988).

[9] A. Gemant, A method of analyzing experimental results obtained from elastoviscous bodies, *Physics* **7**, 311 (1936).

[10] I. S. Gradshteyn and I. M. Ryzhik, *Table of Integrals, Series, and Products*, corrected and enlarged edition, Academic, New York (1980).

[11] K. M. Kolwankar, Studies of fractal structures and processes using methods of the fractional calculus, unpublished thesis, University of Pune (1997).

[12] K. M. Kolwankar and A. D. Gangal, Fractional differentiability of nowhere differentiable functions and dimensions, *Chaos* **6**, 505 (1996).

[13] K. M. Kolwankar and A. D. Gangal, Hölder exponents of irregular signals and local fractional derivatives, *Pramana J. Phys.* **48**, 49 (1997).

[14] B. B. Mandelbrot and J. W. van Ness, Fractional Brownian motions, fractional noise and applications, *SIAM Rev.* **10**, 422 (1968).

[15] B. B. Mandelbrot, *The Fractal Geometry of Nature*, W.H. Freeman, San Francisco (1982).

[16] B. B. Mandelbrot, *Fractals, form, chance and dimension*, W.H. Freeman, San Francisco (1977).

[17] R. D. Mauldin and S. C. Williams, On the Hausdorff dimension of some graphs, *Trans. Am. Math. Soc.* **298**, 793 (1986).

[18] P. Meakin, *Fractals, scaling and growth far from equilibrium*, Cambridge Nonlinear Science Series 5, Cambridge University Press, Cambridge, MA (1998).

[19] E. Ott, *Chaos in Dynamical Systems*, Cambridge University Press, New York (1993).

[20] K. B. Oldham and J. Spanier, *The Fractional Calculus*, Academic, New York (1974).

[21] J. Perrin, Mouvement brownien et réalité moléculaire, *Annales de chimie et de physique* VIII 18, 5-114: Translated by F. Soddy as *Brownian Movement and Molecular Reality*, Taylor and Francis, London.

[22] L. F. Richardson, Atmospheric diffusion shown on a distance-neighbour graph, *Proc. Roy. Soc. London* A **110**, 709-737 (1926).

[23] G. F. Roach, *Green's Functions*, second edition, Cambridge University Press, Cambridge, MA (1982).

[24] A. Rocco and B. J. West, Fractional calculus and the evolution of fractal phenomena, *Physica A* **265**, 535 (1999).

[25] M. Schroeder, *Fractals, Chaos, Power Laws*, W.H. Freeman, San Francisco (1991).

[26] G. W. Scott Blair, B. C. Veinoglou and J. E. Caffyn, Limitations of the Newtonian time scale in relation to non-equilibrium rheological states and a theory of quasi-properties, *Proc. Roy. Soc. Ser. A* **187**, 69 (1947).

[27] D. Sornette, Discrete scale invariance and complex dimensions, *Phys. Rep.* **297**, 239-270 (1994).

[28] B. J. West, *Fractal Physiology and Chaos in Medicine*, Studies of Nonlinear Phenomena in the Life Sciences Vol. **1**, World Scientific, Singapore (1990).

[29] B. J. West and W. Deering, Fractal Physiology for Physicists : Lévy Statistics, *Phys. Rep.* **246**, 1-100 (1994).

[30] B. J. West, *Physiology, Promiscuity and Prophecy at the Millennium : A Tale of Tails*, Studies of Nonlinear Phenomena in the Life Sciences vol. **7**, World Scientific, Singapore (1999).

[31] B. J. West, M. Bargava and A. L. Goldberger, Beyond the principle of similitude: renormalization in the bronchial tree, *J. Appl. Physiol.* **60**, 189 (1986).

[32] A. Zygmund, *Trigonometric Series, vols. I and II* combined, second edition, Cambridge University Press, Cambridge, MA (1977).

Chapter 3

Fractional Dynamics

Everything is Granular   At the turn of the century there were two opposing points of view in physics, the atomists and the antiatomists. The latter camp believed in the continuity of nature and saw no reason why matter should stop being divisible at the level of the atom and should, they reasoned, continue indefinitely to smaller and smaller scales. The atomists, on the other hand, with the successes of the periodic table and the kinetic theory of gases, had Boltzmann as their chief proponent. Boltzmann was such an extreme atomist that he did not even accept the continuity of time. In his St. Louis lecture in 1904 he stated [1]:

> Perhaps our equations are only very close approximations to av-
> erage values that are made up of much finer elements and are not
> strictly differentiable.

It is the atomists' view of the classical microscopic world and its influence on the macroscopic world that we endorse in these lectures.

Time Scales and the Continuum   In earlier discussions we explored some of the reasons why Hamilton's equations of motion may not be adequate to describe complex physical phenomena. In particular, in those systems where there is no clear time-scale separation between the microscopic and macroscopic dynamics, there can be no macroscopic equations of motion determined by a variational equation. Such variational equations have been used to determine the equations of motion for independent particle trajectories, as well as in the construction of the evolution equations for field variables. In fluid mechanics, for example, the individual particle displacement is replaced by the field amplitude and the individual particle momentum is replaced by the momentum density, both field quantities depending on space as well as time. Thus, by relating the Hamiltonian to a Hamiltonian density integrated over space, Hamilton's equations of motion for the individual particles are replaced by variational equations involving the canonical field variables that yield dynamical field equations. This is a straight-forward generalization of particle mechanics to large aggregates of relatively short-range interacting particles. In this way a simple physical model, such as a chain of linear harmonic oscillators, can be used to model the wave equation in the continuum limit [2]. Furthermore, the smoothing of free particle collisions gives rise to the Euler equations in fluid mechanics, that is, the conservation of momentum in the continuum. In these examples, as elsewhere, we assume the separation of time scales that allows us to smooth the microscopic fluctuations and construct a differentiable representation of the dynamics on large space scales and long time scales.

Fluid Phenomena Are Not Necessarily Smooth   There have always been a number of physical phenomena that would not lend themselves to the above methods of smoothing. In fluid flow there is turbulence, the statistical fluctuations in the flow field generated by the internal nonlinear dynamical structure of hydrodynamic systems. Scientists have spent over a century attempting to predict the turbulent properties of hydrodynamic systems starting from the Navier-Stokes equations, that is, from the fluid mechanical equations for viscous fluid flow. All these efforts have met with only limited success. The most successful theories of turbulence have centered around the notion of scaling, that is, the transfer of energy from large-scale to small-scale eddies in a cascade [5] in three spatial dimensions, as first articulated by Kolmogorov.

Viscoelastic Materials Have Memory   Another example of physical phenomena where smoothing has failed is in viscoelastic materials such as plastics and rubber. For these materials the relaxation of stress is described by an integral equation rather than a differential equation. In such phenomena there is no characteristic relaxation time, so the process of relaxation cannot be described by an exponential function. Instead the relaxation has been modeled using either a stretched exponential function or an inverse power law [8]. The relaxation integral explicitly shows the influence of memory on the system dynamics [6].

Phase Transitions Have Long Ranges   Finally, there are phase transitions: the transformation of gases to liquids and liquids to solids [9] and back again. We now understand that as the temperature is lowered in physical systems the range of the particle particle interaction lengthens, and this lengthening is described by renormalization group theory [3]. These concepts of scaling, memory and renormalization group theory are all to be found in the physical applications of the fractional calculus that we discuss.

New Modeling Strategies   Metaphorically, these complex phenomena, whose evolution cannot be described by ordinary differential equations of motion; leap and jump in unexpected ways to obtain food; they unpredictably twist and turn to avoid capture, and suddenly change the direction of a gambit to avoid checkmate. To understand these and other analogous processes we find that we must adopt a new type of modeling, one that is not in terms of ordinary or partial differential equations of motion. It is clear that the fundamental elements of complex physical phenomena, such as phase transitions, the deformation of plastics and the stress relaxation of polymers, satisfy Newton's laws. In these phenomena the evolution of individual particles are described by ordinary differential equations that control the dynamics of individual particle trajectories. It is equally clear that the connection between the fundamental laws of motion controlling the individual particle dynamics and the observed large-scale dynamics can not be made in any straightforward way. In fact the renormalization group approach mentioned above seems to put the doctrine of reductionism into crisis.

A Complex Calculus We have reviewed a number of concepts regarding fractional integrals and fractional derivatives in previous lectures. But the review has been sporadic rather than systematic. Let us begin again, this time starting from a review of the traditional properties of derivatives and build up a fractional calculus that we can then apply in an organized way to some of the complex phenomena mentioned above.

3.1 Elementary Properties: Fractional Derivatives

Standard Derivative We consider the general properties of the derivative operator D_t^n for $n \in \mathcal{N}$; that is, n is an integer. This operator is, in fact, defined to have the following properties, all of which we would like the fractional derivative to share. The first property of interest is that of association

$$D_t^n [Cf(t)] = CD_t^n [f(t)], \tag{3.1}$$

where C is a constant. The second property we would like to incorporate into the fractional calculus is the distributive law

$$D_t^n [f(t) \pm g(t)] = D_t^n [f(t)] \pm D_t^n [g(t)]. \tag{3.2}$$

The final property is that the operator obeys the Leibniz rule for taking the derivative of the product of two functions

$$D_t^n [f(t) g(t)] = \sum_{k=0}^{n} \binom{n}{k} D_t^{n-k} [f(t)] D_t^k [g(t)]$$

$$= \sum_{k=0}^{n} \binom{n}{k} D_t^{n-k} [g(t)] D_t^k [f(t)]; \tag{3.3}$$

where we have changed the order of the two functions on the right-hand side of (3.3) in the second equality to emphasize that the order in which the functions $f(t)$ and $g(t)$ appear is of no consequence in taking the derivative, due to the symmetry of the binomial coefficient:

$$\begin{pmatrix} \alpha \\ k \end{pmatrix} = \frac{\Gamma(\alpha+1)}{\Gamma(k+1)\Gamma(\alpha+1-k)}.$$

The above properties are certainly retained for the nth derivative of a monomial t^m with $m \in \mathcal{N}$, so that

$$D_t^n[t^m] = m(m-1)(m-2)\cdots(m-n+1)t^{m-n}$$

$$= \frac{m!}{(m-n)!}t^{m-n} = \frac{\Gamma(m+1)}{\Gamma(m+1-n)}t^{m-n} \qquad (3.4)$$

for $m > n$. Properties (3.1) and (3.2) establish that the operator D_t^n is linear and (3.4) enables us to compute the nth derivative of any analytic function expressed in terms of Taylor's series. We could, of course, interpret the above operator as the time derivative $D_t \equiv d/dt$.

Fractional Derivatives We now extend these considerations to fractional derivatives. By analogy with (3.4) we define a real-indexed derivative, or more generally, a complex-indexed derivative D_t^α with $\alpha \in \mathcal{R}$ (or $\alpha \in \mathcal{C}$), of a monomial t^β, as

$$\frac{d^\alpha}{dt^\alpha}[t^\beta] \equiv D_t^\alpha[t^\beta] = \frac{\Gamma(\beta+1)}{\Gamma(\beta+1-\alpha)}t^{\beta-\alpha}, \qquad (3.5)$$

where $\beta+1 \neq 0, -1, \cdots, -n$; see, for example, (2.86). For the moment we assume (we show the proof later) that the Leibniz rule for the derivative of the product of functions (3.3) can be generalized to fractional derivatives as

$$D_t^\alpha[f(t)g(t)] = \sum_{k=0}^\infty \begin{pmatrix} \alpha \\ k \end{pmatrix} D_t^{\alpha-k}[f(t)]D_t^k[g(t)]$$

$$= \sum_{k=0}^\infty \begin{pmatrix} \alpha \\ k \end{pmatrix} D_t^{\alpha-k}[g(t)]D_t^k[f(t)], \qquad (3.6)$$

and since α is not integer the upper limit of the sum in (3.6) is infinite. If one of the functions in the product is a constant, say, $g(t) = C$, then (3.6) reduces to

$$D_t^\alpha[f(t)C] = \sum_{k=0}^\infty \begin{pmatrix} \alpha \\ k \end{pmatrix} D_t^{\alpha-k}[f(t)]D_t^k[C] = D_t^\alpha[f(t)]C \qquad (3.7)$$

since only the $k = 0$ term survives in the series because the integer derivatives of the constant vanish. Thus, property (3.1) is retained by the generalized Leibniz rule (3.6).

Associative Property If we now write the function $f(t)$ in (3.7) as the sum of two functions, $h(t) + g(t)$, set the constant C equal to unity, and rewrite the equation in the form

$$D_t^\alpha [f(t) \cdot 1] = D_t^\alpha [f(t) \cdot t^0]$$

we obtain using (3.6),

$$
\begin{aligned}
D_t^\alpha [h(t) + g(t)] &= \sum_{k=0}^{\infty} \binom{\alpha}{k} D_t^{\alpha-k} [t^0] \, D_t^k [h(t) + g(t)] \\
&= D_t^\alpha [h(t)] + D_t^\alpha [g(t)], \qquad (3.8)
\end{aligned}
$$

where we used property (3.2) in going from the first to the second line of (3.8). Thus, we see that the associative property is also true for the fractional derivative D_t^α. Furthermore, equations (3.7) and (3.8), taken together, establish that the fractional derivative is a linear operator.

Fractional Derivative of Negative Monomial Now let us consider the case where the index of the monomial is negative integer valued: $\beta + 1 = 0, -1, \cdots, -n$ and we operate with the ordinary integer derivative. Consider the monomial function $f(t) = t^{-m}$ with m a positive definite integer, from which we obtain

$$D_t^n [t^{-m}] = (-1)^n \frac{m(m+1)\cdots(m+n-1)}{t^{m+n}}$$

or using the properties of gamma functions

$$D_t^n [t^{-m}] = (-1)^n \frac{\Gamma(m+n)}{\Gamma(m)} t^{-(m+n)} \qquad (3.9)$$

with $n \in \mathcal{N}$. If we restrict ourselves to real indices, then again proceeding by analogy we write for $0 < \alpha < 1$,

$$D_t^\alpha [t^{-m}] = (-1)^\alpha \frac{\Gamma(m+\alpha)}{\Gamma(m)} t^{-(m+\alpha)}, \qquad (3.10)$$

but we have to change the definition of the gamma functions when the argument in the numerator is a negative integer. This new definition transforms real functions into complex functions and vice versa, because there is the complex coefficient $(-1)^\alpha = e^{i\alpha\pi}$. We shall have occasion to use (3.10).

3.1.1 Constant Functions

When Derivatives Vanish We define $\mathcal{A}(\alpha)$ to be the set of constant functions under the real indexed derivative D_t^α and $\mathcal{C}(\alpha)$ is the generic constant of index α. So, for example, we consider the two functions: $f(t) = t^{-1/2}$ and $f(t) = C$ and use the derivative of the monomial (3.5) to obtain:

$$D_t^{1/2}\left[t^{-1/2}\right] = \frac{\Gamma(1/2)}{\Gamma(0)}t^{-1} = 0 \tag{3.11}$$

since $\Gamma(0) = \infty$. Thus, a particular function is effectively a constant with regard to a certain fractional derivative. In the second example

$$D_t^{1/2}[C] = C\frac{\Gamma(1)}{\Gamma(1/2)}t^{-1/2} = \frac{C}{\sqrt{\pi t}}, \tag{3.12}$$

where we see that a constant is not constant with regard to fractional derivatives. These two examples demonstrate that there are functions that, under real-indexed derivatives, are additive constants and there are additive constants that, under real-indexed derivatives, are functions. In general a function will be a $C(\alpha)$ constant if it is a linear combination of powers of the independent variable, t^{β_k}, with $\beta_k + 1 - \alpha = 0, -1, \cdots, -n$ with $k = 0, 1, \cdots, n$, since the gamma function in the denominator of (3.10) would diverge. In the example, the more general function

$$f(t) = \sum_{k=0}^{\infty} C_k t^{-k-1/2}, \tag{3.13}$$

would be a $C(1/2)$ constant.

Now for Some Theorems We can now enunciate one of the few theorems that we present in these lectures:

Theorem 1: *Let $f(t)$ be a function having a power series representation and assume that there exist derivatives $D_t^\mu[f(t)]$, $D_t^\nu[f(t)]$, and $D_t^\alpha[f(t)]$ with $\alpha = \mu+\nu$. If $f(t)$ is not a $C(\mu)$ and $C(\nu)$ constant then we have the semi-group property:*

$$D_t^\alpha[f(t)] = D_t^{\mu+\nu}[f(t)] = D_t^\mu[D_t^\nu[f(t)]] = D_t^\nu[D_t^\mu[f(t)]]. \tag{3.14}$$

Demonstration: We first consider a monomial and establish the form of (3.14) for this power of t and then extend the argument to a series of such terms. We obtain the following chain of equalities, by using (3.5) sequentially

$$D_t^\nu\left[D_t^\mu\left[t^\beta\right]\right] = D_t^\nu\left[\frac{\Gamma(\beta+1)}{\Gamma(\beta+1-\mu)}t^{\beta-\mu}\right]$$

$$= \frac{\Gamma(\beta+1)}{\Gamma(\beta+1-\mu-\nu)}t^{\beta-\mu-\nu} = \frac{\Gamma(\beta+1)}{\Gamma(\beta+1-\alpha)}t^{\beta-\alpha} = D_t^\alpha\left[t^\beta\right] \tag{3.15}$$

since by hypothesis t^β is not a $C(\mu)$ or a $C(\nu)$ constant, which is to say that we have excluded $\beta - \alpha + 1 = 0, -1, \cdots$. We can repeat the above argument with the indices in reverse order to obtain

$$D_t^\mu\left[D_t^\nu\left[t^\beta\right]\right] = D_t^\alpha\left[t^\beta\right].$$

Now because the fractional derivative is a linear operator we can write a function as a linear superposition of monomials in the form

$$F(t) = \sum_{k=0}^{\infty} B_k t^{\beta_k}, \tag{3.16}$$

where the B_k are constants and by assumption the monomials are not individually $C(\mu)$ or $C(\nu)$ constants. Thus, we may write by using (3.15) on each of the terms in (3.16),

$$D_t^{\alpha}[F(t)] = D_t^{\nu}[D_t^{\mu}[F(t)]] = D_t^{\mu}[D_t^{\nu}[F(t)]] = D_t^{\mu+\nu}[F(t)] \tag{3.17}$$

which is what we had intended to prove.

Order Is Important   A consequence of the above demonstration is that if the function is a constant function, say, for example, a $C(\mu)$ constant, then we need to keep track of the order of operation of the derivatives

$$D_t^{\alpha}[f(t)] = D_t^{\mu}[D_t^{\nu}[f(t)]]$$

since by assumption

$$D_t^{\mu}[D_t^{\nu}[f(t)]] \neq D_t^{\nu}[D_t^{\mu}[f(t)]] = 0. \tag{3.18}$$

The right-hand side of (3.18) is zero because $f(t)$ is, by assumption, a $C(\mu)$ constant, whereas the left-hand side of the equation is not zero. We infer from this inequality that a constant function under a real derivative destroys the commutativity of the operators D_t^{μ} and D_t^{ν}.

Theorem 2: Let $f(t)$ be a function with fractional derivatives; then we can write

$$D_t^{\alpha}[f(at)] = a^{\alpha} D_x^{\alpha}[f(x)] \ , \ x = at. \tag{3.19}$$

Demonstration: By assumption the function under consideration has fractional derivatives so that using (3.6) we obtain the series

$$D_t^{\alpha}[f(t)] = \sum_{k=0}^{\infty} \binom{\alpha}{k} D_t^{\alpha-k}[t^0] D_t^k[f(t)].$$

We can rewrite this series as

$$D_t^{\alpha}[f(t)] = \sum_{k=0}^{\infty} \binom{\alpha}{k} \frac{t^{k-\alpha}}{\Gamma(k+1-\alpha)} D_t^k[f(t)] \tag{3.20}$$

and inserting the change of variables $x = at$ into the right-hand side of (3.20) we obtain

$$D_t^\alpha \left[f\left(at \right) \right] = \sum_{k=0}^{\infty} \binom{\alpha}{k} \frac{t^{k-\alpha}}{\Gamma\left(k+1-\alpha \right)} D_t^k \left[f\left(at \right) \right]$$

$$= \sum_{k=0}^{\infty} \binom{\alpha}{k} \frac{t^{k-\alpha}}{\Gamma\left(k+1-\alpha \right)} a^k \left[D_x^k \left[f\left(x \right) \right] \right]_{x=at} . \qquad (3.21)$$

Thus, rearranging terms in (3.21), we have

$$D_t^\alpha \left[f\left(at \right) \right] = \sum_{k=0}^{\infty} \binom{\alpha}{k} \frac{a^\alpha \left(at \right)^{k-\alpha}}{\Gamma\left(k+1-\alpha \right)} \left[D_x^k \left[f\left(x \right) \right] \right]_{x=at} = a^\alpha D_x^\alpha \left[f\left(x \right) \right]_{x=at}$$

$$(3.22)$$

which was to be proved.

Convolution Representation   We examine the $\mathcal{C}\left(-n \right)$ constant and in particular the $\mathcal{C}\left(-1 \right)$ constant. We first examine the operator D_t^{-1} applied to a monomial

$$D_t^{-1} \left[t^\beta \right] = \frac{\Gamma\left(\beta+1 \right)}{\Gamma\left(\beta+1+1 \right)} t^{\beta+1} = \frac{t^{\beta+1}}{\beta+1} \qquad (3.23)$$

from which we see that this is the integration operator. Now again using the linearity property of the operator we know that we can take a sum of infinitesimals to obtain the standard definition of the integral and therefore in general we can write

$$D_t^{-1} \left[f\left(t \right) \right] = \int_0^t f\left(\tau \right) d\tau. \qquad (3.24)$$

So in this real indexed fractional derivative formalism the integral is only a particular case, with a negative integer value. By extension we can define

$$g_n\left(t \right) = D_t^{-1} \left[g_{n-1}\left(t \right) \right]$$

so that

$$g_n\left(t \right) = \int_0^t g_{n-1}\left(\tau \right) d\tau = \int_0^t D_\tau^{-1} \left[g_{n-2}\left(\tau \right) \right] d\tau = \int_0^t d\tau_2 \int_0^{\tau_2} g_{n-2}\left(\tau_1 \right) d\tau_1$$

and if $g_0\left(t \right) = f\left(t \right)$, then by induction we can write

$$g_n\left(t \right) \equiv D_t^{-n} \left[f\left(t \right) \right] = \int_0^t \cdots \int_0^{\tau_3} d\tau_2 \int_0^{\tau_2} f\left(\tau_1 \right) d\tau_1 \qquad (3.25)$$

just as we obtained for the $\mathcal{RL}$-fractional operators in an earlier lecture.

Theorem 3: *Let $f\left(t \right)$ be a $\mathcal{C}\left(-1 \right)$ constant function. In this case we have*

$$f(t) \equiv 0.$$

Demonstration: Using the definition of the integral given by (3.24) we know that $D_t^{-1}[f(t)] = \int_0^t f(\tau) \, d\tau$. But by assumption this function is a $C(-1)$ constant so that

$$D_t^{-1}[f(t)] = 0 = \int_0^t f(\tau) \, d\tau \ , \ \forall t \in \mathcal{R}$$

and the only function that can satisfy this integral condition is one that is identically zero for all values of the independent variable t; that is, $f(t) = 0$.

The Leibniz Rule Finally, by using (3.3), we obtain the generalization for the integration by parts equation from the traditional calculus, given by

$$D_t^{-1}[f(t) \, g(t)] = \sum_{k=0}^{\infty} \binom{-1}{k} D_t^{-1-k}[f(t)] \, D_t^k[g(t)]$$

$$= \sum_{k=0}^{\infty} (-1)^k \, D_t^{-1-k}[f(t)] \, D_t^k[g(t)], \qquad (3.26)$$

where we have used

$$(1+t)^{-1} = \sum_{k=0}^{\infty} \binom{-1}{k} t^k = \sum_{k=0}^{\infty} (-1)^k \, t^k \qquad (3.27)$$

thereby establishing the equivalence of the binomial coefficient and the phase factor in (3.26). More formally we could write

$$\binom{-\alpha}{k} = \frac{\Gamma(1-\alpha)}{\Gamma(k+1)\Gamma(1-k-\alpha)}$$

$$= (-1)^k \frac{\Gamma(k+\alpha)}{\Gamma(k+1)\Gamma(\alpha)} = (-1)^k \binom{k+\alpha-1}{k} \qquad (3.28)$$

so that with $\alpha = 1$ the binomial coefficient reduces to a phase, as we saw in (3.27). We have used the generalized Leibniz rule a number of times in our discussion. It is past time that we proved it to be true for fractional derivatives.

3.1.2 The Generalized Leibniz Rule

When the Index Is Not an Integer The classical rule of Leibniz for taking the derivative of the product of two functions is given by the binomial expansion

$$D_t^\alpha[fg] = \sum_{k=0}^{\alpha} \binom{\alpha}{k} D_t^{\alpha-k}[f] \, D_t^k[g], \qquad (3.29)$$

where α is an integer. Therefore we can write

$$
\begin{aligned}
D_t \left[fg \right] &= f D_t \left[g \right] + g D_t \left[f \right] \\
D_t^2 \left[fg \right] &= f D_t^2 \left[g \right] + 2 D_t \left[f \right] D_t \left[g \right] + g D_t^2 \left[f \right] \\
D_t^3 \left[fg \right] &= f D_t^3 \left[g \right] + 3 D_t^2 \left[f \right] D_t \left[g \right] \\
&\quad + 3 D_t \left[f \right] D_t^2 \left[g \right] + g D_t^3 \left[f \right]
\end{aligned}
$$

and the binomial coefficient keeps track of how many factors of each kind contribute to a derivative of a given order. Of course, it is not so simple to write out the derivatives when α is not an integer. Therefore we want to generalize (3.29) to the fractional-α case, that is, to both the fractional integral and fractional derivative cases. For the moment let us assume that we could directly generalize (3.29) to the noninteger α case:

$$
D_t^\alpha \left[fg \right] = \sum_{k=0}^{\infty} \binom{\alpha}{k} D_t^{\alpha-k} \left[f \right] D_t^k \left[g \right] . \tag{3.30}
$$

Note that only the upper limit of the sum is changed in going from (3.29) to (3.30).

Some Confusion as to Order   What is wrong with expression (3.30)? For one thing the left-hand side does not depend on the order of the functions f and g, whereas on the right hand side we have a fractional derivative of f, but an integer derivative of g. Therefore on the right-hand side it appears that we can not interchange the two functions. But this is suspicious because we could just as easily have changed the role of the two functions on the left and thereby obtained the fractional derivative of g and the integer derivative of f on the right. This ambiguity should be a warning flag regarding the interpretation of problems we may encounter in fractional calculus calculations. We need to show that the order on the right-hand side of (3.30) is immaterial, just as it is on the left-hand side.

Series Representation of Function   We want to interpret the generalization of the Leibniz rule to fractional order by considering the functions $f\left(t\right)$ and $g\left(t\right)$ to be analytic on an interval (a, b). Therefore, we can represent either function as a convergent power series of the form

$$
f\left(t\right) = \sum_{k=0}^{\infty} \frac{(-1)^k}{\Gamma\left(k+1\right)} \left(x - t\right)^k f^{(k)}\left(x\right), \tag{3.31}
$$

where $f^{(k)}\left(x\right)$ is the integer k-derivative of the function $f\left(t\right)$ evaluated at $t = x$. We now wish to consider the fractional derivative of $f\left(t\right)$,

$$D_t^\alpha \left[f\left(t\right) \right], \ \alpha > 0, \tag{3.32}$$

which we can also write as

$$D_t^\alpha \left[f\left(t\right) \right] = \left(\frac{d}{dt} \right)^{\left([\alpha]+1 \right)} D_t^{\{\alpha\}-1} \left[f\left(t\right) \right], \tag{3.33}$$

where $[\alpha]$ is the integer part of α and $\{\alpha\}$ is the noninteger part of α, such that

$$\alpha = [\alpha] + \{\alpha\}. \tag{3.34}$$

In (3.33) we see that the first operation is an integer-valued derivative, whereas the second operation is a fractional integral since $\{\alpha\} < 1$. This separation is necessary in order to carry out the proof below.

Series Representation of Fractional Derivative Samko et al. [7] show that a self-consistent definition of the operation of the fractional derivative on an analytic function is given by

$$D_t^\beta \left[f\left(t\right) \right] = \sum_{k=0}^\infty \binom{\beta}{k} \frac{t^{k-\beta}}{\Gamma\left(k+1-\beta\right)} f^{(k)}\left(t\right), \ t \in (a,b). \tag{3.35}$$

Using (3.35) in (3.33) results in

$$D_t^\alpha \left[f\left(t\right) \right] = \left(\frac{d}{dt} \right)^{\left([\alpha]+1 \right)} \sum_{k=0}^\infty \binom{\{\alpha\}-1}{k} \frac{t^{k-\{\alpha\}+1}}{\Gamma\left(k+2-\{\alpha\}\right)} f^{(k)}\left(t\right). \tag{3.36}$$

Samko et al.[7] now carry out a term by term integer-derivative of the factors in the series and therefore the derivative commutes with the sum, yielding

$$D_t^\alpha \left[f\left(t\right) \right] = \sum_{k=0}^\infty \binom{\{\alpha\}-1}{k} \left(\frac{d}{dt} \right)^{\left([\alpha]+1 \right)} \frac{t^{k-\{\alpha\}+1}}{\Gamma\left(k+2-\{\alpha\}\right)} f^{(k)}\left(t\right). \tag{3.37}$$

But we can use the standard Leibniz rule for the product of factors inside the sum to write

$$\begin{aligned}
D_t^\alpha \left[f\left(t\right) \right] &= \sum_{k=0}^\infty \binom{\{\alpha\}-1}{k} \sum_{n=0}^{[\alpha]+1} \binom{[\alpha]+1}{n} \frac{t^{k+n-\alpha} f^{(k+n)}\left(t\right)}{\Gamma\left(k+n-\alpha+1\right)} \\
&= \sum_{k,n=0}^\infty \binom{\{\alpha\}-1}{k} \binom{[\alpha]+1}{n} \frac{t^{k+n-\alpha} f^{(k+n)}\left(t\right)}{\Gamma\left(k+n-\alpha+1\right)}.
\end{aligned} \tag{3.38}$$

We now introduce the dummy variable $j = n + k$ and changing the order of summations obtain

$$D_t^\alpha [f(t)] = \sum_{j=0}^{\infty} \left[\sum_{n=0}^{j} \binom{\{\alpha\}-1}{n} \binom{[\alpha]+1}{j-n} \right] \frac{t^{j-\alpha}}{\Gamma(j-\alpha+1)} f^{(j)}(t) \quad (3.39)$$

and using the sum over gamma functions

$$\sum_{n=0}^{j} \binom{\{\alpha\}-1}{n} \binom{[\alpha]+1}{j-n} = \binom{\{\alpha\}+[\alpha]}{j} = \binom{\alpha}{j},$$

(3.39) reduces to

$$D_t^\alpha [f(t)] = \sum_{j=0}^{\infty} \binom{\alpha}{j} \frac{t^{j-\alpha}}{\Gamma(j-\alpha+1)} f^{(j)}(t). \quad (3.40)$$

Note that (3.40) is just relation (3.35) so we have established self-consistency of this expression for the fractional derivative.

Fractional Derivatives of a Product Now let us turn our attention back to the Leibniz rule and make use of (3.40). Consider the fractional derivative of the product of two functions

$$D_t^\alpha [fg] = \sum_{j=0}^{\infty} \binom{\alpha}{j} \frac{t^{j-\alpha}}{\Gamma(j-\alpha+1)} (fg)^{(j)}, \quad (3.41)$$

where the (j)-superscript denotes the derivative of the product of the functions. Now apply the classical Leibniz rule to the right-hand side of (3.41),

$$D_t^\alpha [fg] = \sum_{k=0}^{\infty} \binom{\alpha}{k} \frac{t^{k-\alpha}}{\Gamma(k-\alpha+1)} \sum_{j=0}^{\infty} \binom{k}{j} f^{(k-j)} g^{(j)} \quad (3.42)$$

and introduce an auxiliary index $m = k - j$ so that

$$D_t^\alpha [fg] = \sum_{j=0}^{\infty} \sum_{m=0}^{\infty} \binom{\alpha}{m+j} \binom{m+j}{j} \frac{t^{m+j-\alpha}}{\Gamma(m+j-\alpha+1)} f^{(m)} g^{(j)}. \quad (3.43)$$

Now we can use the relation between binomial coefficients

$$\binom{\alpha}{m+j} \binom{m+j}{j}$$

$$= \frac{\Gamma(\alpha+1)\Gamma(m+j+1)}{\Gamma(m+j+1)\Gamma(\alpha-m-j+1)} \cdot \frac{1}{\Gamma(j+1)\Gamma(m+1)}$$

$$= \frac{\Gamma(\alpha+1)}{\Gamma(m+1)\Gamma(\alpha-m-j+1)\Gamma(j+1)} \cdot \frac{1}{}$$

$$= \frac{\Gamma(\alpha+1)}{\Gamma(m+1)\Gamma(\alpha-m)} \frac{\Gamma(\alpha-m)}{\Gamma(\alpha-m-j+1)\Gamma(j+1)}$$

$$= \left(\begin{array}{c} \alpha \\ m \end{array} \right) \left(\begin{array}{c} \alpha-m \\ j \end{array} \right)$$

so that (3.43) becomes

$$D_t^\alpha [fg] = \sum_{j=0}^\infty \sum_{m=0}^\infty \left(\begin{array}{c} \alpha \\ m \end{array} \right) \left(\begin{array}{c} \alpha-m \\ j \end{array} \right) \frac{t^{m+j-\alpha}}{\Gamma(m+j-\alpha+1)} D_t^m [f] D_t^j [g].$$

$$(3.44)$$

Using (3.40) we can define

$$D_t^{\alpha-m} [g] = \sum_{j=0}^\infty \left(\begin{array}{c} \alpha-m \\ j \end{array} \right) \frac{t^{j-(\alpha-m)}}{\Gamma(j-(\alpha-m)+1)} D_t^j [g]$$

which when substituted into (3.44) yields

$$D_t^\alpha [fg] = \sum_{m=0}^\infty \left(\begin{array}{c} \alpha \\ m \end{array} \right) D_t^m [f] D_t^{\alpha-m} [g], \qquad (3.45)$$

the generalized Leibniz rule. However, we again point out the lack of symmetry in the derivatives of the two functions in (3.45), but it can be shown that the two functions can be interchanged without changing the value of the fractional derivative.

An Algebraic Function In the general situation the fractional derivative of a function is a series. However, there are some cases where it is possible to express the result in terms of elementary functions. It is not our purpose here to provide an exhaustive list of the fractional derivatives of functions, but it may be useful to see how such expressions are constructed from the definitions provided. An example is given by the function $f(t) = t^\alpha (a+bt)^\beta$, where applying the generalized Leibniz rule (3.45) we obtain

$$D_t^\mu \left[t^\alpha (a+bt)^\beta \right] = \sum_{k=0}^\infty \left(\begin{array}{c} \mu \\ k \end{array} \right) D_t^{\mu-k} [t^\alpha] D_t^k \left[(a+bt)^\beta \right] \qquad (3.46)$$

so that in terms of the fractional derivative of a constant and the integer k derivative of the function we have

$$D_t^\mu \left[t^\alpha (a+bt)^\beta \right] = \sum_{k=0}^\infty \left(\begin{array}{c} \mu \\ k \end{array} \right) \frac{\Gamma(\alpha+1) t^{\alpha+k-\mu}}{\Gamma(\alpha+k-\mu+1)} \beta(\beta-1) \cdots$$

$$\times (\beta-k-1) b^k (a+bt)^{\beta-k}. \qquad (3.47)$$

The multiplicative factors in (3.47) may be expressed in terms of gamma functions

$$\Gamma(\beta+1) = \beta\Gamma(\beta) = \beta(\beta-1)\cdots(\beta-k-1)\Gamma(\beta-k)$$

so that

$$D_t^\mu\left[t^\alpha(a+bt)^\beta\right] = b^{\mu-\alpha}\sum_{k=0}^\infty \left(\begin{array}{c}\mu\\k\end{array}\right)\frac{(bt)^{\alpha+k-\mu}\,\Gamma(\alpha+1)}{\Gamma(\alpha+k-\mu+1)}$$
$$\times\frac{\Gamma(\beta+1)}{\Gamma(\beta+1-k)}(a+bt)^{\beta-k}. \tag{3.48}$$

Now if we set $\mu-\alpha-k = \beta+1-k$, which implies $\mu = \alpha+\beta+1$ as a constraint on the indices, then we use the expression between gamma function

$$\Gamma(\beta+1-k)\,\Gamma(k-\beta) = \frac{\pi}{\sin\pi(\beta-k)}$$

to obtain, using $\sin\pi(\beta-k) = (-1)^k\sin\pi\beta$,

$$D_t^\mu\left[t^\alpha(a+bt)^\beta\right] = \frac{\Gamma(\beta+1)\,\Gamma(\alpha+1)\,[a+bt]^\beta\sin\pi(\mu-\alpha)}{\pi}\frac{}{t^{\mu-\alpha}}$$
$$\times\sum_{k=0}^\infty\left(\begin{array}{c}\mu\\k\end{array}\right)\frac{(-bt)^k}{(a+bt)^k}. \tag{3.49}$$

We sum the geometric series using the binomial relation

$$\sum_{k=0}^\infty\left(\begin{array}{c}\mu\\k\end{array}\right)z^k = (1+z)^\mu,$$

where $z = -bt(a+bt)^{-1}$, to obtain from (3.49)

$$D_t^\mu\left[t^\alpha(a+bt)^\beta\right] = \frac{\Gamma(\beta+1)\,\Gamma(\alpha+1)\,[a+bt]^\beta\sin\pi\mu}{\pi}\frac{}{t^{\mu-\alpha}}\left[1-\frac{bt}{a+bt}\right]^\mu$$
$$= \frac{\Gamma(\beta+1)\,\Gamma(\alpha+1)\,a^\mu[a+bt]^{\beta-\mu}\sin\pi(\mu-\alpha)}{\pi}\frac{}{t^{\mu-\alpha}}. \tag{3.50}$$

In a similar way we can use $\beta = \mu-\alpha-1$ to obtain

$$D_t^\mu\left[t^\alpha(a+bt)^\beta\right] = \frac{\Gamma(\mu-\alpha)\,\Gamma(\alpha+1)}{\pi}\frac{a^\mu\sin\pi(\mu-\alpha)}{[a+bt]^{\alpha+1}\,t^{\mu-\alpha}} \tag{3.51}$$

which reduces to

$$D_t^\mu\left[t^\alpha(a+bt)^\beta\right] = \frac{\Gamma(\alpha+1)}{\Gamma(\mu-\alpha+1)}\frac{a^\mu t^{\alpha-\mu}}{[a+bt]^{\alpha+1}}. \tag{3.52}$$

In the Appendices we have included other fractional derivatives of functions that have been obtained in similar ways.

3.2 The Generalized Exponential Function

The Formalism Is a Shorthand Representation The fractional calculus would not be interesting as a modeling tool for physical scientists unless it provided a shorthand for the complexity encountered in naturally occurring phenomena. The first function that is of interest in this regard is the generalization of the exponential function. We show that the generalized exponential function is to the fractional calculus what the ordinary exponential is to the calculus of Newton, which is to say, that the generalized exponential appears repeatedly in the solutions to the fractional equations of motion for complex systems.

The Fractional Exponential We now turn our attention to the fractional derivative of the exponential function e^t, which when expressed in terms of an infinite series, yields

$$D_t^\mu \left[e^t \right] = D_t^\mu \left[\sum_{k=0}^{\infty} \frac{t^k}{k!} \right] = \sum_{k=0}^{\infty} \frac{t^{k-\mu}}{\Gamma(k+1-\mu)} \equiv E_\mu^t, \tag{3.53}$$

where we define the generalized exponential function, E_μ^t, by the series in (3.53). Note that this definition of the generalized exponential is closely related to, but different from, the definition of the generalized exponential function used by Miller and Ross [4]

$$E_t(-\mu, a) = t^{-\mu} \sum_{k=0}^{\infty} \frac{(at)^k}{\Gamma(k-\mu+1)}, \tag{3.54}$$

which we discuss subsequently. For the time being let us explore the meaning of the generalized exponential given by (3.53). Introducing a constant a into the exponential we obtain from (3.53),

$$D_t^\mu \left[e^{at} \right] = D_t^\mu \left[\sum_{k=0}^{\infty} \frac{(at)^k}{k!} \right] = \sum_{k=0}^{\infty} \frac{a^k t^{k-\mu}}{\Gamma(k+1-\mu)} = a^\mu E_\mu^{at}. \tag{3.55}$$

Thus, we can relate our generalized exponential to that of Miller and Ross by means of the equation

$$E_\mu^{at} = a^{-\mu} E_t(-\mu, a). \tag{3.56}$$

Note that when $\mu = 1$ the fractional derivative in (3.53) becomes the ordinary first derivative and the generalized exponential is the usual exponential.

Series Representation of the Generalized Exponential A second series expansion for the generalized exponential can be obtained using (3.45) as follows

$$D_t^\mu \left[e^t \right] = \sum_{k=0}^{\infty} \binom{\mu}{k} D_t^{\mu-k} \left[t^0 \right] D_t^k \left[e^t \right] = e^t \sum_{k=0}^{\infty} \binom{\mu}{k} \frac{t^{k-\mu}}{\Gamma(k+1-\mu)} \tag{3.57}$$

so that we have

$$e^{-t} E_\mu^t = \sum_{k=0}^{\infty} \binom{\mu}{k} \frac{t^{k-\mu}}{\Gamma(k+1-\mu)}. \tag{3.58}$$

If the real indexed derivative is integer valued, that is, $\mu \in \mathcal{N}$, then we have the coincidence of the generalized exponential and the Napierian exponential, $E_\mu^t = e^t$. In fact if μ is an integer the gamma function in Equation (3.58) gives

$$\Gamma(k+1-\mu) = \Gamma(-m) \text{ with } m = 0, 1, 2, \cdots, \mu - 1$$

which diverges and therefore gives a zero on the right-hand side of (3.58). Only when $k = \mu$ do we have a finite value for the gamma function, so that

$$e^{-t} E_\mu^t = \sum_{k=0}^{\mu} \binom{\mu}{k} \frac{t^{k-\mu}}{\Gamma(k+1-\mu)} = \binom{\mu}{\mu} = 1, \ \mu = \text{integer}, \tag{3.59}$$

as we said.

Negative Indexed Generalized Exponential Now consider the situation when the real-valued index in (3.58) is a negative integer $\mu = -1, -2, \cdots$. Starting from the definition (3.53) we have for $\mu = -1$,

$$E_{-1}^t = D_t^{-1}[e^t] = \sum_{k=0}^{\infty} \frac{t^{k+1}}{\Gamma(k+2)}$$

so that reindexing the series we have

$$E_{-1}^t = \sum_{j=1}^{\infty} \frac{t^j}{\Gamma(j+1)} = e^t - 1. \tag{3.60}$$

Of course, we can also write the negatively indexed generalized exponential as the first-order integral

$$E_{-1}^t = D_t^{-1}[e^t] = \int_0^t e^\tau d\tau = e^t - 1. \tag{3.61}$$

For the case $\mu = -2$ we can use the definition, (3.53), to write

$$E_{-2}^t = D_t^{-2}[e^t] = \sum_{k=0}^{\infty} \frac{t^{k+2}}{\Gamma(k+3)}$$

so that again reindexing the series we can write

$$E_{-2}^t = \sum_{j=2}^{\infty} \frac{t^j}{\Gamma(j+1)} = e^t - 1 - t.$$

Here again we can also write the negatively indexed generalized exponential as the second-order integral

$$E^t_{-2} = D^{-2}_t \left[e^t \right] = \int_0^t d\tau_2 \int_0^{\tau_2} e^{\tau_1} d\tau_1 = \int_0^t \left(e^{\tau_2} - 1 \right) d\tau_2 = e^t - 1 - t. \quad (3.62)$$

We can continue this process, using either the series definition or the integral representation, to the nth integral to obtain

$$E^t_{-n} = D^{-n}_t \left[e^t \right] = e^t - \sum_{k=0}^{n-1} \frac{t^k}{k!}, \quad (3.63)$$

and taking the limit $n \to \infty$ we have

$$\lim_{n \to \infty} D^{-n}_t \left[e^t \right] = 0. \quad (3.64)$$

Equation (3.64) means that the exponential function e^t is a $C(\infty)$ constant.

A Second Generalized Exponential Now let us consider the fractional derivative of the negative exponential function e^{-t}. We do this by considering the fractional derivative

$$D^\mu_t \left[e^{at} \right] = D^\mu_t \left[\sum_{k=0}^\infty \frac{(at)^k}{k!} \right] = a^\mu \sum_{k=0}^\infty \frac{(at)^{k-\mu}}{\Gamma(k+1-\mu)} \equiv a^\mu E^{at}_\mu, \quad (3.65)$$

where a is an arbitrary constant. If we choose $a = -1$ we can use (3.65) to write

$$D^\mu_t \left[e^{-t} \right] = (-1)^\mu E^{-t}_\mu = e^{i\pi\mu} E^{-t}_\mu \quad (3.66)$$

which we can further use to define another generalized exponential function

$${}^* E^{-t}_\mu \equiv e^{i\pi\mu} E^{-t}_\mu. \quad (3.67)$$

In series form we write this new generalized exponential function as

$$D^\mu_t \left[e^{-t} \right] = {}^* E^{-t}_\mu = \sum_{k=0}^\infty \frac{(-1)^k t^{k-\mu}}{\Gamma(k+1-\mu)}. \quad (3.68)$$

Both (3.67) and (3.68) make it abundantly clear that the function ${}^* E^{-t}_\mu$ is not E^t_μ calculated with $-t$; the new function differs from the old by the phase factor $e^{i\pi\mu}$. We can use Theorem 2 to write

$$D^\mu_t \left[e^{-t} \right] = (-1)^\mu D^\mu_x \left[e^x \right]_{x=-t} = e^{i\pi\mu} E^{-t}_\mu \quad (3.69)$$

just as we obtained in (3.66) and here E^{-t}_μ is a function in the complex field. For real functions it is convenient to define ${}^* E^{-t}_\mu$ as E^t_μ calculated with $-t$, but in order to do this we need to define the generalized exponential as

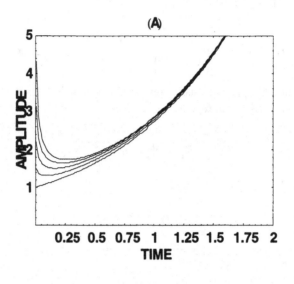

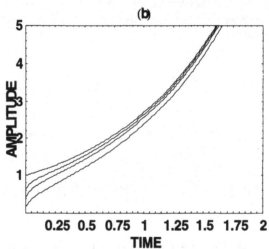

Figure 3.1: *The generalized exponential function is plotted for both positive and negative values of the fractional index. In the top panel (a) we plot E_μ^t for $\mu = 0.0$ to 0.4 in steps of 0.1, with the top curve having the largest value. We see that all these values asymptotically approach the ordinary exponential function. In the lower panel (b) we plot E_μ^t for $\mu = 0.0$ and -0.4 in steps of -0.1 and again we have asymptotic exponential growth.*

$$E_\mu^t = |t|^{-\mu} \sum_{k=0}^{\infty} \frac{t^k}{\Gamma(k+1-\mu)}, \tag{3.70}$$

where it is possible to evaluate this function for both positive and negative values of the independent variable. In general, however, when we are dealing with complex functions we use the first definition of the generalized exponential given by (3.53).

Comparing the Two Definitions To emphasize this last observation concerning the difference between $^*E_{-n}^{-t}$ and E_{-n}^{-t} consider the case $\mu = -n = -1$. We can write the generalized exponential function as

$$E_{-1}^{-t} = D_{-t}^{-1}\left[e^{-t}\right] = t\sum_{k=0}^{\infty}\frac{(-t)^k}{\Gamma(k+2)} = e^{-t} - 1 \qquad (3.71)$$

and the related function by

$$^*E_{-1}^{-t} = D_t^{-1}\left[e^{-t}\right] = \sum_{k=0}^{\infty}\frac{(-t)^{k+1}}{\Gamma(k+2)} = -e^{-t} + 1 = -E_{-1}^{-t}. \qquad (3.72)$$

We can see that the two functions are therefore related by the phase factor $e^{i\pi} = -1$.

Generalized Logarithm For completeness we define the inverse of the generalized exponential function as the generalized logarithm, which is to say, the function that satisfies the relation

$$\ln_\mu E_\mu^t = t. \qquad (3.73)$$

We plot the generalized exponential function for both positive and negative values of the index in Figure 3.1 and the generalized logarithm in Figure 3.2. The global behavior of the generalized exponential function is clear from Figure 3.1. For $\mu > 0$ we see that the function is concave upward and asymptotically diverges

$$\lim_{t\to\infty} E_{\mu>0}^t = \infty. \qquad (3.74)$$

Asymptotic Values For $\mu < 0$ we see that the curvature of the function changes from very small values of the independent variables to intermediate values to large values. Finally, the generalized exponential, again, asymptotically diverges

$$\lim_{t\to\infty} E_{\mu<0}^t = \infty \qquad (3.75)$$

but at essentially the same rate as in the $\mu > 0$ case.

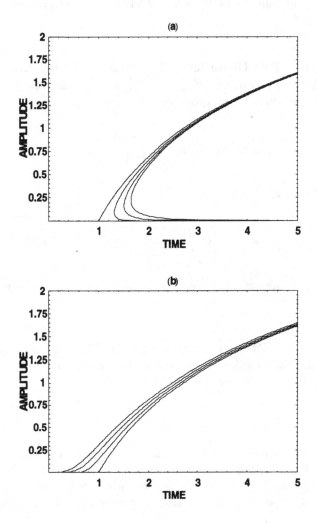

Figure 3.2: *The generalized logarithm is plotted for both positive and negative values of the fractional index. In (a) we plot the generalized logarithm of t for $\mu = 0, 0.1, 0.2$ and 0.3. In (b) we plot the generalized logarithm for $\mu = 0, -0.1, -0.2$ and -0.3.*

Generalized Logarithms and How They Look In Figure 3.2 we graph the generalized logarithmic function for both positive and negative values of the real index of the fractional derivative compared with the $\mu = 0$ function. In the top panel we see that for $\mu > 0$ the generalized logarithm is not necessarily a single valued function, but can be almost hyperbolic in shape. In the lower panel we see that for $\mu < 0$ the generalized logarithm is single-valued and is a monotonically increasing function of the independent variable.

Rate Equation   We now have some idea of the behavior of the generalized exponential, but not much insight into the kind of physical phenomena such functions can be used to model. After all is said and done, what we want to understand are the properties of complex phenomena, not just the mathematics of exotic functions. To gain some perspective let us review one of the first non-trivial functions we study in the ordinary calculus, that being the exponential. In a physical system we obtain the exponential function $e^{\lambda t}$ with λ real, when the percentage change in a dynamical variable is constant over time. If we define the function $X(t) = e^{\lambda t}$, then its time derivative yields

$$\frac{D_t X(t)}{X(t)} = \lambda, \tag{3.76}$$

so that λ is the constant growth rate. Now what happens when we take the time derivative of the generalized exponential?

Generalized Rate Equation   To answer this question we first calculate the D_t^p derivative of E_μ^t, with $p = 0, 1, 2, ...$ Thus

$$D_t^p \left[E_\mu^{\lambda t} \right] = \sum_{j=0}^{\infty} \frac{\lambda^{j-\mu}}{\Gamma(1+j-\mu)} D_t^p \left[t^{j-\mu} \right]$$

$$= (\lambda t)^{-\mu} \lambda^p \sum_{j=0}^{\infty} \frac{(\lambda t)^{j-p}}{\Gamma(1+j-\mu-p)} \tag{3.77}$$

after some rearrangement of terms. Equation(3.77) can also be written as

$$D_t^p \left[E_\mu^{\lambda t} \right] = \lambda^p E_{\mu+p}^{\lambda t}, \tag{3.78}$$

where the integer derivative shifts the index of the generalized exponential and scales the overall amplitude. Now consider the fractional operator $D_t^{\mu-1}$ acting on the exponential

$$D_t^{\mu-1} \left[\lambda e^{\lambda t} \right] = D_t^\mu \left[e^{\lambda t} - 1 \right], \tag{3.79}$$

where the integration D_t^{-1} has been carried out on the right-hand side of (3.79). On the left-hand side of (3.79) we can use (3.78) to write

$$\lambda^\mu E_{\mu-1}^{\lambda t} = \lambda^\mu E_\mu^{\lambda t} - \frac{t^{-\mu}}{\Gamma(1-\mu)}. \tag{3.80}$$

Now taking the time derivative of (3.80) yields

$$D_t \left[E_{\mu-1}^{\lambda t} \right] = D_t \left[E_\mu^{\lambda t} \right] + \frac{\lambda \mu}{\Gamma(1-\mu)(\lambda t)^{\mu+1}} \tag{3.81}$$

but again using (3.78) on the left-hand side of (3.81), after some rearrangement, we obtain

$$\frac{D_t\left[E_\mu^{\lambda t}\right]}{E_\mu^{\lambda t}} = \lambda - \frac{\lambda\mu}{\Gamma\left(1-\mu\right)E_\mu^{\lambda t}\left(\lambda t\right)^{\mu+1}}. \qquad (3.82)$$

Thus, rather than the constant rate of change (see, for example, (3.76)) we have an additional time-dependent term that decreases as the product of an inverse power law and a generalized exponential with increasing time and diverges to infinity as $t \to 0$.

Inhomogeneous Rate Equation   We could also write (3.82) as the inhomogeneous rate equation

$$D_t\left[X\left(t\right)\right] + \frac{\lambda\mu}{\Gamma\left(1-\mu\right)\left(\lambda t\right)^{\mu+1}} = \lambda X\left(t\right), \qquad (3.83)$$

where $X\left(t\right)$ is an unknown function. In the ordinary calculus we would protest such a differential equation, because with $|\mu| < 1$, it has an essential singularity at the origin, indicating that as we approach $t = 0$ the derivative of the function diverges to infinity. However, in later lectures we review how to solve equations such as (3.83) and generalizations of it using Fourier and Laplace transforms.

Asymptotic Solution   Of course, we can solve (3.83) near the origin in a straightforward way. As $t \to 0$ the right-hand side of (3.83) becomes negligible, compared with the inhomogeneous term, so that

$$D_t\left[X\left(t\right)\right] + \frac{\lambda\mu}{\Gamma\left(1-\mu\right)\left(\lambda t\right)^{\mu+1}} \approx 0. \qquad (3.84)$$

Equation (3.84) can be directly integrated to yield

$$X\left(t\right) \approx \frac{1}{\Gamma\left(1-\mu\right)\left(\lambda t\right)^{\mu}}, \qquad (3.85)$$

as it should for the generalized exponential in the vicinity of the origin. One might argue that the singularity could be handled by multiplying the solution by a sufficiently large monomial t^α. To examine this conjecture consider the operator D_t^p acting on the product $t^\alpha E_\mu^{\lambda t}$ for $p = 0, 1, 2, \ldots$ and α noninteger:

$$D_t^p\left[t^\alpha E_\mu^{\lambda t}\right] = \sum_{j=0}^{p} \binom{p}{j} D_t^j\left[t^\alpha\right] D_t^{p-j}\left[E_\mu^{\lambda t}\right]$$

and rearranging terms on the right-hand side of this equation yields

$$D_t^p\left[t^\alpha E_\mu^{\lambda t}\right] = \sum_{j=0}^{p} \binom{p}{j} \frac{\Gamma\left(1+\alpha\right)t^{\alpha-j}\lambda^{p-j}}{\Gamma\left(1+\alpha-j\right)} E_{\mu+p-j}^{\lambda t}. \qquad (3.86)$$

From this general expression with $p = 1$ we obtain

$$D_t \left[t^\alpha E_\mu^{\lambda t} \right] = \lambda t^\alpha E_{\mu+1}^{\lambda t} + \alpha t^{\alpha-1} E_\mu^{\lambda t} \tag{3.87}$$

so that introducing $Y(t) = t^\alpha E_\mu^{\lambda t}$ and using (3.80) this reduces to

$$D_t \left[Y(t) \right] = \left(\lambda + \frac{\alpha}{t} \right) Y(t) + \frac{\lambda}{\Gamma(-\mu)(\lambda t)^{\mu+1}}. \tag{3.88}$$

The rate constant, the coefficient of the dependent variable on the right-hand side of (3.87), still diverges to infinity as $t \to 0$. Near the origin we can neglect $\lambda Y(t)$ relative to the diverging terms and replace (3.88) with

$$D_t \left[e^{-\alpha \ln t} Y(t) \right] \approx \frac{\lambda}{\Gamma(-\mu)(\lambda t)^{\mu+1}}$$

which is immediately solved to give

$$Y(t) \approx \frac{t^{\alpha-\mu}}{\Gamma(1-\mu)\lambda^\mu}. \tag{3.89}$$

Thus, if $\alpha \geq \mu$, the divergence at the origin is quenched, as one might expect. Note that this quenching is a balance between the diverging growth rate and the diverging inhomogeneity in (3.88).

Time Varying Rate Coefficient Thus, rather than a process that increases at a constant rate, the phenomenon described by $E_\mu^{\lambda t}$ grows rapidly at early times in a way dependent on the parameter μ, approaching the constant growth rate of the ordinary exponential asymptotically. An example of the former process is simple Malthusian population increase, or physically the average number of radioactive fragments obtained from a source in a given interval of time. It seems, however, that most processes do not have such regular behavior throughout time. Or at least not those described by the above equations.

Bursting Initial Driver The apparently exponentially growing process described by the generalized exponential may, in fact, start out with an initial burst of activity triggered by some event that exceeds the exponential rate of growth. After some time the process settles down to the more familiar pattern. The smaller the parameter μ, the shorter the time interval over which the generalized exponential deviates from the ordinary exponential, and the shorter the time over which it takes the phenomena to adjust to the regularity of a constant rate of growth.

Complex Generalized Exponential Now let us consider the analogue of the negative exponential. Of course, E_μ^{-t} is complex since it contains a factor $(-1)^\mu$ which has both a real and imaginary part. The function, $^*E_\mu^{-t}$, on the other hand, is defined to be real as given by (3.68). If we restrict our analysis to real fields then the exponential is defined by (3.70) so that its time derivative is given by

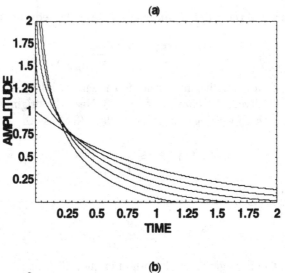

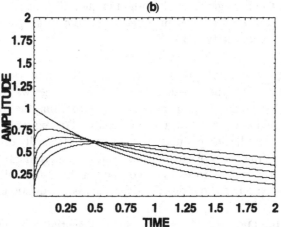

Figure 3.3: *The generalized negative exponential function* $^*E_\mu^{-t}$ *is plotted for positive and negative values of the fractional index. In (a) the* $\mu = 0$ *curve is the ordinary exponential and the generalized exponential is graphed for* $\mu = 0.0$ *to 0.4 in steps of 0.1. In (b) the ordinary exponential is compared with the curves for* $\mu = 0.0$ *to* -0.4 *in steps of* -0.1.

$$D_t \left[E_\mu^{-\lambda t} \right] = \sum_{j=0}^{\infty} \frac{(-1)^j \, \lambda^{j-\mu}}{\Gamma(j-\mu)} t^{j-1-\mu} \qquad (3.90)$$

and changing indices in the sum and reordering terms gives us

$$D_t \left[E_\mu^{-\lambda t} \right] = -\lambda E_\mu^{-\lambda t} + \frac{\lambda}{\Gamma(-\mu)(\lambda t)^{\mu+1}} \qquad (3.91)$$

as the equation of evolution. As $t \to 0$ the second term on the right-hand side of (3.91) dominates and we again obtain (3.85). This behavior is observed in Figure 3.3 with the upward turn of the solution for $\mu > 0$ for increasing time and the downward turn of the solution for $\mu < 0$.

Properties of Negative Generalized Exponentials We observe in Figure 3.3 that $^*E_\mu^{-t}$ decays more rapidly than the exponential e^{-t} for $\mu > 0$, and has a persistent inverse power-law tail above the decaying exponential for $\mu < 0$. Furthermore, unlike E_μ^t, with $\mu, t > 0$ which decays rapidly at early times then increases exponentially as $t \to \infty$, the generalized exponential $^*E_\mu^{-t}$ with $\mu, t > 0$, monotonically decreases from a large (divergent) initial value to zero. Also, unlike the negative exponential that does not achieve zero in finite time, the generalized negative exponential does achieve a zero value in finite time. For $\mu < 0$ and $t > 0$ the negative generalized exponential is nonmonotonic as we see in Figure 3.3, whereas for $\mu < 0$ and $t > 0$ the positive generalized exponential is monotonically increasing.

Thus, we see that the positive generalized exponential can be mistaken for the ordinary exponential asymptotically, whereas no such confusion can arise in the case of the negative exponential.

3.2.1 Generalized Trigonometric Functions

We Can Generalize Many of Our Concepts Now that we have a generalization of the complex exponential function, it should, of course, be possible to construct a generalization of the Euler relation, that being,

$$E_\mu^{it} = \cos_\mu t + i \sin_\mu t. \tag{3.92}$$

From the real part of (3.92) we obtain the equation for the generalized cosine function

$$\cos_\mu t = \frac{1}{2} \left(E_\mu^{it} + E_\mu^{-it} \right) \tag{3.93}$$

and from the imaginary part of (3.92) we obtain the equation for the generalized sine function

$$\sin_\mu t = \frac{1}{2i} \left(E_\mu^{it} - E_\mu^{-it} \right). \tag{3.94}$$

We can then extend these definitions even further and construct the generalized tangent function as well

$$\tan_\mu t \equiv \frac{\sin_\mu t}{\cos_\mu t}. \tag{3.95}$$

Series Expansions   We can also express the generalized sine and generalized cosine functions in series form using the series definition of the generalized exponential. The generalized cosine function is given by

$$\cos_\mu t = \frac{1}{2}\left(E_\mu^{it} + E_\mu^{-it}\right) = \sum_{k=0}^{\infty} \frac{t^{k-\mu}\cos\left[(k-\mu)\,\pi/2\right]}{\Gamma\left(k+1-\mu\right)} \qquad (3.96)$$

and the generalized sine function is given by

$$\sin_\mu t = \frac{1}{2i}\left(E_\mu^{it} - E_\mu^{-it}\right) = \sum_{k=0}^{\infty} \frac{t^{k-\mu}\sin\left[(k-\mu)\,\pi/2\right]}{\Gamma\left(k+1-\mu\right)}. \qquad (3.97)$$

We note in passing that these are not the generalized trigonometric functions defined by Miller and Ross [4]. The present definition of a generalized trigonometric function differs from these earlier definitions by the inclusion of the μ-dependent phase factors.

Negative Parameters   From (3.96) and (3.97) we can see that for integer μ the generalized trigonometric series $\sin_\mu t$ and $\cos_\mu t$ become the ordinary trigonometric functions $\sin t$ and $\cos t$. For negative integer μ we can write, considering, for example, $\cos_\mu t$,

$$\cos_{-m} t = \sum_{k=0}^{\infty} \frac{t^{k+m}}{\Gamma\left(k+1+m\right)}\cos\left[(k+m)\,\pi/2\right], \qquad (3.98)$$

where we have set $\mu = -m$ and m is a positive integer. Making a variable change in the sum, $n = k + m$, (3.98) simplifies to

$$\cos_{-m} t = \sum_{n=m}^{\infty} \frac{t^n}{\Gamma\left(n+1\right)}\cos\left[n\pi/2\right] = \cos t - \sum_{n=0}^{m-1} \frac{t^n\cos\left[n\pi/2\right]}{\Gamma\left(n+1\right)} \qquad (3.99)$$

from which we can infer that for $\mu > -1$ the generalized trigonometric function goes over to the ordinary trigonometric function with the difference between the two being a decreasing power law. The inverse power law arises in (3.99) because $\mu > -1$, so that $m < 1$ and therefore $n \leq 0$. Starting from $\mu = -1$ the generalized trigonometric functions differ from ordinary trigonometric functions by increasing power laws. In particular we have

$$\cos_{-1} t = \cos\, t - 1.$$

The Initial State Is Important   In Figure 3.4 we plot the generalized cosine function for three different values of the μ-index, the $\mu = 0$ being, of course, the cosine function. In this figure we increase μ from 0.1 to 1.5 to try to get a feel for how the various index values change the character of the function. It is clear from the graphs that the predominant effects of all positive μ values are near the origin with the generalized cosine approaching the cosine function as

$t \to \infty$. The generalized cosine diverges at the origin to $+\infty$ for noninteger μ. At the integer value of the index the cosine function is recaptured. The singularity changes sign at $\mu = 2$, but this is not shown here. We find that there is a surge of activity initially, either positive or negative, and out of this formless event emerges the asymptotic oscillatory behavior. This is very different from the classical oscillator that oscillates from its initial excitation such as the plucked string of a harp. To see this, consider the second derivative of the generalized cosine with frequency ω_0

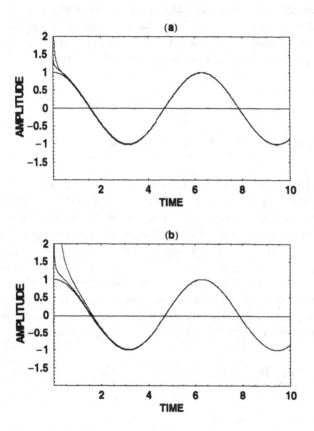

Figure 3.4: *We graph the generalized cosine function and compare it with the ordinary cosine function for a number of positive values of the fractional index. In the upper graph, the lowest curve has $\mu = 0$, the middle curve has $\mu = 0.1$, and the top curve has $\mu = 0.5$. In the lower graph the lowest curve has $\mu = 1$, the middle curve has $\mu = 1.1$, and the upper curve has $\mu = 1.5$.*

$$D_t^2 [\cos_\mu \omega_0 t] = \sum_{k=0}^{\infty} \frac{\omega_0^2 (\omega_0 t)^{k-2-\mu} \cos [(k-\mu) \pi/2]}{\Gamma (k-1-\mu)} \tag{3.100}$$

which after some redefining of terms and introducing $X_\mu (t) \equiv \cos_\mu (\omega_0 t)$, for $\mu > 0$, can be rewritten as

$$D_t^2\left[X_\mu\left(t\right)\right] + \omega_0^2 X_\mu\left(t\right) = \frac{\omega_0^2 \sin\left(\mu\pi/2\right)}{\Gamma\left(-\mu\right)\left(\omega_0 t\right)^{2+\mu}} + \frac{\omega_0^2 \cos\left(\mu\pi/2\right)}{\Gamma\left(-1-\mu\right)\left(\omega_0 t\right)^{2+\mu}}. \qquad (3.101)$$

Driven Harmonic Oscillator   It is clear that (3.101) is the differential equation of motion for a driven harmonic oscillator. However, rather than an isolated pulse on the right-hand side, we have an inverse power-law driving force. The oscillatory response therefore rides atop a slowly decreasing driving force. The single oscillator frequency ω_0 can be replaced by the frequency ω_n, the eigenvalue of the homogeneous oscillator equation, and the solution to the resulting differential equation is the linear superposition of solutions

$$X\left(t\right) = \sum_{n=0}^{\infty} C_n \cos_\mu\left(\omega_n t\right), \qquad (3.102)$$

where the C_n are determined by the initial conditions.

Generalized Cosine   In Figure 3.5 we depict the behavior of the generalized cosine function for negative μ values. Again for small negative values of the index the difference between the cosine and the generalized cosine is most significant in the vicinity of the origin. After a few oscillations the generalized cosine converges back on the cosine. For larger negative values, but still $\mu > -1$, the generalized cosine is strongly distorted for an oscillation, but it converges back to the cosine as $t \to \infty$. On the other hand, when the index is more strongly negative, $-1 \geq \mu \geq -2$ the oscillations appear to ride on a linear ramp of negative slope, and to diverge asymptotically. The $\mu = -1$ curve is tangent to the horizontal axis and the $\mu = -2$ curve is exactly the same.

Inhomogeneous Harmonic Oscillator   In the case $\mu < 0$, say, $\mu = -|\mu|$, we can write (3.101) as

$$D_t^2\left[X_\mu\left(t\right)\right] + \omega_0^2 X_\mu\left(t\right) = -\frac{\omega_0^2 \sin\left(|\mu|\,\pi/2\right)}{\Gamma\left(|\mu|\right)\left(\omega_0 t\right)^{1-|\mu|}} + \frac{\omega_0^2 \cos\left(|\mu|\,\pi/2\right)}{\Gamma\left(-1+|\mu|\right) t^{2-|\mu|}}, \qquad (3.103)$$

where we see that the driving force remains the inverse power law, but the two terms have different signs. The sign of the driving force changes from positive to negative at a time given by

$$t_\mu = \frac{\Gamma\left(|\mu|\right)}{\Gamma\left(-1+|\mu|\right)} \cot\left(|\mu|\,\pi/2\right) \qquad (3.104)$$

and this is the crossover point so evident in Figure 3.5. Equation (3.103) clearly shows the ramping up of the oscillations.

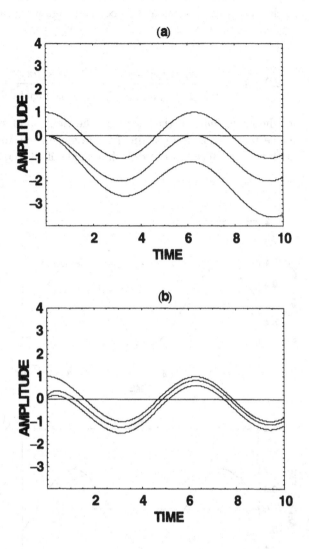

Figure 3.5: *We graph the generalized cosine function and compare it with the ordinary cosine function for different values of the fractional index. In the upper graph the middle curve has $\mu = -1.0$ and the lower curve has $\mu = -1.5$. In the lower graph the middle curve has $\mu = -0.5$ and the lower curve has $\mu = -0.7$.*

Generalized Sine In Figure 3.6 we plot the generalized sine function for a number of different values of the μ-index and compare these with the sine function. In this figure we increase μ from 0.1 to 0.5, from top to bottom, to again develop our intuition for how the various index values change the character of the function. It is clear from the comparisons that as in the case of the generalized cosine for $\mu > 0$, the generalized sine changes most dramatically in the vicinity of the origin, where it diverges, for all positive noninteger values of the index and approaches the sine function as $t \to \infty$. Here again we can construct an inhomogeneous harmonic oscillator equation using $Y_\mu(t) \equiv \sin_\mu(\omega_0 t)$:

$$D_t^2[Y_\mu(t)] + \omega_0^2 Y_\mu(t) = \frac{\omega_0^2 \sin(\mu\pi/2)}{\Gamma(-1-\mu)(\omega_0 t)^{2+\mu}} - \frac{\omega_0^2 \cos(\mu\pi/2)}{\Gamma(-\mu)(\omega_0 t)^{1+\mu}} \qquad (3.105)$$

which aside from the coefficients of the terms on the right-hand side, bears a striking resemblance to (3.101). Here again the inhomogeneous terms diverge at early times, with the first term on the right-hand side of (3.105) dominating for $0 < \mu < 1$.

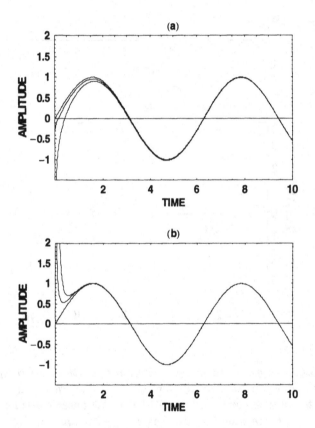

Figure 3.6: *We graph the generalized sine function and compare it with the ordinary sine function for different values of the fractional index. In the upper graph the middle curve has $\mu = 0.1$ are the lower curve has $\mu = 0.5$. In the lower graph the middle curve has $\mu = 1.1$ are the upper curve has $\mu = 1.5$.*

Inhomogeneous Oscillator   In Figure 3.7 we graph the behavior of the generalized sine function for negative values of the index. There is little or no change from the sine function near the origin for all negative values in the region $0 \geq \mu \geq -1.0$. For small negative values of the index there is only a slight shift from the generalized sine to the sine; however, as μ becomes more negative, $\mu < -1.0$, the two functions separate from each nother, not unlike what occurred for the generalized cosine and cosine functions. There is a linear ramp of negative slope leading to a diverging function as $t \longrightarrow \infty$. In the case $\mu < 0$, say, $\mu = -|\mu|$, we can write from (3.105),

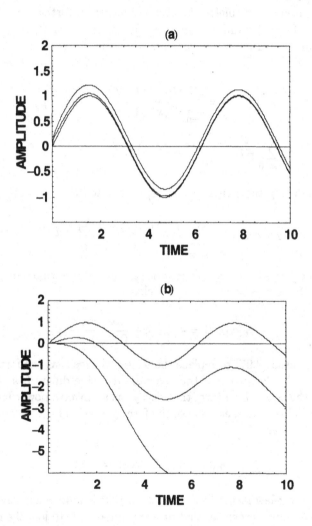

Figure 3.7: *We graph the generalized sine function and compare it with the ordinary sine function for different values of the fractional index. In the upper graph we have for the middle curve $\mu = -0.1$ and for the lower curve $\mu = -0.5$. In the lower graph, note the change in scale. For the middle curve $\mu = -1.5$ and for the lower curve $\mu = -2.0$.*

$$D_t^2 \left[Y_\mu(t)\right] + \omega_0^2 Y_\mu(t) = -\frac{\omega_0^2 \sin\left(|\mu|\,\pi/2\right)}{\Gamma\left(-1+|\mu|\right)\left(\omega_0 t\right)^{2-|\mu|}} - \frac{\omega_0^2 \cos\left(|\mu|\,\pi/2\right)}{\Gamma\left(|\mu|\right)\left(\omega_0 t\right)^{1-|\mu|}}, \quad (3.106)$$

where, here again, the driving force remains the inverse power law.

Derivative of Generalized Cosine   It is also useful to study the derivatives of the generalized trigonometric functions in order to further understand how these periodic functions differ from those in the standard form. Consider the first-order time derivative of the generalized cosine function

$$D_t \left[\cos_\mu t\right] = \sum_{k=0}^{\infty} \frac{(k-\mu)\, t^{k-\mu-1}}{\Gamma(k+1-\mu)} \cos\left[(k-\mu)\,\pi/2\right]$$

$$= \sum_{k=0}^{\infty} \frac{t^{k-\mu-1}}{\Gamma(k-\mu)} \cos\left[(k-\mu)\,\pi/2\right], \quad (3.107)$$

where by reindexing the series, $k = j+1$, we can write

$$D_t \left[\cos_\mu t\right] = \sum_{j=-1}^{\infty} \frac{t^{j-\mu} \cos\left[(j+1-\mu)\,\pi/2\right]}{\Gamma(j+1-\mu)}. \quad (3.108)$$

Separating the $j = -1$ term from the series and using the trigonometric identity $\cos(j+1-\mu)\pi/2 = -\sin(j-\mu)\pi/2$ yields

$$D_t \left[\cos_\mu t\right] = -\sin_\mu t + \frac{\cos\left(\mu\pi/2\right)}{\Gamma(-\mu)\, t^{\mu+1}}, \quad (3.109)$$

where we have used (3.97) to replace the series. We see that the formal relation resulting from the derivative of the generalized cosine differs from that of the derivative of the cosine by a term that decays as an inverse power law in the independent variable. Thus, as $t \to \infty$, the formal relation for the two derivatives approach each other:

$$\lim_{t \to \infty} D_t \left[\cos_\mu t\right] = D_t \left[\cos t\right] = -\sin t.$$

The inverse power-law form of the term in (3.109) is quite suggestive, since the memory in dynamical processes that makes it impossible to join the microscopic and macroscopic descriptions of complex phenomena is exactly of this inverse power-law form. This was precisely the form in the inhomogeneous term in the generalized harmonic oscillator equation.

Derivative of Generalized Sine   Let us now examine the derivative of the generalized sine function

$$D_t \left[\sin_\mu t \right] = \sum_{k=0}^{\infty} \frac{(k - \mu) \, t^{k-\mu-1}}{\Gamma (k + 1 - \mu)} \sin \left[(k - \mu) \, \pi/2 \right]$$

$$= \sum_{k=0}^{\infty} \frac{t^{k-\mu-1}}{\Gamma (k - \mu)} \sin \left[(k - \mu) \, \pi/2 \right], \qquad (3.110)$$

where by reindexing the series, $k = j + 1$, we can write

$$D_t \left[\sin_\mu t \right] = \sum_{j=-1}^{\infty} \frac{t^{j-\mu} \sin \left[(j + 1 - \mu) \, \pi/2 \right]}{\Gamma (j + 1 - \mu)}. \qquad (3.111)$$

Separating the $j = -1$ term from the series and using the trigonometric identity $\sin (j + 1 - \mu) \, \pi/2 = \cos (j - \mu) \, \pi/2$ yields

$$D_t \left[\sin_\mu t \right] = \cos_\mu t - \frac{\sin (\mu\pi/2)}{\Gamma (-\mu) \, t^{\mu+1}}, \qquad (3.112)$$

where we have used (3.96) to replace the series. We see that the formal relation resulting from the derivative of the generalized sine differs from the derivative of the sine by a term that decays as an inverse power law in the independent variable, just as it did for the generalized cosine. Thus, as $t \to \infty$ the formal relations for the two derivatives approach each other:

$$\lim_{t \to \infty} D_t \left[\sin_\mu t \right] = D_t \left[\sin t \right] = \cos t.$$

Here again, the inverse power-law form of the term in (3.112) is quite suggestive.

Generalized Euler Relations We have examined what happens to a generalized trigonometric function when we take an ordinary derivative. Now let us examine what happens to an ordinary trigonometric function when we take a fractional derivative. Consider the fractional derivative of the sine function

$$D_t^\mu \left[\sin t \right] = \frac{1}{2i} \left(D_t^\mu \left[e^{it} \right] - D_t^\mu \left[e^{-it} \right] \right) = \frac{1}{2i} \left(e^{i\mu\pi/2} E_\mu^{it} - e^{-i\mu\pi/2} E_\mu^{it} \right)$$

so that using the Euler relations for both the exponential and generalized exponential and combining terms we obtain

$$D_t^\mu \left[\sin t \right] = \sin (\mu\pi/2) \cos_\mu t + \cos (\mu\pi/2) \sin_\mu t. \qquad (3.113)$$

Equation (3.113) is reminiscent of the trigonometric expansion of $\sin (t + \mu\pi/2)$. In the same way we take the fractional derivative of the cosine function

$$D_t^\mu \left[\cos t \right] = \frac{1}{2} \left(D_t^\mu \left[e^{it} \right] + D_t^\mu \left[e^{-it} \right] \right) = \frac{1}{2} \left(e^{i\mu\pi/2} E_\mu^{it} + e^{-i\mu\pi/2} E_\mu^{-it} \right)$$

so again using the Euler relations and combining terms we obtain

$$D_t^\mu \left[\cos t \right] = \cos (\mu\pi/2) \cos_\mu t - \sin (\mu\pi/2) \sin_\mu t. \qquad (3.114)$$

Equation (3.114) is reminiscent of the trigonometric expansion of $\cos (t + \mu\pi/2)$.

Generalized Hyperbolic Functions   A similar kind of analysis can be done for the generalization of the hyperbolic sines and cosines. The generalized hyperbolic cosine function is given by

$$\cosh_\mu t = \frac{1}{2}\left(E_\mu^t +^* E_\mu^{-t}\right),$$ (3.115)

the generalized hyperbolic sine function is given by

$$\sinh_\mu t = \frac{1}{2}\left(E_\mu^t -^* E_\mu^{-t}\right),$$ (3.116)

and finally the generalized hyperbolic tangent is given by

$$\tanh_\mu t = \frac{\sinh_\mu t}{\cosh_\mu t}.$$ (3.117)

Note that we use the generalized exponential for real functions given by $^*E_\mu^{-t}$ and not E_μ^{-t} in the definitions of the generalized hypergeometric functions. Now let us consider the derivatives of the generalized hypergeometric functions.

Derivative of Generalized Exponentials   It is possible to incorporate all the derivatives for the complex exponential into the single relationship

$$D_t\left[E_\mu^{\pm it}\right] = \pm i\left[E_\mu^{\pm it} + \frac{1}{\Gamma(-\mu)}\frac{1}{(\pm i)^\mu\, t^{\mu+1}}\right]$$ (3.118)

from which one can immediately construct (3.109) and (3.112) from the definition of the generalized cosine and generalized sine, respectively. In this same way we can use the definition of the generalized hyperbolic cosine and (3.118) to obtain

$$D_t\left[\cosh_\mu t\right] = \sinh_\mu t + \frac{1}{\Gamma(-\mu)\, t^{\mu+1}}$$ (3.119)

and the definition of the generalized hyperbolic sine and (3.118) to obtain

$$D_t\left[\sinh_\mu t\right] = \cosh_\mu t.$$ (3.120)

It is interesting that only the derivative of the hyperbolic cosine function has the additional inverse power-law term. This is due to the symmetry of the generalized hyperbolic cosine. There is no such correction to the antisymmetric generalized hyperbolic sine function.

Derivative of the Generalized Logarithm   The derivative of the generalized logarithm is

$$D_t\left[\ln_\alpha t\right] = \frac{1}{D_\alpha\left[E_\alpha^t\right]} = \frac{1}{t + \frac{1}{\Gamma(-\alpha)[\ln_\alpha t]^{\alpha+1}}}$$

which again reduces to the familiar result when α is an integer and the gamma function diverges.

3.3   Parametric Derivatives

Derivatives of Parameters and Not Variables   It is convenient to introduce a fractional derivative operator acting on the parameters in our equations rather than on the independent variable. For this purpose we introduce a new operator as follows

$$F(t) = D_\mu^\alpha \left[D_t^\mu \left[f(t) \right] \right], \tag{3.121}$$

where we have written

$$D_t^\alpha \left[\cdot \right] = D_\mu^\alpha \left[D_t^\mu \left[\cdot \right] \right]$$

and the outside fractional derivative is with respect to the parameter μ. In particular we examine the case $\alpha = 1$ and calculate the derivative with respect to the parameter μ in the vicinity of $\mu = 0$, so we may write (3.121) as:

$$F(t) = \frac{\partial}{\partial \mu} \left[D_t^\mu \left[f(t) \right] \right]_{\mu=0} \tag{3.122}$$

Let us use (3.3) to write

$$D_t^\mu \left[f(t) \right] = \sum_{k=0}^\infty \binom{\mu}{k} \frac{t^{k-\mu}}{\Gamma(k+1-\mu)} f^{(k)}(t)$$

so that taking the derivative with respect to the parameter μ we obtain

$$F(t) = \frac{\partial}{\partial \mu} t^{-\mu} \sum_{k=0}^\infty \binom{\mu}{k} \frac{t^k}{\Gamma(k+1-\mu)} f^{(k)}(t) \Big|_{\mu=0}. \tag{3.123}$$

It is clear that the μ-derivatives separate into the factor in front of the series and those μ-dependent terms within the series. After some manipulation of gamma functions we obtain

$$F(t) = -\ln t \; f(t) + \sum_{k=0}^\infty \frac{t^{k-\mu}}{\Gamma(k+1)} D_t^k \left[f(t) \right]$$

$$\times \frac{\partial}{\partial \mu} \left[\frac{\Gamma(\mu+1)(-1)^{k+1} \sin \mu\pi}{(k-\mu)\pi} \right]_{\mu=0}, \tag{3.124}$$

where the last factor yields

$$\frac{\partial}{\partial \mu} \left[\frac{\Gamma(\mu+1) \sin \mu\pi}{(k-\mu)\pi} \right]_{\mu=0} = \delta_{k,0} \left[\frac{\partial \Gamma(\mu+1)}{\partial \mu} \right]_{\mu=0} + \frac{1}{k}$$

which when inserted into (3.124) gives us

$$F(t) = -\left[\ln t + \gamma\right] f(t) + \sum_{k=0}^{\infty} \frac{(-1)^{k+1} t^k}{k\Gamma(k+1)} f^{(k)}(t), \qquad (3.125)$$

where the constant γ is given by the value of the psi function $\psi(1) = -\gamma$, obtained from the derivative of the gamma function.

An application of the idea   Let us now consider an application of the above parametric differentiation. Consider the following integral

$$
\begin{aligned}
F(t) &= \int_0^t e^{-\tau} \ln \tau \, d\tau \\
&= \frac{\partial}{\partial \mu} \left[\int_0^t e^{-\tau} \tau^{\mu} d\tau \right]_{\mu=0}. \qquad (3.126)
\end{aligned}
$$

But we know the value of the integral in the square brackets, it is given by (??), so we can rewrite (3.126) as

$$F(t) = \frac{\partial}{\partial \mu} \left[\left\{ \Gamma(\mu+1) e^{-\tau} E_{-(\mu+1)}^{\tau} \right\}_0^t \right]_{\mu=0}$$

which upon differentiating yields

$$
\begin{aligned}
F(t) &= \frac{\partial}{\partial \mu} \left[\Gamma(\mu+1)\right]_{\mu=0} e^{-t} E_{-1}^t + e^{-t} \frac{\partial}{\partial \mu} \left\{ D_t^{-\mu} \left[E_{-1}^t \right] \right\}_{\mu=0} \\
&= \Gamma'(1) e^{-t} (e^t - 1) + e^{-t} \frac{\partial}{\partial \mu} \left\{ D_t^{-\mu} \left[e^t - 1 \right] \right\}_{\mu=0}. \qquad (3.127)
\end{aligned}
$$

Thus, by applying (3.125) to the second term on the right-hand side of (3.127) we obtain

$$F(t) = \ln t \left[1 - e^{-t} \right] - B(t) \qquad (3.128)$$

with the newly defined function

$$B(t) = \sum_{k=0}^{\infty} \frac{(-1)^{k+1} t^k}{k\Gamma(k+1)}. \qquad (3.129)$$

However to evaluate the integral at long times we note that the parametric derivative yields

$$\frac{\partial}{\partial \mu} D_t^{\mu} \left[e^t \right] = \frac{\partial E_{\mu}^t}{\partial \mu}$$

so that for $t \to \infty$ and recalling that $E_{\mu}^t \to e^t$ in this limit we have

$$\lim_{t \to \infty} \frac{\partial E_\mu^t}{\partial \mu} = 0,$$

which is a consequence of the fact that the asymptotic value of the generalized exponential is independent of μ. Thus, using

$$\frac{\partial E_\mu^t}{\partial \mu} = [B(t) - (\gamma + \ln t)] e^t$$

we can write

$$\lim_{t \to \infty} B(t) = \gamma + \ln t. \tag{3.130}$$

Equation (3.130) allows us to write, using (3.126) and (3.128), the definite integral

$$\int_0^\infty e^{-t} \ln t \, dt = -\gamma. \tag{3.131}$$

Using an Infinitesimal Generator   It is now possible to use the above evaluation of the integral to suggest a procedure for simplifying our formalism. Let us define an operator in terms of the parametric derivative of a fractional derivative

$$\mathcal{L} \equiv \frac{\partial D_t^\mu}{\partial \mu}. \tag{3.132}$$

Thus, $\mathcal{L}$ is the infinitesimal generator of the fractional derivative operator

$$D_t^\mu \sim D_t^0 + \mu \mathcal{L} \equiv 1 + \mu \mathcal{L} , \ |\mu| << 1 \tag{3.133}$$

where from (3.133) it is clear that $\mathcal{L}$ is a linear operator with the properties

$$\mathcal{L}[af(t) + bg(t)] = a\mathcal{L}[f(t)] + b\mathcal{L}[g(t)] \tag{3.134}$$

and

$$\frac{d}{dt}[\mathcal{L}f(t)] = \mathcal{L}\left[\frac{df}{dt}\right] - \frac{f(0)}{t}. \tag{3.135}$$

To prove (3.135) it is sufficient to take the derivative of the operator (3.132) applied to the function $f(t)$ from which we obtain

$$\frac{d}{dt}[\mathcal{L}f(t)] = \mathcal{L}\left[\frac{df}{dt}\right] - \frac{f(t)}{t} + \sum_{k=1}^\infty (-1)^{k+1} \frac{t^{k-1}}{\Gamma(k+1)} f^{(k)} \tag{3.136}$$

and taking cognizance of Taylor's theorem

$$\sum_{k=1}^{\infty}(-1)^{k+1}\frac{t^{k-1}}{\Gamma(k+1)}f^{(k)} = \frac{f(t)-f(0)}{t}$$

reduces (3.136) to (3.135).

Using a Commutator   Furthermore, we can integrate (3.135) to obtain

$$\mathcal{L}[f(t)] - \mathcal{L}[f(t)]_{t=a} = \int_{a}^{t}\mathcal{L}f'(\tau)\,d\tau - f(0)(\ln t - \ln a) \qquad (3.137)$$

and using

$$f(t) = f(a) + \int_{a}^{t}f'(\tau)\,d\tau \qquad (3.138)$$

we have from (3.137)

$$\mathcal{L}\left[f(a) + \int_{a}^{t}f'(\tau)\,d\tau\right] = \mathcal{L}[f(t)]_{t=a} + \int_{a}^{t}\mathcal{L}f'(\tau)\,d\tau - f(0)(\ln t - \ln a).$$

But as we saw above

$$\mathcal{L}[f(a)] = -(\gamma + \ln t)f(a)$$

so that

$$\mathcal{L}\left[\int_{a}^{t}f'(\tau)\,d\tau\right] = \int_{a}^{t}\mathcal{L}f'(\tau)\,d\tau - f(0)\ln t + f(a)\ln t + c, \qquad (3.139)$$

where the integration constant is given by

$$c = \gamma f(a) + f(0)\ln a + [\mathcal{L}f(t)]_{t=a}. \qquad (3.140)$$

In a similar way we can use (3.135) to obtain commutation rules for the operator $\mathcal{L}$, the derivative d/dt and the integral $\int_{a}^{t}$:

$$\left[\mathcal{L}, \frac{\partial}{\partial t}\right]f(t) = \frac{f(0)}{t} \qquad (3.141)$$

$$\left[\mathcal{L}, \int_{a}^{t}\right]f'(t) = [f(a) - f(0)]\ln t + c \qquad (3.142)$$

$$[\mathcal{L}, t]f(t) = \int_{0}^{t}f(t')\,dt'. \qquad (3.143)$$

This last commutator, (3.143), is a consequence of the definition

$$\mathcal{L}f(t) \equiv -(\gamma + \ln t) f(t) + \sum_{k=1}^{\infty} \frac{(-1)^{k-1} t^k}{k\Gamma(k+1)} f^{(k)}(t) \qquad (3.144)$$

so that the product function $tf(t)$ yields

$$\begin{aligned}
\mathcal{L}[tf(t)] &= -(\gamma + \ln t) tf(t) + \sum_{k=1}^{\infty} \frac{(-1)^{k-1} t^k}{k\Gamma(k+1)} \frac{d^k}{dt^k} [tf(t)] \\
&= -(\gamma + \ln t) tf(t) + \sum_{k=1}^{\infty} \frac{(-1)^{k-1} t^k}{k\Gamma(k+1)} \left[kf^{(k-1)} + tf^{(k)} \right].
\end{aligned}$$

$$(3.145)$$

So we have after some rearrangement of terms

$$\mathcal{L}[tf(t)] = t - (\gamma + \ln t) tf(t) + \sum_{k=1}^{\infty} \frac{(-1)^{k-1} t^k}{k\Gamma(k+1)} f^{(k)} + \sum_{k=1}^{\infty} \frac{(-1)^{k-1} t^k}{k\Gamma(k+1)} f^{(k-1)}$$

$$(3.146)$$

which simplifies to

$$\mathcal{L}[tf(t)] = t\mathcal{L}[f(t)] + \sum_{k=1}^{\infty} \frac{(-1)^{k-1} t^k}{k\Gamma(k+1)} f^{(k-1)}(t). \qquad (3.147)$$

But the series in (3.147) is actually an expansion of the definite integral

$$\sum_{k=1}^{\infty} \frac{(-1)^{k-1} t^k}{k\Gamma(k+1)} f^{(k-1)} = \sum_{n=0}^{\infty} \frac{(-1)^n t^{n+1}}{\Gamma(n+1)} f^{(n)} \equiv \int_0^t f(t')\, dt'$$

which when substituted into (3.147) yields the commutation relation (3.143).

Some General Results   Consider a function of n variables $f(x_1, x_2, \cdots, x_n)$. The mth order, integer, partial derivative of this function has the symmetry

$$\frac{\partial^m f(x_1, x_2, \cdots, x_n)}{\partial x_1 \cdots \partial x_j \cdots \partial x_k} = \frac{\partial^m f(x_1, x_2, \cdots, x_n)}{\partial x_j \cdots \partial x_1 \cdots \partial x_k} = \cdots \qquad (3.148)$$

so that the value of the multiple derivative is independent of the order in which the derivatives are taken. We can rewrite (3.148) for fractional partial derivatives as

$$\frac{\partial^{\alpha_1}}{\partial x_1^{\alpha_1}} \cdots \frac{\partial^{\alpha_j}}{\partial x_j^{\alpha_j}} \cdots \frac{\partial^{\alpha_k}}{\partial x_k^{\alpha_k}} f(x_1, x_2, \cdots, x_n) = \frac{\partial^{\alpha_j}}{\partial x_j^{\alpha_j}} \cdots \frac{\partial^{\alpha_1}}{\partial x_1^{\alpha_1}} \cdots \frac{\partial^{\alpha_k}}{\partial x_k^{\alpha_k}} f(x_1, x_2, \cdots, x_n)$$

$$(3.149)$$

where the operator $\partial^{\alpha_j}/\partial x_j^{\alpha_j}$ is the fractional partial derivative $D_{x_j}^{\alpha_j}$. In this sense the multiple integral and the parametric derivative under the integral are particular cases of (3.149). Take, for example, the double integral

$$\frac{\partial^{-1}}{\partial y^{-1}}\frac{\partial^{-1}}{\partial x^{-1}}f(x,y) = \int f(x',y')\,dx'dy' = \frac{\partial^{-1}}{\partial x^{-1}}\frac{\partial^{-1}}{\partial y^{-1}}f(x,y)$$

so that we see that the order of the integration operation does not matter. From this relation we can also write that the parametric derivative and the integration commute with each other

$$\frac{\partial}{\partial \lambda}\frac{\partial^{-1}}{\partial x^{-1}}f(x,\lambda) = \frac{\partial^{-1}}{\partial x^{-1}}\frac{\partial}{\partial \lambda}f(x,\lambda)$$

which means

$$\frac{\partial}{\partial \lambda}\frac{\partial^{-1}}{\partial x^{-1}}f(x,\lambda) = \int \frac{\partial}{\partial \lambda}f(x',\lambda)\,dx'. \tag{3.150}$$

We can use the fractional derivative with respect to a parameter to produce the possibility of new transformations. Consider, for example, the integral

$$I(a,b) = \int_0^\infty t^\alpha \left(a + bt^\beta\right)^{\gamma-1} dt \tag{3.151}$$

which is convergent for $\alpha > -1$ and $\gamma + (\alpha + 1)/\beta < 1$ (with $a, b \neq 0$). We rewrite (3.151) in terms of parametric derivatives

$$\begin{aligned}
I(a,b) &= b^{\gamma-1}\int_0^\infty t^\alpha \left(a/b + t^\beta\right)^{\gamma-1} dt \\
&= b^{\gamma-1}D_{\frac{a}{b}}^{-\gamma}\left[D_{\frac{a}{b}}^\gamma \int_0^\infty t^\alpha \left(a/b + t^\beta\right)^{\gamma-1} dt\right] \tag{3.152}
\end{aligned}$$

in order to simplify its evaluation. We can use (3.50) in terms of the parametric fractional derivative to obtain from (3.152)

$$I(a,b) = b^{\gamma-1}\frac{\sin \pi\gamma}{\pi}\Gamma(\gamma)D_\lambda^{-\gamma}\left[\lambda^{-\gamma}\int_0^\infty t^{\alpha+\gamma\beta}\left(\lambda + t^\beta\right)^{-1} dt\right], \tag{3.153a}$$

where $\lambda = a/b$. Making the further substitution $z = t^\beta$ in (3.153a) we have:

$$I(a,b) = b^{\gamma-1}\frac{\sin \pi\gamma}{\pi}\Gamma(\gamma)D_\lambda^{-\gamma}\left[\lambda^{-\gamma}\beta^{-1}\int_0^\infty \frac{z^{\frac{\alpha+1}{\beta}+\gamma-1}}{z+\lambda}\,dz\right] \tag{3.154}$$

and using the calculus of residues to evaluate the simple pole in the remaining integral we have

$$I\left(a,b\right) = b^{\gamma-1}\frac{\Gamma\left(\gamma\right)\sin\pi\gamma}{\pi}\frac{\pi}{\sin\left[\left(\frac{\alpha+1}{\beta}+\gamma\right)\pi\right]}D_{\lambda}^{-\gamma}\left[\lambda^{\frac{\alpha+1}{\beta}-1}\right]$$

so that we finally obtain

$$I\left(a,b\right) = \frac{b^{\gamma-1}\sin\pi\gamma}{\beta\sin\left[\left(\frac{\alpha+1}{\beta}+\gamma\right)\pi\right]}\frac{\Gamma\left(\gamma\right)\Gamma\left(\frac{\alpha+1}{\beta}\right)}{\Gamma\left(\frac{\alpha+1}{\beta}+\gamma\right)}\left(\frac{a}{b}\right)^{\frac{\alpha+1}{\beta}+\gamma-1}. \tag{3.155}$$

This was a relatively painless way (as these things go) to evaluate a complicated integral.

3.4 Commentary

Fractal Operators Modeling Complex Phenomena In this chapter we have viewed the tip of the fractional calculus iceberg. We have seen that even the simplest of fractional derivatives may have deep physical implications when modeling a complex phenomenon. Fractional time derivatives have diverging initial values and long-time memory, whereas spatial fractional derivatives have long-range interactions. In general the underlying phenomena modeled by such fractal operators violate the traditional assumptions of differentiability, as well as locality in space and time.

The Generalized Functions In these last few lectures we have constructed a basis for a fractional calculus using only the definition of the fractional derivative and the fractional integral operating on series. This definition has allowed us to generalize the definitions of the exponential function, the trigonometric functions, and the hyperbolic functions. We noted that the generalized functions contain an inverse power-law part that is not present in the traditional definitions of these functions. One interpretation of the meaning of the inverse power-law part of the generalized functions is that the physical concepts of relaxation and oscillation generalize, and due to the complexity of the phenomena being considered, these new functions contain a memory that is not present in simpler phenomena.

Interpretation of Generalized Harmonic Functions The generalized exponential can be real or complex and therefore it was used to construct generalizations of both the trigonometric and hyperbolic functions. The generalized harmonic functions were shown to be solutions to inhomogeneous linear harmonic oscillator equations of motion, where the inhomogeneity is the inverse power law in time. Thus, if we interpret the independent variable to be space rather than time, the second derivative can be interpreted as the limit of nearest neighbor interactions on a spatial lattice. The field amplitude is asymptotically harmonic as it should be, but locally the field is distorted, even singular, due to the long-range interactions driving the system. This is not unlike the distortion of a regular lattice that is introduced by a local defect. However, this is not the only distortion to which the field is subjected. For certain parameter values the long-range interactions essentially provide a ramp on which the oscillations ride. It is of interest to determine if some phenomena manifest these types of behaviors.

Bibliography

[1] E. Broda, *Ludwig Boltzmann, Man-Physicist-Philosopher*, Ox Bow, Woodbridge (1983).

[2] H. Goldstein, *Advanced Classical Mechanics*, John Wiley, New York, first edition (1955).

[3] P. Meakin, *Fractals, scaling and growth far from equilibrium*, Cambridge Nonlinear Science Series 5, Cambridge University Press, Cambridge, MA (1998).

[4] K. S. Miller and B. Ross, *An Introduction to the Fractional Calculus and Fractional Differential Equations*, John Wiley, New York (1993).

[5] D. Ruelle, *Chaotic Evolution and Strange Attractors*, Cambridge University Press, Cambridge (1989).

[6] Yu. N. Rabotnov, *Elements of Hereditary Solid Mechanics*, MIR , Moscow (1980).

[7] S. G. Samko, A. A. Kilbas and O. I. Marichev, *Fractional Integrals and Derivatives*, Gordon and Breach, New York (1993).

[8] M. F. Shlesinger, Fractal time and 1/f noise in complex systems, *Ann. N. Y. Acad. Sci.* **504**, 214 (1987).

[9] H. E. Stanley, *Introduction to Phase Transitions and Critical Phenomena*, Oxford University Press, Oxford (1979).

Chapter 4

Fractional Fourier Transforms

Linear Transport and Propagation In the next few lectures we provide a brief overview of Fourier analysis and how it has been used to model linear physical phenomena, particularly the reversible propagation of scalar waves in homogeneous media and the irreversible diffusion of one molecular species within another. The purpose of this review is to orient the reader so that the significance of wave propagation in fractal media will be apparent as will anomalous diffusion. These latter topics have emerged in the last two decades as the natural successors of the phenomena examined in the 19th and early 20th centuries.

Generalized Fourier Series It is no longer acceptable to make the approximation that the functions describing a phenomenon are continuous and differentiable, only because of the resulting simplicity in the physical modeling. We have moved beyond such simplifying assumptions in describing complex phenomena, and require that the functions used more completely reflect the richness in structure that is experimentally observed. Thus, we examine some of the most pervasive mathematical tools for the analysis of physical systems, the Fourier series and Fourier transform, and highlight the physical restrictions that are implied when these methods are used to describe a process. Our purpose is, in part, to show that a generalization of Fourier analysis may be developed using the fractional calculus to model complex phenomena.

4.1 A Brief Review of Fourier Analysis

Fourier Series Fourier analysis is one of those areas of mathematics that finds its way into the description of all manner of physical phenomena, including acoustics, heat transport, electricity and magnetism, quantum mechanics, and many others. Dirichlet proved that it is possible to express any single-valued function as a sum of sinusoidal components, of frequencies which are multiples of a fundamental frequency, that is, as a Fourier series. A function $f(t)$ is periodic if there exists a constant $\overline{T}$ for which

$$f\left(t + \overline{T}\right) = f(t) \tag{4.1}$$

for any t in the domain of the definition of the function. The constant $\overline{T}$ is called the period of the function. Let us construct a series consisting of a sum over the harmonics of the fundamental frequency $\omega_0 = 2\pi/\overline{T}$:

$$f_N(t) = \frac{a_0}{2} + \sum_{k=1}^{N} \left[a_k \cos \omega_k t + b_k \sin \omega_k t\right], \tag{4.2}$$

where the harmonics are $\omega_k = k\omega_0 = 2\pi k/\overline{T}$. Equation (4.2) is a trigonometric series of order N and period $\overline{T}$. It is clear that $f_N(t)$ is a sum of various harmonics, but the function itself is much more complicated than one which is simply harmonic, even though it is periodic. Various functions can be represented by (4.2) by selecting different sets of coefficients. The limit $N \to \infty$ can be taken to obtain an infinite trigonometric series (when it converges) for a function with period $\overline{T}$,

$$f(t) \equiv \lim_{N \to \infty} f_N(t). \tag{4.3}$$

Not All Aeries Are Differentiable   How large is the class of functions that can be represented by such infinite trigonometric series? This question was raised in the 18th century by Euler and Bernoulli in connection with Bernoulli's study of a vibrating string. Daniel Bernoulli had previously formulated the principle of superposition and using physical arguments showed that a very large class of continuous function could be expanded in trigonometric series. This view was strongly criticized by Bernoulli's contemporaries, including Euler who had, ironically, shown that the solution to the wave equation could be expressed as such a trigonometric series. The conflict centered on the idea that if a function is represented as an analytic expression (such as a trigonometric series) then it must have appropriate properties of differentiability. However, Bernoulli believed that one could represent functions with point discontinuities, for example, a plucked string, in such a way. It took nearly a century for Bernoulli's physical intuition to be vindicated. Recall that the Weierstrass function, which is nowhere differentiable, is of the form of an infinite-order trigonometric series.

Fourier Series   When the coefficients in (4.2) are given in a prescribed way, as discussed shortly, and $N = \infty$, the series is called a Fourier series representation of the function. Fourier series were first investigated in connection with small oscillations in elastic materials such as the stretched string of a musical instrument. The series

$$f(t) = \frac{a_0}{2} + \sum_{k=1}^{\infty} [a_k \cos \omega_k t + b_k \sin \omega_k t] \tag{4.4}$$

clearly shows that the vibration of any point on a string is a linear superposition of the harmonics of a fundamental frequency ω_0. Thus, one can represent a complicated oscillatory motion as a sum of individual oscillations that are particularly simple. It is also possible to use this idea in the infinite period limit, that is, to represent nonoscillatory phenomena with such series as well.

The Nonoscillatory Part If we introduce a dimensionless time $\tau = \omega_0 t$ then (4.4) becomes

$$f(\tau) = \frac{a_0}{2} + \sum_{k=1}^{\infty} [a_k \cos k\tau + b_k \sin k\tau] \tag{4.5}$$

which is a function of period 2π. To determine the constant coefficients in (4.5) we assume the series is integrable in such a way that

$$\int_0^{2\pi} f(\tau) d\tau = \frac{a_0}{2} \int_0^{2\pi} d\tau + \sum_{k=1}^{\infty} \left[a_k \int_0^{2\pi} d\tau \cos k\tau + b_k \int_0^{2\pi} d\tau \sin k\tau \right].$$

The periodic nature of the terms in the series requires that each integral give zero contribution to the sum so that the integral reduces to

$$a_0 = \frac{1}{\pi} \int_0^{2\pi} f(\tau) d\tau \tag{4.6}$$

and the constant term is just the average value of the function over its domain.

Fourier Coefficients Next, multiply (4.5) by $\cos n\tau$, where n is an integer, and integrate the resulting expression from 0 to 2π:

$$\int_0^{2\pi} f(\tau) \cos n\tau d\tau = \frac{a_0}{2} \int_0^{2\pi} \cos n\tau d\tau$$

$$+ \sum_{k=1}^{\infty} a_k \int_0^{2\pi} d\tau \cos k\tau \cos n\tau + \sum_{k=1}^{\infty} b_k \int_0^{2\pi} d\tau \sin k\tau \cos n\tau. \tag{4.7}$$

The first integral vanishes due to periodicity, and to evaluate the second integral we use the fact that the set of functions $\{e^{-ik\theta}\}$ is complete on the interval $(0, 2\pi)$ and satisfies the orthogonality relation

$$\int_0^{2\pi} e^{-ik\theta} e^{im\theta} d\theta = 2\pi \delta_{k,m}, \tag{4.8}$$

where $\delta_{k,n} = 0$ if $k \neq n$ and $\delta_{k,n} = 1$ if $k = n$, is the Kronecker delta function. Using (4.8) we obtain

$$\int_0^{2\pi} \cos k\tau \cos n\tau = \pi\delta_{k,n} \text{ for } k > 0, n > 0 \tag{4.9}$$

and

$$\int_0^{2\pi} \sin k\tau \sin n\tau = \pi\delta_{k,n} \text{ for } k > 0, n > 0. \tag{4.10}$$

Thus, we obtain

$$a_k = \frac{1}{\pi} \int_0^{2\pi} f(\tau) \cos k\tau d\tau \tag{4.11}$$

and in an exactly equivalent way, multiplying (4.5) by $\sin n\tau$ and integrating yields

$$b_k = \frac{1}{\pi} \int_0^{2\pi} f(\tau) \sin k\tau d\tau. \tag{4.12}$$

The quantities calculated using (4.11) and (4.12) are called the Fourier coefficients for the function $f(t)$, and the trigonometric series with these coefficients is called the Fourier series of $f(t)$.

Complex Coefficients   We can also write the Fourier series in complex form using the Euler expansion

$$e^{ik\tau} = \cos k\tau + i \sin k\tau \tag{4.13}$$

so that (4.5) can be written

$$f(\tau) = \sum_{k=-\infty}^{\infty} c_k e^{-ik\tau}, \tag{4.14}$$

where the coefficients in (4.14) are complex:

$$c_k = \frac{1}{2} [a_k + ib_k], \tag{4.15}$$

$$c_{-k} = \frac{1}{2} [a_k - ib_k], \tag{4.16}$$

$$c_0 = \frac{a_0}{2}. \tag{4.17}$$

The complex Fourier coefficients in the function $f(t)$ are easily shown to satisfy the relations

$$c_k = \frac{1}{2\pi} \int_0^{2\pi} f(\tau) e^{ik\tau} d\tau, \ k = 0, \pm 1, \pm 2, \cdots. \tag{4.18}$$

To establish (4.18) we again use Euler's formula to write (4.18) as

$$c_k = \frac{1}{2\pi} \int_0^{2\pi} f(\tau) \cos k\tau \, d\tau + i \frac{1}{2\pi} \int_0^{2\pi} f(\tau) \sin k\tau \, d\tau \qquad (4.19)$$

and using (4.11) and (4.12) for the two integrals in (4.19) we obtain (4.15), with a similar result for c_{-k}. It is useful to remember that if the function $f(\tau)$ is real, as it would be for a physical observable, then the coefficients c_k and c_{-k} are complex conjugates of each other; that is, $c_k^* = c_{-k}$.

Closure Now let us review what happens in the continuum limit. Inserting (4.18) into (4.14) and changing the order of integration and summation gives

$$f(\tau) = \int_0^{2\pi} f(\tau') \, d\tau' \left[\frac{1}{2\pi} \sum_{k=-\infty}^{\infty} e^{-ik\tau} e^{ik\tau'} \right]. \qquad (4.20)$$

In order for this equation to reduce to a tautology the term in square brackets must be identified with the Dirac delta function

$$\delta(\tau - \tau') = \frac{1}{2\pi} \sum_{k=-\infty}^{\infty} e^{-ik(\tau - \tau')}. \qquad (4.21)$$

We can put (4.21) into more familiar form by changing the time index to the fraction of a time interval $\overline{T}$, that is, $\tau = 2\pi t / \overline{T}$ so that

$$f(t) = \sum_{k=-\infty}^{\infty} c_k e^{-ik2\pi t / \overline{T}} \qquad (4.22)$$

with the complex coefficient given by

$$c_k = \frac{1}{\overline{T}} \int_0^{\overline{T}} e^{ik2\pi t / \overline{T}} f(t) \, dt. \qquad (4.23)$$

The closure relation given by (4.21) can then be expressed as

$$\delta(t - t') = \frac{1}{\overline{T}} \sum_{k=-\infty}^{\infty} e^{-ik2\pi(t - t')/\overline{T}} \qquad (4.24)$$

as we identified in (4.21).

Infinite Time As $\overline{T} \to \infty$ the above series become integrals and we obtain the Fourier integral transformations. An indication of how this comes about can be seen by noting that for large $\overline{T}$, the function $c_k e^{-ik2\pi t / \overline{T}}$ in the summand of (4.22) changes little as k changes by unity or by an amount Δk such that $2\pi t \Delta k << \overline{T}$. If the series for $f(t)$ is written as

$$f(t) = \sum_{k=-\infty}^{\infty} \Delta k c_k e^{-ik2\pi t / \overline{T}}$$

for finite $\overline{T}$, where the summand has been replaced by an average over the Δk interval, with the sum running over the Δk intervals, and if we let

$$\omega_k = 2\pi k/\overline{T} \to \Delta\omega_k = 2\pi\Delta k/\overline{T}$$

the series for the function becomes

$$f(t) = \sum_{k=-\infty}^{\infty} \Delta k e^{-i\omega_k t} \frac{1}{\overline{T}} \int_0^T e^{i\omega_k t'} f(t')\, dt'.$$

In the limit $\overline{T} \to \infty$ we have

$$\frac{1}{\overline{T}} \sum_{k=-\infty}^{\infty} \Delta k e^{-i\omega_k t} \to \frac{1}{2\pi} \int_{-\infty}^{\infty} e^{-i\omega t}\, d\omega$$

so that we obtain

$$f(t) = \frac{1}{2\pi} \int_{-\infty}^{\infty} e^{-i\omega t}\, d\omega \int_0^{\infty} e^{i\omega t'} f(t')\, dt'. \qquad (4.25)$$

Thus, we define the Fourier transform of a function by

$$\hat{f}(\omega) = \int_{-\infty}^{\infty} e^{i\omega t} f(t)\, dt \equiv \mathcal{FT}\{f(t); \omega\}, \qquad (4.26)$$

the inverse Fourier transform of the function by

$$f(t) = \frac{1}{2\pi} \int_{-\infty}^{\infty} e^{-i\omega t}\, \hat{f}(\omega)\, d\omega \equiv \mathcal{FT}^{-1}\left\{\hat{f}(\omega); t\right\}, \qquad (4.27)$$

and the closure relation (4.24) becomes the Dirac delta function

$$\delta(t - t') = \frac{1}{2\pi} \int_{-\infty}^{\infty} e^{-i\omega(t-t')}\, d\omega \qquad (4.28)$$

which when inserted into (4.25) yields a tautology. Note that there are other definitions of the Fourier transform and its inverse that are more symmetric; that is, the normalization of the integrals is both $\mathcal{FT}$ and $\mathcal{FT}^{-1}$ is $1/\sqrt{2\pi}$.

4.2   Linear Fields

Continuous Field Phenomena   The previous lectures separated phenomena into those that can be described as discrete Fourier series, and those that are modeled as continuous Fourier integrals. In our earlier discussion of classical mechanics we concentrated on discrete, individual particle, phenomena. We now extend that discussion from systems with a finite number of degrees of freedom to those having an infinite number of degrees of freedom. Such systems are called fields and the dependent variables are continuous rather than discrete. An example of such a system is the surface of a fluid, say, the ocean, on which water waves propagate. Each point on this surface participates in the wave motion and the complete motion can only be described by specifying the position coordinate of all fluid particles in the surface. Similar models can be constructed for scalar waves in solids and gases as well, the difference between the media being characterized by changes in the physical parameters such as the wave speed. One can straightforwardly adopt the techniques of classical mechanics to describe the motion of such systems. The most direct method is to construct a discrete model of the system having a finite number of degrees of freedom and then take the limit of some parameters that carry the discrete model smoothly over into a continuous model.

Elastic Energy A traditional example of making the transition from a discrete to a continuum model is that of a chain of oscillators modeling the atoms in an elastic rod discussed at length in Chapter 12 of Goldstein [6]. To facilitate our investigation we present a truncated version of that discussion here. Consider an infinitely long elastic rod that can undergo only longitudinal vibrations. This model consists of a chain of equal mass oscillators, with equilibrium spacing, whose oscillations are constrained to only move along the direction of the chain. Note that this elastic rod is restricted to one dimension. We can denote the kinetic energy associated with the rod as the sum of the kinetic energies of the individual masses:

$$T = \frac{1}{2} \sum_{j=1}^{N} m_j \dot{\eta}_j^2, \tag{4.29}$$

where $\dot{\eta}_j$ is the velocity of the jth particle. We specify that the springs connecting the masses are themselves massless and only represent the linear displacement force between particles in the solid. Thus, the potential energy is the sum of the potential energies associated with the relative displacements of each particle or, equivalently, the displacement of each particle from its equilibrium position:

$$V = \frac{1}{2} \sum_{j=1}^{N} \kappa \left(\eta_{j+1} - \eta_j \right)^2. \tag{4.30}$$

Here κ is the spring constant for each of the springs, and η_j is the displacement of the jth particle from its equilibrium position.

Lagrange Density The Lagrangian for the discrete model of the elastic rod is given by the difference between the kinetic and potential energies

$$L = T - V = \frac{1}{2} \sum_{j=1}^{N} \left[m_j \dot{\eta}_j^2 - \kappa \left(\eta_{j+1} - \eta_j \right)^2 \right]. \tag{4.31}$$

It is useful in the study of fields to factor the Lagrangian into the form

$$L = \frac{1}{2}\sum_{j=1}^{N}\Delta\left[\frac{m_j}{\Delta}\dot{\eta}_j^2 - \kappa\Delta\left(\frac{\eta_{j+1} - \eta_j}{\Delta}\right)^2\right]$$

$$= \sum_{j=1}^{N}\Delta\mathcal{L}_j, \tag{4.32}$$

where Δ is the equilibrium spacing between masses on the spring lattice and $\mathcal{L}_j$ is the discrete Lagrangian density. If we assume that η_j is a generalized coordinate and $\dot{\eta}_j$ is its conjugate velocity, then the Euler-Lagrange equations for each of the individual particles are

$$\frac{d}{dt}\frac{\partial\mathcal{L}_j}{\partial\dot{\eta}_j} - \frac{\partial\mathcal{L}_j}{\partial\eta_j} = 0 \ , \ j = 1, 2, \cdots, N. \tag{4.33}$$

Substituting the definition of the discrete Lagrangian density into the Euler-Lagrange equations of motion, we obtain for the description of the motion of the jth particle

$$\frac{m_j}{\Delta}\ddot{\eta}_j - \kappa\Delta\left(\frac{2\eta_j - \eta_{j+1} - \eta_{j-1}}{\Delta^2}\right) = 0 \tag{4.34}$$

which for finite Δ is a chain of oscillators with nearest neighbor interactions only. The linear nature of (4.34) makes it possible to solve these equations as a Fourier series with the Fourier amplitudes given by the initial displacements and initial velocities of each of the oscillators. However, we are not interested in the solution to the discrete problem here, but rather we want to consider these equations of motion in the continuous limit, that is, the limit as the equilibrium distance between the particles vanishes as $\Delta \to 0$.

Continuum Wave Equation As $\Delta \to 0$ the ratio of the particle mass to the equilibrium particle separation, m_j/Δ becomes the mass per unit length of the rod, say, ρ, independently of the index j. For an elastic rod obeying Hooke's law one can write the displacement force as the product of Young's modulus (Y) and the elongation per unit length. In the continuum limit $Y = \kappa\Delta$ and the elongation per unit length is the gradient of the mass displacement field:

$$\lim_{\Delta\to 0}\frac{\eta_{j+1} - \eta_j}{\Delta} = \frac{\partial\eta(x,t)}{\partial x} \equiv \eta'(x,t), \tag{4.35}$$

where x denotes the continuous position of the mass as $\Delta \to 0$ and replaces the discrete index j. Therefore the Lagrangian (4.32) takes on the continuous form

$$L = \frac{1}{2}\int\left[\rho\,\dot{\eta}(x,t)^2 - Y\eta'(x,t)^2\right]dx \tag{4.36}$$

so that the Lagrangian becomes an integral over the continuous Lagrangian density

$$\mathcal{L}\left(\eta,\dot{\eta},\eta';x,t\right) = \frac{1}{2}\left[\rho\,\dot{\eta}\,(x,t)^2 - Y\eta'\,(x,t)^2\right], \qquad (4.37)$$

where we have set the mass density to a constant and replaced $\mathcal{L}_j$ in (4.32) by $\mathcal{L}$. We could generalize the variational formalism to include the spatial dependence of the field amplitude, but it is more interesting here to just take the continuous limit of (4.34) and use

$$\lim_{\Delta \to 0}\left(\frac{2\eta_j - \eta_{j+1} - \eta_{j-1}}{\Delta^2}\right) = \frac{\partial^2\eta\,(x,t)}{\partial x^2}$$

so that the equation of motion (4.34) becomes the wave equation for the longitudinal vibrations in the elastic rod with propagation velocity $u = \sqrt{Y/\rho}$:

$$\frac{1}{u^2}\frac{\partial^2\eta\,(x,t)}{\partial t^2} - \frac{\partial^2\eta\,(x,t)}{\partial x^2} = 0. \qquad (4.38)$$

Continuous Wave Field *What have we learned here?* First, we have learned that the position coordinate x is not a generalized coordinate. It is the displacement of the masses $\eta\,(x,t)$, that is, the generalized coordinate and x is just a continuous index in the same way that the discrete index j locates the mass along the chain. Second, we have learned that just as η_j was a different coordinate from η_k, when $j \neq k$, so too $\eta\,(x,t)$ is different from $\eta\,(x',t)$, when $x \neq x'$. Since there is an infinity of values of x there is an infinity of generalized coordinates for each value of the time t. The field amplitude is, of course, labeled by the two continuous parameters x and t. If the above discussion had been carried out in three spatial dimensions rather than one, then the field quantities would be a function of the spatial vector $\mathbf{x} = (x,y,z)$ as well as time and the second derivative in space would be replaced by the Laplacian operator ∇^2.

Fourier Transforms Let us now return to the Fourier transform. The space-time Fourier transform of a function, based on the two-dimensional generalization of (4.26) is denoted by

$$\hat{\hat{\eta}}\,(k,\omega) = \int_{-\infty}^{\infty} e^{ikx}dx \int_{-\infty}^{\infty} e^{i\omega t}dt\,\eta\,(x,t) \qquad (4.39)$$

and the inverse space-time Fourier transform by

$$\eta\,(x,t) = \frac{1}{(2\pi)^2}\int_{-\infty}^{\infty} e^{-ikx}dk \int_{-\infty}^{\infty} e^{-i\omega t}d\omega\,\hat{\hat{\eta}}\,(k,\omega). \qquad (4.40)$$

Thus, if we introduce (4.40) into the wave equation we obtain

$$\frac{1}{(2\pi)^2}\int_{-\infty}^{\infty} e^{-ikx}dk \int_{-\infty}^{\infty} e^{-i\omega t}d\omega\left[\frac{\omega^2}{u^2} - k^2\right]\hat{\hat{\eta}}\,(k,\omega) = 0, \qquad (4.41)$$

where we have integrated the second derivative terms by parts twice to obtain

$$\int_{-\infty}^{\infty} e^{-iz\lambda} \frac{\partial^2 f(\lambda)}{\partial \lambda^2} d\lambda = -z^2 \int_{-\infty}^{\infty} e^{-iz\lambda} f(\lambda) d\lambda,$$

where $z\lambda$ can be either ωt or kx. Because $\hat{\eta}(k,\omega)$ is quite generally nonzero, (4.41) can only be satisfied by the dispersion relation

$$\frac{\omega^2}{u^2} = k^2. \tag{4.42}$$

The dispersion relation gives the frequency of the propagating wave as a function of the wavenumber of that same wave. The solutions to the quadratic equation (4.42) are the frequencies $\omega = \pm ku$ indicating waves that propagate to the right or to the left and are linearly dependent on k. Since the phase speed of the wave u is independent of ω and k, the medium is nondispersive, as is the case of light in a vacuum. For linear, deep water, gravity waves on the surface of a fluid, the dispersion relation is given by $\omega = \sqrt{gk}$, where g is the acceleration of gravity. The phase speed of the gravity wave is $u = \omega/k = \sqrt{g/k}$, so that the phase speed is dependent on wavelength $(2\pi/k)$. The longer the wave, the greater the phase speed, so that longer waves travel faster than shorter waves. The physics of waves is summarized in their dispersion relations.

Fourier Amplitudes   If we just consider the spatial Fourier transform of the displacement field amplitude, leaving the time-dependence unmolested,

$$\eta(x,t) = \frac{1}{2\pi} \int_{-\infty}^{\infty} e^{-ikx} dk \, \hat{\eta}(k,t) \tag{4.43}$$

we obtain from the wave equation

$$\frac{\partial^2 \hat{\eta}(k,t)}{\partial t^2} + \omega_k^2 \, \hat{\eta}(k,t) = 0, \tag{4.44}$$

where $\omega_k = ku$ is the solution to the dispersion relation. Note that (4.44) is the equation for a linear harmonic oscillator in terms of the spatial Fourier mode amplitudes and the mode amplitudes are independent of one another. The solution to the wave equation can be written as the linear superposition of the mode amplitudes

$$\eta(x,t) = \frac{1}{2\pi} \int_{-\infty}^{\infty} \left[A(k) e^{-i(kx - \omega_k t)} + cc \right] dk \tag{4.45}$$

where the mode amplitude function A is determined by the initial conditions

$$A(k) = \int_{-\infty}^{\infty} e^{ikx} \left[\eta(x,0) - \frac{i}{\omega_k} \dot{\eta}(x,0) \right] dx. \tag{4.46}$$

Note that (4.45) denotes the superposition of plane waves propagating to the right and to the left as the general solution to the wave equation; that is, $kx \pm \omega_k t = k(x \pm ut)$, just as found by Euler.

Physical Waves The above discussion is based on longitudinal modes propagating along an elastic rod, but they could just as easily have been the sounds from an orchestra, the vibrations of the floor when a truck rumbles by, white light shining through the library window, or any of a broad array of other oscillatory physical phenomena. Of course, there are details such as whether the field amplitudes refer to longitudinal or transverse modes of oscillation, whether the field is vector or scalar, and so on, but the essential point regarding the dynamics is whether the physical process response is sufficiently smooth to be described by a second-order partial differential equation in space and time. If it is sufficiently smooth then Fourier analysis becomes a powerful tool with which to analyze the phenomenon.

4.2.1 Inhomogeneous Linear Wave Fields

At this point our considerations regarding generalized trigonometric functions may be introduced into the discussion of linear wave fields. For example, if the linear oscillator equation (4.44) is inhomogeneous with terms that are inverse power law in time

$$\frac{\partial^2 \overset{\wedge}{\eta}(k,t)}{\partial t^2} + \omega_k^2 \overset{\wedge}{\eta}(k,t) = \frac{a_\mu}{t^{1+\mu}} + \frac{b_\mu}{t^{2+\mu}} \tag{4.47}$$

then with the proper choices of the constants a_μ and b_μ the solution to (4.47) will be the generalized trigonometric functions. In this way the linear wave field can be constructed from a superposition of generalized harmonic terms or in terms of the generalized exponential functions in time:

$$\eta(x,t) = \frac{1}{2\pi} \int_{-\infty}^{\infty} \left[A(k) E_\mu^{i\omega_k t} e^{-ikx} + cc \right] dk. \tag{4.48}$$

Thus, the solution to the inhomogeneous modal equation yields propagating wave modes asymptotically. However, at early times there can be significant distortion of the wave field, distortion that dampens out after a few oscillations.

Harmonic Oscillator in Space Of course we could have also taken the Fourier transform of the linear wave field amplitude in time to obtain the harmonic oscillator in space

$$\frac{\partial^2 \overset{\wedge}{\eta}(x,\omega)}{\partial x^2} + \frac{\omega^2}{u^2} \overset{\wedge}{\eta}(x,\omega) = 0. \tag{4.49}$$

For ease of notation we define $\psi(x) \equiv \overset{\wedge}{\eta}(x,\omega)$ and using the linear dispersion relation (4.42), we can rewrite (4.49) as

$$\psi''(x) + k^2 \psi(x) = 0, \tag{4.50}$$

where the primes denote the derivative with respect to the spatial coordinate. Of course, (4.50) again gives harmonic solutions in space that depend on the boundary conditions.

Heterogeneous Helmholz Equation   Now let us consider a spatial inhomogeneity that shifts the phase in the wave as it propagates in space, for example, a radio wave penetrating the ionosphere or an acoustic wave stimulating an elastic medium. To model such effects one traditionally replaces the wavenumber k in (4.50) with the product of a fundamental wavenumber k_0 and the spatially dependent index of refraction $n(x)$ to yield the Helmholtz wave equation

$$\psi''(x) + k_0^2 n^2(x)\psi(x) = \delta(x). \tag{4.51}$$

The Dirac delta function on the right-hand side of (4.51) corresponds to a unit amplitude point source, and an arbitrary source of waves may be constructed from a superposition of such point sources with various amplitudes. Thus, in the more general case the harmonic oscillator has a position-dependent frequency. For weak scattering the index of refraction in the interval $(0, L)$ for the one-dimensional case, varies about the value unity

$$n^2(x) = 1 + \lambda(x), \tag{4.52}$$

where typically the variations are spatially random with a mean-square amplitude that is very small, $\langle \lambda^2(x) \rangle << 1$. We separate the wave solution to (4.51) into an incident wave and a scattered wave, $\psi(x) = \psi_0(x) + \psi_s(x)$. The incident wave satisfies the free-space equation

$$\psi_0''(x) + k_0^2 \psi_0(x) = 0. \tag{4.53}$$

The scattered wave is determined by the inhomogeneous equation

$$\psi_s''(x) + k_0^2 \psi_s(x) = \delta(x) - \lambda(x) k_0^2 \psi(x) \tag{4.54}$$

and satisfies the radiation condition at infinity,

$$\psi_s'(x) = ik_0 \psi_s(x), \text{ as } |x| \to \infty. \tag{4.55}$$

Introducing the solution to the free-space Green's function equation

$$G_0''(x) + k_0^2 G_0(x) = \delta(x) \tag{4.56}$$

the total wave solution can be written

$$\psi(x) = \psi_0(x) - k_0^2 \int_0^L G_0(x - x')\lambda(x')\psi(x')\,dx' \tag{4.57}$$

and the integral is across the slab of width L, containing the random variations in the index of refraction. The free-space Green's function is given by the one-dimensional plane wave

$$G_0(x - x') = \frac{e^{ik_0|x-x'|}}{2ik_0}. \tag{4.58}$$

No exact closed form solution to (4.57), with a random variation in the index of refraction, has yet been developed.

Green's Function Solutions The scalar wave equation in one dimension is not a good approximation to many physical phenomena, but it shares a great many properties with all wave equations in complex media. For example, the total Green's function satisfies the integral equation

$$G\left(x,x'\right) = G_0\left(x,x'\right) - k_0^2 \int_0^L G_0\left(x,x''\right)\lambda\left(x''\right)G\left(x'',x'\right)dx'', \qquad (4.59)$$

with the random kernel $\lambda\left(x\right)$. The total solution to the wave equation with a source $J\left(x\right)$ is given by

$$\psi\left(x\right) = \int G\left(x,x'\right)J\left(x'\right)dx' \qquad (4.60)$$

and if $J\left(x\right) = \delta\left(x\right)$, then (4.60) reduces to (4.57).

Infinite Order Series In the situation where the mean-square fluctuations in the medium are small one of the major problems associated with the formal solution (4.57), or the Green's function equation (4.59), is the convergence of the expansion in terms of successively higher orders in the inhomogeneity $\lambda\left(x\right)$. Symbolically, such series have the form

$$G = G_0 - k_0^2 G_0 * \lambda * G_0 + k_0^4 G_0 * \lambda * G_0 * \lambda * G_0 - \cdots,$$

where the products are actually multiple convolutions. In general such series for the wave scattering problem converge too slowly to be of practical value, or they contain secular terms and therefore diverge asymptotically in time [4]. In short, we do not know the general properties of the solutions to the wave equation with a fluctuating index of refraction, even when those fluctuations are relatively weak. A number of techniques have been developed to handle phenomena of this kind including diagrammatic expansions, which we do not pursue, and path integral methods, which we address in due course.

Sometimes Even Linear Equations Cannot Be Solved Our purpose in this lecture has been to emphasize that even in the case of a linear wave field, where we are fairly confident regarding our level of understanding of the basic phenomenon, the simplest complication leads to unsolved and perhaps unsolvable modeling equations. In light of these results, it is not unreasonable to explore a new way of modeling the influence of media irregularities on the generation and propagation of linear waves. We do that subsequently, where instead of using a differential equation with nonconstant coefficients we model the phenomenon of wave propagation in irregular media using fractional wave equations.

4.2.2   Linear Diffusive Fields

Heat Equation   Fourier's contribution to the linear world view began in 1807 in his study of the flow of heat in a solid body. He established the principle of superposition for the heat equation. If $T(x,t)$ is the heat measured at the point x at time t in a one-dimensional solid body, the description of the evolution of heat in the body is given by the transport equation

$$\frac{\partial T(x,t)}{\partial t} = \kappa_T \frac{\partial^2 T(x,t)}{\partial x^2}, \qquad (4.61)$$

where κ_T is the thermal conductivity of the solid. Note that the heat equation describes irreversible dissipative processes unlike the wave equation which describes reversible periodic processes. It is significant that the principle of superposition could be applied to these two fundamentally different physical phenomena. In 1822 Fourier argued, and it was eventually proven in 1829 by Dirichlet, that any piecewise continuous (smooth) function could be expanded as a trigonometric (Fourier) series. Thus, the emerging disciplines of acoustics, heat transport, and electromagnetic theory (Maxwell's equations) in the 18th and 19th centuries could all be treated in a unified way. The linear nature of all these processes ensures that they satisfy the principle of superposition.

Principle of Superposition   The penultimate form of the principle of superposition was expressed in the mathematical formulation of the Sturm-Liouville theory (1836-1838). The authors of this theory demonstrated that a large class of differential equations (and thereby the associated physical processes intended to be modeled by these equations) could be cast in the form of eigenvalue problems. The solution to the original equation could then be represented as a linear superposition of the characteristic or eigenmotions of the physical phenomenon. This was immediately perceived as the systematic way of unraveling complex events in the physical world and giving them relatively simple mathematical descriptions. In the following century this would allow the problems of the microscopic world to be treated in a mathematically rigorous way. All of quantum mechanics would become accessible in investigation due to the existence of this mathematical apparatus.

Einstein Diffusion   One of the more illuminating applications of the above ideas was to the explanation of the irreversible process of diffusion. If we write the density of, say, the aromatic molecules of perfume at location x at time t in a one-dimensional fluid (the air) by $\rho(x,t)$, then the diffusion equation of Einstein is given by

$$\frac{\partial \rho(x,t)}{\partial t} = D \frac{\partial^2 \rho(x,t)}{\partial x^2}, \qquad (4.62)$$

where D is the diffusion coefficient. The diffusion equation for the mass density is a parabolic partial differential equation whose solution can be represented as the eigenfunction expansion

$$\rho(x,t \mid x_0) = \sum_{n=0}^{\infty} \rho_{ss}(x) \frac{\Psi_n(x)\Psi_n(x_0)}{N_n} e^{-\lambda_n t}. \tag{4.63}$$

The steady-state solution to the diffusion equation is given by

$$\frac{\partial \rho_{ss}(x)}{\partial t} = 0 \tag{4.64}$$

and the eigenfunctions Ψ_n are orthogonal with respect to the weighting function given by the steady-state distribution

$$N_n \delta_{mn} = \int_\Omega dx \rho_{ss}(x)\Psi_m(x)\Psi_n(x) \tag{4.65}$$

and Ω is the domain of the process. The eigenfunctions $\Psi_n(x)$ and the eigenvalues λ_n satisfy the differential equation

$$D\Psi_n''(x) + D\frac{d[\ln\rho_{ss}(x)]}{dx}\Psi_n'(x) + \lambda_n\Psi_n(x) = 0 \tag{4.66}$$

subject to the appropriate boundary conditions. Equation (4.66) can be written in the self-adjoint form

$$\frac{d}{dx}\left[D\rho_{ss}(x)\frac{d}{dx}\Psi_n(x)\right] + \lambda_n\rho_{ss}(x)\Psi_n(x) = 0. \tag{4.67}$$

Since $\rho_{ss}(x)$ is a positive function with a finite number of poles the differential equation (4.67) is of the Sturm-Liouville type and the eigenvalues λ_n are real, nondegenerate, nonnegative, and may be ordered such that $\lambda_0 < \lambda_1 < \cdots$. [13]

Continuity and Differentiability The above formalism is based on the mathematical ideas of continuity and differentiability, which to the physicists of the 19th century meant more or less the same thing. Cantor and Weierstrass, among a select few, however, were able to show that this was not always true. In fact we now know that this is almost never true in physical phenomena: *continuity - yes; differentiability - almost never*. In previous lectures we saw that Fourier series could be used to represent fractal functions, for example, those of Weierstrass. We also saw that complex phenomena, those described by fractional derivatives, can be represented by generalized exponential functions. Is it therefore possible to synthesize these two ideas and express fractal functions as generalized Fourier series, that is, series of the form (4.5) but with the trigonometric functions replaced with generalized trigonometric functions? Furthermore, can the Dirac delta function in (4.28) be extended to include the generalized exponential function and thereby obtain a fractional Dirac delta function? In the next few lectures we discuss these questions and find that the answer to both is a tentative yes.

4.3   Fourier Transforms in the Fractional Calculus

Properties of Functions   In the usual Fourier analysis we would restrict our considerations to functions $f(t)$ with finite integrated norm on the open interval

$$\int_{-\infty}^{\infty} |f(t)|\, dt < \infty. \tag{4.68}$$

To establish notation we define the Fourier transform of the functions by (4.26)

$$\hat{f}(\omega) = \mathcal{FT}\left[f(t);\omega\right] = \int_{-\infty}^{\infty} e^{i\omega t} f(t)\, dt$$

and the inverse Fourier transform by

$$f(t) = \mathcal{FT}^{-1}\left[\hat{f}(\omega);t\right] = \frac{1}{2\pi} \int_{-\infty}^{\infty} e^{-i\omega t}\, \hat{f}(\omega)\, d\omega.$$

Note that to extend our ideas concerning Fourier analysis we consider functions that are not $\mathcal{L}_2$ integrable. For the moment let us consider some additional properties of Fourier transforms that we shall find useful in the subsequent generalizations. The integer frequency derivative of the Fourier transform of the function $f(t)$ is given by

$$\frac{d^n \hat{f}(\omega)}{d\omega^n} = \int_{-\infty}^{\infty} (it)^n\, e^{i\omega t} f(t)\, dt \tag{4.69}$$

which can be expressed as

$$\frac{d^n \hat{f}(\omega)}{d\omega^n} = i^n \mathcal{FT}\left[t^n f(t);\omega\right], \tag{4.70}$$

the Fourier transform of the product $t^n f(t)$. In the same way we can write for the Fourier transform of the time derivative of the function $f(t)$:

$$\mathcal{FT}\left[\frac{d^n f(t)}{dt^n};\omega\right] = \int_{-\infty}^{\infty} e^{i\omega t} \frac{d^n f(t)}{dt^n} dt$$

which when integrated by parts n times, each integration by parts providing a factor $(-i\omega)$, yields

$$\mathcal{FT}\left[\frac{d^n f(t)}{dt^n};\omega\right] = \int_{-\infty}^{\infty} (-i\omega)^n\, e^{i\omega t} f(t)\, dt$$

provided that the function and its first n derivatives vanish as $t \to \pm\infty$, so there are no contributions from the boundaries after each of the integrations by parts. Now we obtain

$$FT\left[\frac{d^n f(t)}{dt^n};\omega\right] = (-i\omega)^n\, FT\,[f(t)\,;\omega] = (-i\omega)^n\,\hat{f}(\omega) \tag{4.71}$$

which taking the inverse Fourier transform yields

$$\frac{d^n f(t)}{dt^n} = (-i)^n\, FT^{-1}\left[\omega^n\,\hat{f}(\omega)\,;t\right]. \tag{4.72}$$

Note the complementary aspect of (4.70) and (4.72).

4.3.1 Fourier Transforms of Fractional Derivatives

Generalized Transform Let us now turn our attention to the study of the behavior of the fractional derivative operator and the generalized exponential function under the Fourier transform operation. Consider the definition of the fractional derivative of the function $f(t)$:

$$\begin{aligned}
D_t^\alpha\,[f(t)] &= \sum_{k=0}^{\infty}\binom{\alpha}{k} D_t^{\alpha-k}\,[t^0]\, D_t^k\,[f(t)]\\
&= \sum_{k=0}^{\infty}\binom{\alpha}{k}\frac{t^{k-\alpha}}{\Gamma(k-\alpha+1)}D_t^k\,[f(t)] \tag{4.73}
\end{aligned}$$

where we now introduce the Fourier transform of the derivatives of $f(t)$ given by (4.71);

$$D_t^\alpha\,[f(t)] = \sum_{k=0}^{\infty}\binom{\alpha}{k}\frac{t^{k-\alpha}}{\Gamma(k-\alpha+1)}\int_{-\infty}^{\infty} d\omega\, e^{-i\omega t}\,(-i\omega)^k\,\hat{f}(\omega). \tag{4.74}$$

We can write the series in the form of the generalized exponential

$$E_\alpha^{-i\omega t} = \sum_{k=0}^{\infty}\binom{\alpha}{k}\frac{(-i\omega t)^{k-\alpha}}{\Gamma(k-\alpha+1)}e^{-i\omega t} \tag{4.75}$$

so that (4.74) reduces to

$$D_t^\alpha\,[f(t)] = \frac{1}{2\pi}\int_{-\infty}^{\infty} d\omega\, E_\alpha^{-i\omega t}\,(-i\omega)^\alpha\,\hat{f}(\omega). \tag{4.76}$$

Equation (4.76) is the fractional derivative of the function $f(t)$ expressed as a Fourier transform (4.27) with respect to the independent variable ω; it provides a way to calculate the fractional derivative of a function without using the series expansion. Equation (4.76) is analogous to the integer derivative of the function given by (4.72) with the exponential function $e^{-i\omega t}$ replaced with the generalized exponential function $E_\alpha^{-i\omega t}$.

Fractional Derivatives of Fourier Amplitudes In a similar way we take the fractional derivative of the Fourier transform with respect to the transform variable

$$D_\omega^\alpha \left[\hat{f}(\omega) \right] = \sum_{k=0}^\infty \left(\begin{array}{c} \alpha \\ k \end{array} \right) D_\omega^{\alpha-k} [\omega^0] \, D_\omega^k \left[\hat{f}(\omega) \right]$$

$$= \sum_{k=0}^\infty \frac{\left(\begin{array}{c} \alpha \\ k \end{array} \right) \omega^{k-\alpha}}{\Gamma(k-\alpha+1)} D_\omega^k \left[\hat{f}(\omega) \right], \tag{4.77}$$

where we now introduce the inverse Fourier transform of $\hat{f}(\omega)$ given by (4.69),

$$D_\omega^\alpha \left[\hat{f}(\omega) \right] = \sum_{k=0}^\infty \left(\begin{array}{c} \alpha \\ k \end{array} \right) \frac{\omega^{k-\alpha}}{\Gamma(k-\alpha+1)} \int_{-\infty}^\infty (it)^k \, e^{i\omega t} f(t) \, dt. \tag{4.78}$$

We can write the series in (4.78) in the form of the generalized exponential

$$E_\alpha^{i\omega t} = \sum_{k=0}^\infty \left(\begin{array}{c} \alpha \\ k \end{array} \right) \frac{(i\omega t)^{k-\alpha}}{\Gamma(k-\alpha+1)} e^{i\omega t} \tag{4.79}$$

which when reinserted into (4.78) yields

$$D_\omega^\alpha \left[\hat{f}(\omega) \right] = \int_{-\infty}^\infty (it)^\alpha \, E_\alpha^{i\omega t} f(t) \, dt. \tag{4.80}$$

Generalized Fourier Transform Thus, we see that (4.76) and (4.80) are analogous to the properties of Fourier transforms given by (4.70) and (4.72). The relation between the derivatives of the function and its Fourier amplitude is the same as the fractional derivative of the function and its Fourier transform with the exponential replaced with the generalized exponential. Also the relations between the derivatives of the Fourier transform of the function and the function is the same as the fractional derivative of the Fourier transform of the function and the function with the exponential replaced with the generalized exponential.

Exponential Example Let us consider the example $f(t) = e^{at}$, whose Fourier transform is given by

$$\hat{f}(\omega) = \int_{-\infty}^\infty e^{i\omega t} e^{at} \, dt \tag{4.81}$$

from which we obtain the Dirac delta function

$$\hat{f}(\omega) = 2\pi \delta(\omega - ia). \tag{4.82}$$

Inserting the Fourier amplitude (4.82) into (4.76) we have for the fractional derivative of the exponential

$$D_t^\alpha \left[e^{at} \right] = \frac{1}{2\pi} \int_{-\infty}^{\infty} d\omega \, E_\alpha^{-i\omega t} \left(-i\omega \right)^\alpha 2\pi \delta \left(\omega - ia \right) = a^\alpha E_\alpha^{at} \qquad (4.83)$$

a result we obtained earlier and which was also obtained by Bologna [1].

4.3.2 Weyl's Fractional Operators

A Fourier-based Fractional Calculus Here we discuss the similarity between the Fourier transform of the fractional derivative as defined by (4.76) and the definition of the fractional integro-differential operators developed by Weyl. The latter definition was developed for the treatment of periodic functions so that the operation of fractional derivatives or fractional integrals acting on periodic functions would yield periodic functions. The Riemann-Liouville fractional integro-differentiation does not do this and therefore was thought to be unsatisfactory.

Derivatives of Fourier Series We construct the definition of the Weyl fractional operators by defining the actions of the operators on the periodic functions expressed as the Fourier series (4.4) and/or (4.14). We follow Samko et al. [16], in part, and set the constant term in these series to zero; that is, $c_0 = 0$. In this way we can write the nth derivative, n an integer, of a periodic function using (4.14) as

$$\frac{d^n f \left(\tau \right)}{d\tau^n} = \sum_{k=-\infty}^{\infty} \left(-ik \right)^n c_k e^{-ik\tau}. \qquad (4.84)$$

Equation (4.84) suggests the immediate generalization to the Weyl fractional operator

$$\mathcal{W}_\tau^\alpha \left[f \left(\tau \right) \right] = \sum_{k=-\infty}^{\infty} \left(-ik \right)^\alpha c_k e^{-ik\tau}, \qquad (4.85)$$

where $1 > \alpha > 0$ indicates a Weyl fractional derivative and $0 > \alpha > -1$ denotes a Weyl fractional integral. Equation (4.85) meets the requirement that the integro-derivative of a periodic function is again a periodic function. A further generalization of this operator includes the relation with generalized exponential functions. Recall the fractional derivative of the generalized exponential function given by (3.55) which will allow us to write (4.85) as

$$\mathcal{W}_\tau^\alpha \left[f \left(\tau \right) \right] = D_\tau^\alpha \left[\sum_{k=-\infty}^{\infty} c_k E_{-\alpha}^{-ik\tau} \right] \qquad (4.86)$$

from which we can define the fractional function by a generalized Fourier series

$$f(\tau) = \sum_{k=-\infty}^{\infty} c_k E_{-\alpha}^{-ik\tau}, \tag{4.87}$$

where we note that this expansion depends on the fractional index α.

Convolutions Using Fourier Series We define the convolution of two 2π-periodic functions $f(t)$ and $g(t)$ by the relation

$$(f * g)(\tau) = \frac{1}{2\pi} \int_0^{2\pi} f(\tau - \tau') g(\tau') d\tau' \tag{4.88}$$

so that inserting Fourier series under the integral for each of these functions

$$f(\tau) = \sum_{k=-\infty}^{\infty} c_k e^{-ik\tau},$$

and

$$g(\tau) = \sum_{k=-\infty}^{\infty} g_k e^{-ik\tau},$$

and using the orthogonality of the exponentials given by (4.8) we obtain

$$(f * g)(\tau) = \sum_{k=-\infty}^{\infty} c_k g_{-k} e^{-ik\tau}, \tag{4.89}$$

where c_k and g_k are the Fourier coefficients of the 2π-periodic functions $f(\tau)$ and $g(\tau)$. Using the convolution property (4.89) we can write the definition of the Weyl fractional integral (4.85) as

$$W_\tau^{-\beta}[f(\tau)] = \frac{1}{2\pi} \int_0^{2\pi} f(\tau - \tau') \phi_\beta(\tau') d\tau', \tag{4.90}$$

where we have introduced the function

$$\phi_\beta(\tau) = \sum_{k=-\infty, \neq 0}^{\infty} \frac{e^{-ik\tau}}{(-ik)^\beta} \tag{4.91}$$

and we explicitly denote the exclusion of the $k = 0$ term from the sum. Here again we could do the analysis with the generalized exponential and write (4.91) as

$$\phi_\beta(\tau) = D_\tau^{-\beta} \left[\sum_{k=-\infty, \neq 0}^{\infty} E_\beta^{-ik\tau} \right], \tag{4.92}$$

but since we are only using the Weyl calculus as an example we do not explore that option here. We can write an alternate expression for (4.91) by noting that

$$(-ik)^\beta = |k|^\beta e^{-i\beta\pi k/2|k|}$$

which when substituted into (4.91) yields

$$\phi_\beta(\tau) = 2 \sum_{k=1}^{\infty} \frac{\cos[k\tau + \beta\pi/2]}{k^\beta}. \tag{4.93}$$

It is a simple matter to show by direct substitution of (4.93) into (4.90) that we again have the Weyl fractional derivative (4.85) with $\alpha = -\beta > 0$. Zygmund ([19], page 201) has shown that (4.90) converges for all $\tau \in (0, 2\pi)$ if $\beta > 0$.

Weyl's Form of a Fractional Derivative   It is possible to prove using the Poisson summation formula for the function (4.91) that the region of integration for the Weyl fractional integral can be extended from the interval $(0, 2\pi)$ to the interval $(0, \infty)$ such that (4.90) is replaced with [16]

$$\mathcal{W}_t^{-\beta}[f(t)] = \frac{1}{\Gamma(\beta)} \int_0^\infty f(t - t') t'^{\beta-1} dt' \tag{4.94}$$

so that changing the integration variable we have

$$\mathcal{W}_t^{-\beta}[f(t)] = \frac{1}{\Gamma(\beta)} \int_{-\infty}^t f(t')(t - t')^{\beta-1} dt'. \tag{4.95}$$

Thus, the definition of the Weyl fractional integral is in complete agreement with the Riemann-Liouville definition of the fractional derivative with the lower bound of the integral being $-\infty$. We emphasize, however, that the convergence of the integral (4.95) is strongly dependent on the integral of the function $f(t)$ vanishing over one period. We refer the reader to Samko et al. [16] for a complete discussion.

Comparison of Different Fractional Operators   To establish the equivalence of (4.85) and an integral form of the Weyl fractional derivative for $1 > \alpha > 0$ we define,

$$D_t^\alpha[f(t)] = \frac{d}{dt} \mathcal{W}_t^{1-\alpha}[f(t)] \tag{4.96}$$

and using the integral (4.90)

$$\mathcal{W}_t^{1-\alpha}[f(t)] = \frac{1}{2\pi} \int_0^{2\pi} f(t - t') \phi_{1-\alpha}(t') dt'$$

we obtain for the right-hand side of (4.96)

$$\frac{d}{dt} \mathcal{W}_t^{1-\alpha}[f(t)] = \frac{1}{2\pi} \int_0^{2\pi} \frac{df(t - t')}{dt} \phi_{1-\alpha}(t') dt'. \tag{4.97}$$

Therefore from the identity

$$\frac{df\,(t - t')}{dt} = -\frac{df\,(t - t')}{dt'}$$

we can integrate (4.97) by parts and obtain

$$\frac{d}{dt}\mathcal{W}_t^{1-\alpha}\left[f\,(t)\right] = \frac{1}{2\pi}\int_0^{2\pi} f\,(t - t')\,\frac{d\phi_{1-\alpha}\,(t')}{dt'}dt' \qquad (4.98)$$

so that using the series expansion for the kernel we have

$$\frac{d\phi_{1-\alpha}\,(t')}{dt'} = \sum_{k=-\infty,\neq 0}^{\infty} \frac{-ike^{-ikt'}}{(-ik)^{1-\alpha}}$$

and (4.97) reduces to

$$\frac{d}{dt}\mathcal{W}_t^{1-\alpha}\left[f\,(t)\right] = \frac{1}{2\pi}\sum_{k=-\infty,\neq 0}^{\infty} (-ik)^{\alpha}\int_0^{2\pi} e^{-ikt'}f\,(t - t')\,dt'. \qquad (4.99)$$

Thus, replacing the integral in (4.99) with the Fourier transform of $f\,(t)$ we can write

$$D_t^{\alpha}\left[f\,(t)\right] = \sum_{k=-\infty,\neq 0}^{\infty} (-ik)^{\alpha}\,c_k e^{-ikt} \qquad (4.100)$$

which agrees with the definition of the Weyl fractional derivative; that is,

$$D_t^{\alpha}\left[\cdot\right] = \mathcal{W}_t^{\alpha}\left[\cdot\right] \qquad (4.101)$$

comparing (4.85) with (4.100). A relatively complete discussion of the Weyl calculus is given in Miller and Ross [14].

In Terms of Fourier Integrals Now we consider the continuum limit of the time series (4.85) using the device implemented earlier where we introduced the period of the function as $\overline{T}$, and then took the limit as $\overline{T} \to \infty$. Recall from (4.23) that

$$c_k = \frac{1}{\overline{T}}\int_0^{\overline{T}} e^{ik2\pi t/\overline{T}}f\,(t)\,dt$$

so that we can rewrite the Fourier series as

$$f\,(t) = \sum_{k=-\infty}^{\infty} \Delta k c_k e^{-ik2\pi t/\overline{T}}$$

with $2\pi t\Delta k \ll \overline{T}$. Then in the limit $\overline{T} \to \infty$ we obtain the associations

$$c_k \to \hat{f}(\omega)$$

and

$$\frac{1}{T} \sum_{k=-\infty}^{\infty} \Delta k e^{-i\omega_k t} \to \frac{1}{2\pi} \int_{-\infty}^{\infty} e^{-i\omega t} d\omega$$

so that (4.85) can be written as

$$W_t^\alpha [f(t)] = \frac{1}{2\pi} \int_{-\infty}^{\infty} (-i\omega)^\alpha e^{-i\omega t} \hat{f}(\omega) d\omega. \tag{4.102}$$

Equation (4.102) is the continuum representation of the Weyl fractional derivative (4.100). Note that this fractional derivative is quite different from the one given by (4.76) in that the former has the generalized exponential function under the integral, and the latter has the usual exponential function. However the form of the two relations is the same. It may be of interest to consider the fractional derivative given by (4.76) as a generalization of the Weyl fractional derivative (4.102) to nonperiodic or infinitely periodic functions.

A Different Way to Get the Equation   To complete our brief comparison we calculate the Fourier transform of the fractional derivative of a function

$$\mathcal{FT} \{D_t^\alpha [f(t)]; \omega\} = \int_{-\infty}^{\infty} e^{i\omega t} D_t^\alpha [f(t)] dt. \tag{4.103}$$

Using the $\mathcal{RL}$-form of the fractional derivative we have

$$D_t^\alpha [f(t)] = \frac{1}{\Gamma(-\alpha)} \int_{-\infty}^{t} \frac{f(t') dt'}{(t-t')^{1+\alpha}}$$

$$= \frac{1}{\Gamma(-\alpha)} \int_0^{\infty} \xi^{-(1+\alpha)} f(t-\xi) d\xi, \tag{4.104}$$

so that substituting the Fourier transform of the function

$$f(t-\xi) = \frac{1}{2\pi} \int_{-\infty}^{\infty} e^{-i\omega(t-\xi)} \hat{f}(\omega) d\omega$$

into (4.104) yields

$$D_t^\alpha [f(t)] = \frac{1}{\Gamma(-\alpha)} \int_0^{\infty} \xi^{-(1+\alpha)} d\xi \int_{-\infty}^{\infty} e^{-i\omega(t-\xi)} \hat{f}(\omega) \frac{d\omega}{2\pi}. \tag{4.105}$$

The Fourier transform of (4.105) then simplifies to

$$\mathcal{FT}\left\{D_t^\alpha\left[f\left(t\right)\right];\omega\right\} = \frac{\hat{f}\left(\omega\right)}{\Gamma\left(-\alpha\right)}\int_0^\infty e^{i\omega\xi}\xi^{-(1+\alpha)}d\xi. \qquad (4.106)$$

Evaluating the integral in (4.106) using the sine and cosine transforms separately we obtain from Gradshteyn and Ryzhik ([7], pgs. 1146-1152)

$$\int_0^\infty e^{i\omega\xi}\xi^{-(1+\alpha)}d\xi = \Gamma\left(-\alpha\right)\omega^\alpha\left[\cos\left(\alpha\pi/2\right) - i\sin\left(\alpha\pi/2\right)\right] = \Gamma\left(-\alpha\right)\left(-i\omega\right)^\alpha$$
$$(4.107)$$

so that (4.106) becomes

$$\mathcal{FT}\left\{D_t^\alpha\left[f\left(t\right)\right];\omega\right\} = \left(-i\omega\right)^\alpha \hat{f}\left(\omega\right), \qquad (4.108)$$

the product of the Fourier transform of the operator and the Fourier transform function. Thus, taking the inverse Fourier transform of (4.108) we obtain

$$D_t^\alpha\left[f\left(t\right)\right] = \frac{1}{2\pi}\int_{-\infty}^\infty e^{-i\omega t}\left(-i\omega\right)^\alpha \hat{f}\left(\omega\right)d\omega \qquad (4.109)$$

which is of the same form obtained using the Weyl derivative, that is, (4.102).

4.3.3   The Dirac Delta Function

The Impulse Function   The Dirac delta function is another example of a useful quantity that may be generalized to the fractional calculus[1]. It is often stated that the delta function was first introduced into physics by Dirac [3] in his treatment of relativistic quantum phenomena. However, the idea was developed more than a century earlier by Cauchy [2] and independently by Poisson [15], in their derivation of the Fourier integral theorem. Their arguments were not only independent of each other, but independent of the work of Fourier as well. Some 85 years later, Hermite [8] referred to their work when he also made use of what they called the impulse function.

Many Functional Forms   It was not only mathematicians that made use of the delta function; Kirchoff and Kon [11] used it in their formulation of Huygen's principle in the wave theory of light, and Lord Kelvin [18] showed that a heat source could also be characterized by such a function. In each of these cases the impulse function was considered as a limiting form of certain types of sequence functions. For example, four types of sequence functions are [5]

$$\delta_1\left(t,\varepsilon\right) = \begin{cases} \frac{1}{2\varepsilon}, & |t| < \varepsilon \\ 0, & |t| > \varepsilon \end{cases}, \qquad (4.110)$$

$$\delta_2\left(t,\varepsilon\right) = \frac{1}{\pi}\frac{\varepsilon}{\varepsilon^2 + t^2}, \qquad (4.111)$$

---

[1]The historical references and analyses in this lecture benefitted a great deal from Giri [5].

$$\delta_3 (t, \varepsilon) = \frac{1}{\pi \varepsilon} \frac{\sin (t/\varepsilon)}{t/\varepsilon}, \tag{4.112}$$

$$\delta_4 (t, \varepsilon) = \frac{1}{\varepsilon \sqrt{\pi}} exp \left[-\frac{t^2}{\varepsilon^2} \right] \tag{4.113}$$

so that in general

$$\delta (t) = \lim_{\varepsilon \to 0} \delta_j (t, \varepsilon). \tag{4.114}$$

Each of the four functions used in (4.114), $j = 1, 2, 3, 4$, becomes increasingly peaked at the origin as $\varepsilon \to 0$ and tends to vanish everywhere else. Of course, these four functions are by no means exhaustive of those that can be used to characterize the impulse function asymptotically.

Heaviside's Representation Heaviside [9], in his study of electromagnetic waves, found the infinite series representation of the delta function

$$\delta (x - y) = \frac{1}{\overline{T}} \sum_{n=1}^{\infty} \sin \left(\frac{n \pi x}{\overline{T}} \right) \sin \left(\frac{n \pi y}{\overline{T}} \right) \tag{4.115}$$

for real nonzero x and y such that $|x|, |y| < \overline{T}$. Note that (4.115) is a divergent series, which it must be in order for the delta function to be an improper function and not a regular function. A regular function is one that has a definite value for each point in its domain. It was to provide a mathematical foundation for such objects as the delta function that generalized function theory [12] was invented. Thus, we should, in fact, refer to $\delta (t)$ as a delta distribution, but the nomenclature of the Dirac delta function has become so ingrained in the physics literature that it would be nearly impossible to change. From the physical perspective it is the utility of $\delta (t)$ and its derivatives that are important, and as long as no contradictions arise with its applications, we are justified in its use.

Fourier Series Representation We can use the sequence function $\delta_j (t)$ in conjunction with the finite Fourier transform (4.5) on a unit interval to write the Fourier coefficient as

$$
\begin{aligned}
a_k &= \frac{1}{\pi} \int_{-\pi}^{\pi} dt \delta_1 (t, \varepsilon) \cos kt = \frac{1}{\pi} \int_{-\varepsilon}^{\varepsilon} \frac{dt}{2\varepsilon} \cos kt \\
&= \frac{\sin k\varepsilon}{k \pi \varepsilon}, \text{ with } a_0 = \frac{1}{\pi}.
\end{aligned}
\tag{4.116}
$$

Using (4.116) the Fourier series becomes

$$\delta_1 (t, \varepsilon) = \frac{1}{2\pi} + \frac{1}{\pi} \sum_{k=1}^{\infty} \frac{1}{k\varepsilon} \sin (k\varepsilon) \cos (kt) \tag{4.117}$$

and using (4.114), we obtain

$$\delta\left(t\right)=\frac{1}{2\pi}+\frac{1}{\pi}\sum_{k=1}^{\infty}\cos\left(kt\right)=\frac{1}{2\pi}\sum_{k=-\infty}^{\infty}e^{ikt}\text{ , for }\left|t\right|\leq\pi. \tag{4.118}$$

Thus, by a suitable choice of variables, $t\rightarrow\left(x-y\right)\pi/\overline{T}$, it is possible to obtain Heaviside's formula (4.115) in the limit $\overline{T}\rightarrow\infty$. Note that to extend (4.118) to all values of the independent variable, it is only necessary to invoke periodic boundary conditions to obtain

$$\sum_{k=-\infty}^{\infty}\delta\left(t-2k\pi\right)=\frac{1}{2\pi}+\frac{1}{\pi}\sum_{k=1}^{\infty}\cos kt=\frac{1}{2\pi}\sum_{k=-\infty}^{\infty}e^{ikt}. \tag{4.119}$$

This argument can, of course, be constructed for any of the sequence functions yielding different Fourier series

$$\delta\left(t\right)=\lim_{\varepsilon\rightarrow0}\left[\frac{1}{2\pi}+\frac{1}{\pi}\sum_{k=1}^{\infty}a_{k}\left(j,\varepsilon\right)\cos kt\right] \tag{4.120}$$

where the Fourier coefficients, $a_{k}\left(j,\varepsilon\right)$, are given by the integrals

$$a_{k}\left(j,\varepsilon\right)=\frac{1}{\pi}\int_{-\pi}^{\pi}dt\delta_{j}\left(t,\varepsilon\right)\cos kt. \tag{4.121}$$

Fourier Integral Representation   The continuous form of the Dirac delta function is obtained by inserting one of the sequence functions into the definition of the Fourier transform. Consider the Fourier transform of a representation of the delta function

$$\begin{aligned}\hat{f}\left(\omega\right) &= \int_{-\infty}^{\infty}e^{i\omega t}f\left(t\right)dt=\int_{-\infty}^{\infty}e^{i\omega t}dt\delta\left(t\right)\\ &= \lim_{\varepsilon\rightarrow0}\int_{-\infty}^{\infty}e^{i\omega t}dt\delta_{j}\left(t,\varepsilon\right).\end{aligned} \tag{4.122}$$

As an example substitute the $j=2$ function into (4.122) to obtain

$$\hat{f}\left(\omega\right)=\lim_{\varepsilon\rightarrow0}\int_{-\infty}^{\infty}e^{-i\omega t}\frac{dt}{\pi}\frac{\varepsilon}{\varepsilon^{2}+t^{2}}=\lim_{\varepsilon\rightarrow0}\left[e^{-\left|\omega\right|\varepsilon}\right]. \tag{4.123}$$

Now taking the inverse Fourier transform of the Dirac delta function in (4.122) and using (4.123) yields

$$\delta\left(t\right)=\frac{1}{2\pi}\int_{-\infty}^{\infty}e^{-i\omega t}\lim_{\varepsilon\rightarrow0}\left[e^{-\left|\omega\right|\varepsilon}\right]d\omega=\frac{1}{2\pi}\int_{-\infty}^{\infty}e^{-i\omega t}d\omega. \tag{4.124}$$

The delta function is only rigorously defined by its integration over an interval such that for a test function $f\left(t\right)$:

$$\int_{-\infty}^{\infty} f(t)\, \delta(t-a)\, dt = f(a) \tag{4.125}$$

putting all the weight of the function at the value where the delta function diverges, $t = a$. In addition to the integral over the Dirac delta function we also have integrals over the derivatives of the delta function.

Derivatives of Delta Functions Suppose we have the first derivative of the delta function denoted by $\delta'(t-a)$; then

$$
\begin{aligned}
\int_{-\infty}^{\infty} f(t)\, \delta'(t-a)\, dt &= f(t)\, \delta(t-a)\, |_{-\infty}^{\infty} - \int_{-\infty}^{\infty} f'(t)\, \delta(t-a)\, dt \\
&= -f'(a),
\end{aligned}
\tag{4.126}
$$

where we have integrated by parts and $f'(a)$ is the derivative of the test function evaluated at $t = a$. In the same way we have for the nth order derivative of the delta function $\delta^{(n)}(t-a)$ the result

$$\int_{-\infty}^{\infty} f(t)\, \delta^{(n)}(t-a)\, dt = (-1)^n f^{(n)}(a), \tag{4.127}$$

where the phase arises due to integrations by parts. Equations (4.125) to (4.127) are the relations we wish to generalize.

Fractional Derivatives of Delta Functions We want to establish a connection among the above relations concerning the Dirac delta function and the generalized exponential. Consider the fractional derivative of an exponential given by (4.83) from which we can define the integral over the frequency involving the fractional time derivative

$$\int_{-\infty}^{\infty} D_t^\alpha \left[e^{i\omega t} \right] d\omega = \int_{-\infty}^{\infty} (i\omega)^\alpha E_\alpha^{i\omega t}\, d\omega. \tag{4.128}$$

Also the integral over time involving the fractional frequency derivative is

$$\int_{-\infty}^{\infty} D_\omega^\alpha \left[e^{i\omega t} \right] dt = \int_{-\infty}^{\infty} (it)^\alpha E_\alpha^{i\omega t}\, dt. \tag{4.129}$$

From (4.129) we can factor out the fractional frequency derivative

$$D_\omega^\alpha \left[\int_{-\infty}^{\infty} e^{i\omega t}\, dt \right] = \int_{-\infty}^{\infty} (it)^\alpha E_\alpha^{i\omega t}\, dt,$$

but the integral in the square brackets is just the Dirac delta function so that we can write for the fractional derivative of the delta function

$$\delta^{(\alpha)}(\omega) \equiv D_\omega^\alpha \left[\delta(\omega) \right] = \frac{1}{2\pi} \int_{-\infty}^{\infty} (it)^\alpha E_\alpha^{i\omega t}\, dt. \tag{4.130}$$

Thus, the αth derivative of the delta function can be expressed in terms of an integral over the generalized exponential function.

Fourier Integral Representation   In the same way we can factor the fractional time derivative from (4.128) to obtain

$$D_t^\alpha \left[\int_{-\infty}^\infty e^{i\omega t} d\omega \right] = \int_{-\infty}^\infty (i\omega)^\alpha E_\alpha^{i\omega t} d\omega,$$

where again the term in brackets is a Dirac delta function, but now in the time so we can write

$$\delta^{(\alpha)}(t) \equiv D_t^\alpha \left[\delta(t) \right] = \frac{1}{2\pi} \int_{-\infty}^\infty (i\omega)^\alpha E_\alpha^{i\omega t} d\omega, \qquad (4.131)$$

a completely equivalent definition to (4.130).

Using the Fractional Derivative of the Delta Function   Now let us consider the convolution of the fractional derivative of the Dirac delta function, $\delta^{(\alpha)}(t - t')$, and a test function $f(t')$ integrated over $-\infty \leq t \leq \infty$,

$$I_\alpha(t) = \int_{-\infty}^\infty \delta^{(\alpha)}(t - t') f(t') dt'. \qquad (4.132)$$

Of course, since (4.132) has the form of a convolution we can write for the Fourier transform of the integral as the product of the Fourier amplitudes

$$\mathcal{FT} \left\{ I_\alpha(t) ; \omega \right\} = \overset{\wedge(\alpha)}{\delta}(\omega) \, \hat{f}(\omega). \qquad (4.133)$$

Furthermore, the Fourier transform of the fractional delta function is

$$\overset{\wedge(\alpha)}{\delta}(\omega) = \int_{-\infty}^\infty e^{-i\omega t} D_t^\alpha \left[e^{i\omega t} \right] dt \qquad (4.134)$$

which can readily be evaluated from (4.108) to be

$$\overset{\wedge(\alpha)}{\delta}(\omega) = (-i\omega)^\alpha. \qquad (4.135)$$

Thus, we have

$$\mathcal{FT} \left\{ I_\alpha(t) ; \omega \right\} = (-i\omega)^\alpha \, \hat{f}(\omega), \qquad (4.136)$$

whose inverse Fourier transform is

$$\int_{-\infty}^\infty \delta^{(\alpha)}(t - t') f(t') dt' = (-1)^\alpha D_t^\alpha \left[f(t) \right] \qquad (4.137)$$

which is the direct generalization of the integer derivative of the Dirac delta function given by (4.126).

4.4 Generalized Fourier Transforms

Generalized Fourier Integral The analysis in the preceding section could have been facilitated by introducing the generalized Fourier transform of a function by

$$\hat{f}_\alpha(\omega) \equiv \int_{-\infty}^{\infty} E_\alpha^{i\omega t} f(t)\, dt. \tag{4.138}$$

We construct the inverse of this transform by introducing the function

$$g(t) = \frac{f(t)}{(it)^\alpha} \tag{4.139}$$

so that we can rewrite the fractional Fourier transform as

$$\hat{f}_\alpha(\omega) = \int_{-\infty}^{\infty} (it)^\alpha\, E_\alpha^{i\omega t} g(t)\, dt. \tag{4.140}$$

In terms of the fractional derivative with respect to frequency we have

$$\hat{f}_\alpha(\omega) = D_\omega^\alpha \left[\int_{-\infty}^{\infty} e^{i\omega t} g(t)\, dt \right], \tag{4.141}$$

since

$$D_\omega^\alpha \left[e^{i\omega t} \right] = (it)^\alpha\, E_\alpha^{i\omega t}.$$

The integral in (4.141) yields the ordinary Fourier transform of $g(t)$, that is, $\hat{g}(\omega)$, so that (4.141) can be formally written as

$$\hat{f}_\alpha(\omega) = D_\omega^\alpha \left[\hat{g}(\omega) \right]. \tag{4.142}$$

Equation (4.142) may be inverted to obtain

$$\hat{g}(\omega) = D_\omega^{-\alpha} \left[\hat{f}_\alpha(\omega) \right] \tag{4.143}$$

so that taking the inverse Fourier transform of (4.143) yields

$$g(t) = \int_{-\infty}^{\infty} \frac{d\omega}{2\pi} e^{-i\omega t} D_\omega^{-\alpha} \left[\hat{f}_\alpha(\omega) \right]$$

or in terms of the original function, using (4.139), we have

$$f(t) = (it)^\alpha \int_{-\infty}^{\infty} \frac{d\omega}{2\pi} e^{-i\omega t} D_\omega^{-\alpha} \left[\hat{f}_\alpha(\omega) \right]. \tag{4.144}$$

Equation (4.144) defines the inverse of the generalized Fourier transform given by (4.138).

Inversion of Gneralized Transform The inverse generalized Fourier transform relation is easily verified by substituting the definition of the generalized Fourier transform into (4.144) to obtain

$$f(t) = (it)^\alpha \int_{-\infty}^{\infty} \frac{d\omega}{2\pi} e^{-i\omega t} D_\omega^{-\alpha} \left[D_\omega^\alpha \left[\hat{g}(\omega) \right] \right]$$

but

$$D_\omega^{-\alpha} D_\omega^\alpha = 1$$

so that

$$f(t) = (it)^\alpha \int_{-\infty}^{\infty} \frac{d\omega}{2\pi} e^{-i\omega t} \hat{g}(\omega)$$

which, using the Fourier transform of (4.139) and the delta function (4.124), is a tautology.

4.4.1   Examples of Generalized Transforms

Transform of Monomial As an example of the consistent nature of the generalized Fourier transform and its inverse, consider the generalized Fourier transform of the monomial $t^{-\beta}$,

$$\mathcal{FT}_\alpha \left\{ t^{-\beta}; \omega \right\} \equiv \int_{-\infty}^{\infty} E_\alpha^{i\omega t} t^{-\beta} dt, \tag{4.145}$$

where the subscript denotes the generalized Fourier transform. Inserting the definition of the generalized exponential given by (4.83) into (4.145) yields

$$\mathcal{FT}_\alpha \left\{ t^{-\beta}; \omega \right\} = \int_{-\infty}^{\infty} t^{-\beta} D_\omega^\alpha \left[\frac{e^{i\omega t}}{(it)^\alpha} \right] dt$$
$$= e^{-i\alpha\pi/2} D_\omega^\alpha \left[\int_{-\infty}^{\infty} t^{-\beta-\alpha} e^{i\omega t} dt \right] \tag{4.146}$$

and evaluating the integral using the usual Fourier transform gives, after some algebra,

$$\mathcal{FT}_\alpha \left\{ t^{-\beta}; \omega \right\} = 2e^{-i\alpha\pi/2} \Gamma(1-\alpha-\beta) \cos\left[\pi(\alpha+\beta)/2 \right] D_\omega^\alpha \left[\omega^{\alpha+\beta-1} \right]. \tag{4.147}$$

The fractional derivative of the monomial is

$$D_\omega^\alpha \left[\omega^{\alpha+\beta-1} \right] = \frac{\Gamma(\alpha+\beta)}{\Gamma(\beta)} \omega^{\beta-1} \tag{4.148}$$

so that (4.147) becomes

$$FT_\alpha \left\{ t^{-\beta}; \omega \right\} = 2e^{-i\alpha\pi/2} \frac{\Gamma(\alpha+\beta)\,\Gamma(1-\alpha-\beta)}{\Gamma(\beta)} \cos\left[\pi(\alpha+\beta)/2\right] \omega^{\beta-1}$$

$$= \frac{\pi e^{-i\alpha\pi/2}}{\Gamma(\beta)\sin\left[\pi(\alpha+\beta)/2\right]} \omega^{\beta-1}. \tag{4.149}$$

Inversion of Generalized Transform The test now comes in determining the inverse Fourier transform of (4.149). Denote the coefficient in (4.149) by the constant $C(\alpha,\beta)$ so that the inverse generalized Fourier transform of this expression can be written, using (4.144), as

$$FT_\alpha^{-1}\left\{FT_\alpha\left\{t^{-\beta};\omega\right\};t\right\} \equiv \int_{-\infty}^\infty (it)^\alpha\,e^{-i\omega t} D_\omega^{-\alpha}\left[FT_\alpha\left\{t^{-\beta};\omega\right\}\right] \frac{d\omega}{2\pi}. \tag{4.150}$$

Inserting the generalized Fourier transform (4.149) into (4.150) yields

$$FT_\alpha^{-1}\left\{FT_\alpha\left\{t^{-\beta};\omega\right\};t\right\} = C(\alpha,\beta)\int_{-\infty}^\infty (it)^\alpha\,e^{-i\omega t} D_\omega^{-\alpha}\left[\omega^{\beta-1}\right] \frac{d\omega}{2\pi} \tag{4.151}$$

so that using

$$D_\omega^{-\alpha}\left[\omega^{\beta-1}\right] = \frac{\Gamma(\beta)}{\Gamma(\alpha+\beta)} \omega^{\alpha+\beta-1}$$

we obtain

$$FT_\alpha^{-1}\left\{FT_\alpha\left\{t^{-\beta};\omega\right\};t\right\} = \frac{\Gamma(\beta)}{\Gamma(\alpha+\beta)} C(\alpha,\beta)\,(it)^\alpha \int_{-\infty}^\infty \omega^{\alpha+\beta-1} e^{-i\omega t} \frac{d\omega}{2\pi}.$$

Evaluating the integral, after a little algebra, yields

$$FT_\alpha^{-1}\left\{FT_\alpha\left\{t^{-\beta};\omega\right\};t\right\} = C(\alpha,\beta)\,e^{i\alpha\pi/2} \frac{\Gamma(\beta)}{\pi} \sin\left[\pi(\alpha+\beta)/2\right] t^{-\beta}, \tag{4.152}$$

so that inserting the value for the overall coefficient into (4.152)

$$FT_\alpha^{-1}\left\{FT_\alpha\left\{t^{-\beta};\omega\right\};t\right\} = \frac{e^{-i\alpha\pi/2}\pi}{\Gamma(\beta)\sin\left[\pi(\alpha+\beta)/2\right]} \frac{e^{i\alpha\pi/2}\Gamma(\beta)}{\pi}$$

$$\times \sin\left[\pi(\alpha+\beta)/2\right] t^{-\beta} = t^{-\beta}, \tag{4.153}$$

which was to be shown. Thus, since the generalized Fourier transform works for monomials, by induction it should also work for analytic functions, since such functions have Maclaurin series expansions.

4.4.2   Generalized Convolutions

Parseval's Theorem   Consider the convolution

$$I = \int_{-\infty}^{\infty} G\left(t - t'\right) Q\left(t'\right) dt' \qquad (4.154)$$

and assuming that both functions have generalized Fourier transforms we obtain

$$I = \int_{-\infty}^{\infty} dt' \int_{-\infty}^{\infty} \frac{d\omega}{2\pi} E_{-\alpha}^{-i\omega(t-t')} \hat{G}_{\alpha}(\omega) \int_{-\infty}^{\infty} \frac{d\omega'}{2\pi} E_{\alpha}^{-i\omega't'} \hat{Q}_{-\alpha}(\omega'). \qquad (4.155)$$

To put (4.155) in its simplest form we must first evaluate

$$I^* = \int_{-\infty}^{\infty} dt' E_{-\alpha}^{-i\omega(t-t')} E_{\alpha}^{-i\omega't'} \qquad (4.156)$$

which can immediately be written

$$I^* = \int_{-\infty}^{\infty} dt' D_{t'}^{\alpha} \left[e^{-i\omega't'}\right] (-i\omega')^{-\alpha} D_{t-t'}^{-\alpha} \left[e^{-i\omega(t-t')}\right] (-i\omega)^{\alpha}. \qquad (4.157)$$

Using the integral definition of the fractional derivative we have

$$D_{t-t'}^{-\alpha} \left[e^{-i\omega(t-t')}\right] = \frac{1}{\Gamma(\alpha)} \int_{-\infty}^{t-t'} \frac{e^{i\omega\xi} d\xi}{(t - [\xi + t'])^{1-\alpha}}$$

so that changing variables $y = t' + \xi$ yields

$$D_{t-t'}^{-\alpha} \left[e^{-i\omega(t-t')}\right] = \frac{e^{-i\omega t}}{\Gamma(\alpha)} \int_{-\infty}^{t} \frac{e^{i\omega y} dy}{(t - y)^{1-\alpha}} = e^{-i\omega t'} D_{t}^{-\alpha} \left[e^{i\omega t}\right]. \qquad (4.158)$$

Substituting (4.158) into (4.157) allows us to write

$$\begin{aligned} I^* &= \int_{-\infty}^{\infty} dt' D_{t'}^{\alpha} \left[e^{-i\omega't'}\right] (-i\omega')^{-\alpha} e^{-i\omega t'} D_{t}^{-\alpha} \left[e^{i\omega t}\right] (-i\omega)^{\alpha} \\ &= E_{-\alpha}^{i\omega t} \mathcal{FT} \left\{ D_{t'}^{\alpha} \left[e^{-i\omega't'}\right] (-i\omega')^{-\alpha}; \omega \right\} = E_{-\alpha}^{i\omega t} \delta\left(\omega - \omega'\right) \quad (4.159) \end{aligned}$$

where we have used

$$\mathcal{FT} \left\{ D_{t}^{\alpha} \left[f\left(t\right)\right]; \omega \right\} = (-i\omega)^{\alpha} \hat{f}(\omega),$$

so that when (4.159) is inserted into (4.155) the latter reduces to

$$I = \frac{1}{2\pi} \int_{-\infty}^{\infty} d\omega \hat{G}_{\alpha}(\omega) \hat{Q}_{-\alpha}(\omega) E_{-\alpha}^{i\omega t}. \qquad (4.160)$$

Equation (4.160) is a generalization of Parseval's theorem to these generalized Fourier transforms.

4.5 Commentary

Telegrapher's Equation We have considered how Fourier series and Fourier transforms can be used to model the physical observables in linear fields. The simple physical model of matter consists of a lattice of coupled harmonic oscillators, such that in the continuum limit, the initial displacement, like the plucked string of a violin, produces a superposition of traveling waves. In an elastic material there is little or no damping, to first-order, and the motion is described by a wave equation. In a gas, where molecular collisions dominate, small initial disturbances are themselves dissipated and the motion is described by a diffusion equation. A combination of the two kinds of motion is given by the telegrapher's equation

$$\frac{1}{u^2}\frac{\partial^2 \rho(x,t)}{\partial t^2} + \lambda \frac{\partial \rho(x,t)}{\partial t} = D'\frac{\partial^2 \rho(x,t)}{\partial x^2}, \qquad (4.161)$$

where λ is the dissipation rate, u is the phase velocity of the wave, and the diffusion coefficient is $D = D'/\lambda$. The telegrapher's equation was developed to describe traveling waves along long distance cables. The solutions showed the wave motion at early times and dissipative, or diffusive motion, at late times.

Diffusion Equation Is Unphysical The equation of heat transport (4.61) has an infinite propagation speed and therefore it cannot completely describe the dynamics of physical phenomena. According to the heat equation when one end of a material body is heated, the other end, no matter how distant, immediately begins to feel the influence of the heat. This is, of course, unphysical and bothered Maxwell a great deal. In fact, starting from the kinetic theory of gases and incorporating a ballistic term into the equation of heat conduction, he obtained the telegrapher's equation (4.161) instead of the diffusion equation. Maxwell dropped the second derivative term after concluding [10] that it :"...may be neglected, as the rate of conduction will rapidly establish itself."

Finite Propagation Speed As Scales and Sneider [17] point out, Maxwell's conclusion was valid according to what was experimentally known over a century ago, but not today. We now know that heat is not a macroscopic fluid, but is the macroscopic manifestation of microscopic motion. In a solid, heat is the linear superposition of lattice vibrations called phonons. Lattice vibrations are responsible for the transport of heat, which we know is a diffusive phenomenon. The higher the temperature the more the phonons scatter on the lattice and the more nearly the transport process is purely diffusive. As the temperature is lowered there is less and less scattering of the phonons and the transport becomes more and more coherent, which is to say, wave-like. This is also the content of the fluctuation-dissipation relation, which implies that as the temperature goes to zero $\lambda \to 0$, and the telegrapher's equation reduces to the wave equation.

Scattering in Heterogeneous Media The linear wave field in media with erratic fluctuations is traditionally modeled by partial differential equations with nonconstant coefficients. When the fluctuations are time-independent, the scalar wave equation can be reexpressed as a Helmholtz equation (4.51). The total wave solution given by (4.57) expresses the total wave field in terms of the scatterings from the irregularities in the medium. Note that here the irregularities are assumed to be frozen in time since they are time-independent. These multiple scatterings are expressed as perturbation terms in an infinite order expansion of the total Green's function in (4.59).

Superposition of Fractal Functions   The derivative of an analytic function has an expression in terms of a weighted Fourier integral and we found that a fractional derivative has an analogous integral in terms of the generalized exponential function. In effect the traditional analysis in terms of Fourier transforms appears to be generalizable into transforms in terms of the generalized exponential. Thus, we can extend the ideas for linear superposition using analytic functions that are solutions to differential equations of motion, to the linear superposition of generalized exponential functions that are solutions to fractional differential equations of motion. This then becomes the natural basis for expanding the physical observables of complex linear phenomena. This is discussed, subsequently, in connection with fractional propagation and fractional transport equations.

Bibliography

[1] M. Bologna, *Derivata ad Indice Reale*, Ets Editrice, Italy (1990).

[2] A. L. Cauchy, Theorie de la propagation des ondes (Prix d'analyses mathematique), *Concours de* 1815 et de 1816, 140-2, 281-2.

[3] P. A. M. Dirac, *The Principles of Quantum Mechanics*, Oxford University Press, Oxford (1953).

[4] U. Frisch and G. Parisi, in *Turbulence and Predictability in Geophysical Fluid Dynamics and Climate Dynamics*, M. Ghil, R. Benzi and G. Parisi, eds., North-Holland, Amsterdam (1985).

[5] D. V. Giri, *Dirac Delta Functions*, unpublished report (1979).

[6] H. Goldstein, *Advanced Classical Mechanics*, first edition, John Wiley, New York (1955).

[7] I. S. Gradshteyn and I. M. Ryzhik, *Table of Integrals, Series, and Products*, corrected and enlarged edition, Academic, New York (1980).

[8] M. Hermite, (*Faculte des Sciences de Paris*), IV edition, 155, Paris (1891).

[9] O. Heaviside, *Electromagnetic Theory*, vol. II, 54-55 and 92, First issue 1893 and reissued London 1922.

[10] D. D. Joseph and L. Preziosi, *Rev. Mod. Phys.* **61**, 41 (1989).

[11] G. R. Kirchoff and S. B. Kon, *Adad. Wiss, Berlin Vom.* 22, 641, juni (1992); *Wied. Ann* XVIII, 663 (1883).

[12] M. J. Lighthill, *Introduction to Fourier Analysis and Generalized Functions*, Cambridge University Press, Cambridge (1964).

[13] K. Lindenberg and B.J. West, *The first, the biggest, and other such considerations*, J. Stat. Phys. **42**, 201 (1986).

[14] K. S. Miller and B. Ross, *An Introduction to the Fractional Calculus and Fractional Differential Equations*, John Wiley, New York (1993).

[15] S. D. Poisson, *Memoire sur la theorie des ondes* (lu le 2 octobre et le 18 decembre 1815), 85-86.

[16] S. G. Samko, A. A. Kilbas and O. I. Marichev, *Fractional Integrals and Derivatives*, Gordon and Breach, New York (1993).

[17] J. A. Scales and R. Snieder, What is a wave?, Nature **401**, 739 (1999).

[18] W. Thomson, *Trans. Roy. Soc. of Edinburgh, XXI*, (May) (1854).

[19] A. Zygmund, *Trigonometric Series, vols. I and II* combined, second edition, Cambridge University Press, Cambridge, MA (1977).

Chapter 5

Fractional Laplace Transforms

Solutions to Differential Equations The Laplace transform and its inverse constitute a powerful technique for solving linear differential equations. This approach to solving such equations, when combined with complex analysis, provides one of the most formidable tools in the analyst's tool kit. In the next few lectures we review how this approach can be applied to fractional differential equations with constant coefficients having rational and irrational exponents. In this way we find that the solutions to such equations can be written as series of generalized exponentials and in so doing make contact with our earlier lectures.

The Inverse Transform Is More Complex The formal analysis of analytic functions using Laplace and Fourier transforms can be included in a single perspective by replacing the frequency by a complex parameter, $i\omega \rightarrow -s$, so that we define the Laplace transform of the function $f(t)$ by the integral

$$\widetilde{f}(s) = \mathcal{LT}\{f(t); s\} = \int_0^\infty f(t) e^{-st} dt, \ Re\ s > 0. \tag{5.1}$$

Thus, the discussion of the properties of the Laplace transform is not independent of that of the Fourier transform, since the former is a special case of the latter. The inverse in the two cases is quite different, however. In the Fourier transform the inverse is just the mirror image of the direct transform. For the Laplace transform, the situation is quite different due to the dependence of the inverse on the initial conditions. The inverse Laplace transform is given by

$$f(t) = \mathcal{LT}^{-1}\left\{\widetilde{f}(s); t\right\} = \frac{1}{2\pi i} \int_{C-i\infty}^{C+i\infty} e^{st}\, \widetilde{f}(s)\, ds, \tag{5.2}$$

where C is an arbitrary constant that keeps the contour of integration off the imaginary axis and $Ret > 0$. There are many properties of Laplace transforms and their possible use on a large number of functions that might be of interest, but only some of which we discuss below. We are especially concerned with those properties that may guide our understanding of the fractional calculus and how these can be used to model complex phenomena.

Harmonic Function Consider the harmonic function $f(t) = e^{i\omega t}$, whose Laplace transform we obtain from (5.1) to be

$$\widehat{f}(s) = \frac{1}{s - i\omega}. \tag{5.3}$$

Introducing (5.3) into (5.2) we obtain

$$f(t) = \frac{1}{2\pi i} \int_{C-i\infty}^{C+i\infty} e^{st} \frac{1}{s - i\omega} ds \tag{5.4}$$

where the integral may be evaluated using the calculus of residues. For $t > 0$, we close the contour by a semicircle in the left half of the complex plane, capturing the pole at $s = i\omega$, so that $f(t) = e^{i\omega t}$ for $t > 0$. For $t < 0$, the semicircle must be closed in the right half-plane. But in this region the integrand is analytic unless ω is negative, so that $f(t) = e^{i\omega t}\Theta(-\omega)$ for $t < 0$ where $\Theta(\cdot)$ is the Heaviside unit step function. When the Laplace transform of a function has simple poles, the calculus of residues is a great help in determining the function.

Waves Carry Information In our discussion of Fourier series and transforms, the connection with physical phenomena was immediate. Everything we see and hear has an intermediary wave transmitting information from the points in the environment to where the waves are perceived. The light striking our eyes, the sound enveloping our ears, are both wave phenomena, and understanding the physical nature is of immediate interest to us. Thus, the mathematics of trigonometric series and superposition resonate with our experience of the physical world and the methods of Fourier feels right. This same intuition does not apply to the method of Laplace, where the approach is more formal and the physical interpretation only comes after the application, not before. However, the simplicity of the algebraic manipulations makes mastery of the method desirable.

5.1 Solving Differential Equations

Laplace Transform of nth Derivative Let us consider some of the properties of Laplace transforms that are analogous to those of the Fourier transforms of the nth derivatives of the functions. Consider the nth derivative of the Laplace transform of $f(t)$ with respect to the Laplace variable s,

$$\frac{d^n \widehat{f}(s)}{ds^n} = \int_0^\infty (-t)^n f(t) e^{-st} dt. \tag{5.5}$$

Compare (5.5) with the Laplace transform of the nth derivative of $f(t)$,

$$\mathcal{LT}\left\{\frac{d^n f(t)}{dt^n}; s\right\} = \int_0^\infty \frac{d^n f(t)}{dt^n} e^{-st} dt$$

$$= s^n \widetilde{f}(s) - s^{n-1} f(0) - \cdots - f^{(n-1)}(0) \quad (5.6)$$

where the series on the right-hand side of (5.6) depends on the initial value of the function and the initial values of the first $(n-1)$-derivatives of the function. Note that this dependence of the transform on the initial conditions arises because the Laplace transform, unlike the Fourier transform, begins at the initial time $t = 0$. Thus, the Laplace transform is useful for solving initial value problems, whereas the Fourier transform is not.

Driven Harmonic Oscillator Consider, for example, the simple harmonic oscillator driven by a time-dependent function $g(t)$,

$$\ddot{X}(t) + \omega_0^2 X(t) = g(t). \quad (5.7)$$

The Laplace transform of the equation for the driven harmonic oscillator yields

$$s^2 \widetilde{X}(s) - sX(0) - \dot{X}(0) + \omega_0^2 \widetilde{X}(s) = \widetilde{g}(s) \quad (5.8)$$

so that rearranging terms gives us

$$\widetilde{X}(s) = \frac{sX(0)}{s^2 + \omega_0^2} + \frac{\dot{X}(0)}{s^2 + \omega_0^2} + \frac{\widetilde{g}(s)}{s^2 + \omega_0^2}. \quad (5.9)$$

The roots in (5.9) are, of course, given by $s = \pm i\omega_0$, so that the inverse Laplace transform is determined by the simple poles at these roots. After some algebra, the inverse Laplace transform of (5.9) is given by

$$X(t) = X(0) \cos[\omega_0 t] + \frac{\dot{X}(0)}{\omega_0} \sin[\omega_0 t] + \mathcal{LT}^{-1}\left\{\frac{\widetilde{g}(s)}{s^2 + \omega_0^2}; t\right\}, \quad (5.10)$$

where the contribution of the driver to the solution depends on its functional form, and therefore on the inverse Laplace transform explicitly indicated in the solution.

Convolution and Green's Functions An alternative interpretation of the solution (5.10) can be constructed by noting that the Laplace transform of the convolution of two analytic functions is the product of their respective Laplace transforms

$$\mathcal{LT}\left\{\int_0^t f(t - t') g(t') dt'; s\right\} = \widetilde{f}(s) \widetilde{g}(s). \quad (5.11)$$

Therefore taking the inverse Laplace transform of (5.11) we have

$$\mathcal{LT}^{-1}\left\{\tilde{f}(s)\,\tilde{g}(s)\,;t\right\} = \int_0^t f(t-t')\,g(t')\,dt'. \tag{5.12}$$

This allows us to rewrite (5.10) as the sum of the homogeneous and inhomogeneous solutions

$$X(t) = X(0)\cos[\omega_0 t] + \frac{\dot{X}(0)}{\omega_0}\sin[\omega_0 t] + \int_0^t G(t-t')\,g(t')\,dt', \tag{5.13}$$

where $G(t)$ is a Green's function given by

$$G(t) = \mathcal{LT}^{-1}\left\{\frac{1}{s^2+\omega_0^2}\,;t\right\}. \tag{5.14}$$

The influence of the driving force on the system response, determined by the system dynamics, is manifest in the Green's function. In this simple example the Green's function is just the trigonometric function $\sin[\omega_0 t]$. We show that the method of constructing Green's functions from the solutions to linear ordinary differential equations carries over to linear fractional differential equations.

5.1.1 Extension to Fractional Powers

When Is a Constant Not a Constant? In correspondence with our discussion of Fourier transforms we now extend the definition of the Laplace transform and its inverse to fractional derivatives. To do this we introduce the following theorem.

  Theorem 4: *Let the function $f(t)$ have a power series representation that is valid in the vicinity of $t = 0$. If the fractional derivative $D_t^\alpha[f(t)]$ exists, does not contain a $C(\alpha)$ constant, and the Laplace transform of both $f(t)$ and its fractional derivative also exist then*

$$\mathcal{LT}\left\{D_t^\alpha[f(t)]\,;s\right\} = s^\gamma\left[s^n\,\hat{f}(s) - s^{n-1}f(0) - \cdots - f^{(n-1)}(0)\right] \tag{5.15}$$

and the integer part of α is given by $n = [\alpha]$, when $\alpha \geq 1$ and $\gamma = \alpha - n \geq 0$ so that for $\alpha < 1$ (5.15) reduces to

$$\mathcal{LT}\left\{D_t^\alpha[f(t)]\,;s\right\} = s^\alpha\,\hat{f}(s). \tag{5.16}$$

Transform of Series To demonstrate the validity of Theorem 4 we write the function of interest $f(t)$ as the power series

$$f(t) = \sum_{k=0}^{\infty} C_k t^{\rho_k} \tag{5.17}$$

and using the Laplace transform of a monomial with $\mu > -1$,

$$\mathcal{LT}\left\{t^{\mu}; s\right\} = \frac{\Gamma\left(\mu + 1\right)}{s^{\mu+1}},$$

we obtain for the Laplace transform of the function (5.17),

$$\tilde{f}\left(s\right) = \sum_{k=0}^{\infty} C_k \frac{\Gamma\left(\rho_k + 1\right)}{s^{\rho_k+1}}. \tag{5.18}$$

Now consider the fractional derivative of the function $D_t^{\alpha}\left[f\left(t\right)\right]$ with $\alpha < 1$ and $f\left(t\right)$ given by (5.17), so that

$$D_t^{\alpha}\left[f\left(t\right)\right] = \sum_{k=0}^{\infty} C_k \frac{\Gamma\left(\rho_k\right)}{\Gamma\left(\rho_k + 1 - \alpha\right)} t^{\rho_k - \alpha}. \tag{5.19}$$

The Laplace transform of (5.19) is

$$\begin{aligned}
\mathcal{LT}\left\{D_t^{\alpha}\left[f\left(t\right)\right]; s\right\} &= \sum_{k=0}^{\infty} C_k \frac{\Gamma\left(\rho_k + 1\right)}{s^{\rho_k+1-\alpha}} \\
&= s^{\alpha} \sum_{k=0}^{\infty} C_k \frac{\Gamma\left(\rho_k + 1\right)}{s^{\rho_k+1}} = s^{\alpha}\,\tilde{f}\left(s\right);
\end{aligned} \tag{5.20}$$

that is, the Laplace transform of the fractional derivative of a function is the product of the individual Laplace transforms, since the initial values of the function and its derivatives vanish. Now consider the case $\alpha \geq 1$ so that we can separate out the integer part of the index, $\alpha = n + \gamma$. By assumption $f\left(t\right)$ does not have a $C\left(n\right)$ constant, so we can factor the fractional derivative as

$$D_t^{\alpha} = D_t^{\gamma} D_t^{n}$$

and define a new function $g\left(t\right)$ by the equation

$$g\left(t\right) \equiv D_t^{n}\left[f\left(t\right)\right] \tag{5.21}$$

so that the Laplace transform of the fractional derivative of the function can be written

$$\begin{aligned}
\mathcal{LT}\left\{D_t^{\alpha}\left[f\left(t\right)\right]; s\right\} &= \mathcal{LT}\left\{D_t^{\gamma}\left[g\left(t\right)\right]; s\right\} = s^{\gamma}\,\tilde{g}\left(s\right) \\
&= s^{\gamma} s^{n}\,\tilde{f}\left(s\right) - s^{n-1} f\left(0\right) - \cdots - f^{(n-1)}\left(0\right), \tag{5.22}
\end{aligned}$$

where we have used (5.6) to replace $\tilde{g}\left(s\right)$.

Laplace Tansform of General Function   Now let $f(t)$ contain a $C(n)$ constant, then we must keep track of the order in which we factor the fractional derivative, that is,

$$D_t^\alpha = D_t^n D_t^\gamma.$$

But a $C(n)$ constant function means that it has a piece that behaves as

$$q(t) = \sum_{k=0}^{n-1} A_k t^k \tag{5.23}$$

so that we should write the series form of $f(t)$ as the sum of two functions

$$f(t) = h(t) + q(t). \tag{5.24}$$

The first function in (5.24) has the form considered above, that is, the part of the function that is not a $C(n)$ constant

$$h(t) = \sum_{k=0}^{\infty} C_k t^{\rho_k} \tag{5.25}$$

and the second function is given by (5.23). The fractional derivative of $q(t)$ is given by

$$D_t^\gamma [q(t)] = D_t^\gamma \left[\sum_{k=0}^{n-1} A_k t^k \right] = \sum_{k=0}^{n-1} A_k \frac{\Gamma(k+1)}{\Gamma(k+1-\gamma)} t^{k-\gamma} \equiv q_\gamma(t). \tag{5.26}$$

Inspecting the function $q_\gamma(t)$ it is clear that when we take the integer derivative we shall have terms like $t^{k-\gamma-n}$. But from the upper bound of the series we know that $k \le n-1$ so that $k - \gamma - n < -1$. This constraint on the summation index means that the Laplace transform does not exist since the initial values of this function and its derivatives diverge. However, this is contrary to the hypothesis that the Laplace transform of the function $f(t)$ exists, therefore the coefficients of the $C(n)$ constant function must be identically zero for all k. Thus, the function must have the form (5.17) for which the theorem has been shown to be true.

Transform of Fractional Derivatives   Let us pursue this argument in a bit more detail. Assume a $C(\alpha)$ constant in the series (5.20) where such terms would disappear with application of the fractional derivative. Thus, (5.20) has to be written

$$s^\alpha \sum_{k=0}^{\infty} C_k \frac{\Gamma(\rho_k + 1)}{s^{\rho_k+1}} = s^\alpha \left[\tilde{f}(s) - \sum_{k \in \{\tilde{n}\}} C_k \frac{\Gamma(\rho_k + 1)}{s^{\rho_k+1}} \right], \tag{5.27}$$

where $k \in \{\tilde{n}\}$ means that the sum is over all indices contained in the $C(\alpha)$ constant, because these terms are set to zero after the fractional derivative and therefore must be removed from the definition of the function. But what are these terms? For $\alpha < 0$ there are no $C(\alpha)$ constants because to be a $C(\alpha)$ constant requires that the terms have the form $t^{\rho_{\overline{n}}}$ with $\rho_{\overline{n}} = \alpha - k - 1$, $k = 0, 1, \cdots$. In this case

$$D_t^{\alpha}\left[t^{\rho_{\overline{n}}}\right] = \frac{\Gamma(\rho_{\overline{n}}+1)}{\Gamma(\rho_{\overline{n}}+1-\alpha)}t^{\rho_{\overline{n}}-\alpha} = 0$$

since $\Gamma(-\text{integer}) = \infty$. But also, for $\rho_{\overline{n}} > -1$ there exists a Laplace transform, so $\alpha - k - 1 > -1 \Rightarrow \alpha > k$, but $\alpha < 0$, so we do not have this term in the series.

Initial Values for Fractional Derivatives For $0 < \alpha < 1$ we can have the single term $k = 0$ because we again have $\alpha > k$, which is true only for $k = 0$. In this latter case the series (5.20) is

$$s^{\alpha}\left[\sum_{k=0}^{\infty} C_k \frac{\Gamma(\rho_k+1)}{s^{\rho_k+1}} - C_{\overline{n}} \frac{\Gamma(\rho_{\overline{n}}+1)}{s^{\rho_{\overline{n}}+1}}\right], \tag{5.28}$$

where we have added and subtracted the $C(\alpha)$ constant term. The index for the $C(\alpha)$ constant is given by $\rho_{\overline{n}} = \alpha - 1$ so we may write (5.28) as

$$s^{\alpha}\left[\sum_{k=0}^{\infty} C_k \frac{\Gamma(\rho_k+1)}{s^{\rho_k+1}} - C_{\overline{n}} \frac{\Gamma(\alpha)}{s^{\alpha}}\right]. \tag{5.29}$$

But we have

$$C_{\overline{n}} = \frac{1}{\Gamma(\alpha)} D_t^{\alpha-1}\left[f(t)\right]\Big|_{t=0} \tag{5.30}$$

which can be determined from

$$D_t^{\alpha-1}\left[\sum_{k=0}^{\infty} C_k t^{\rho_k}\right] = \left.\left[\sum_{k=0}^{\infty} C_k \frac{\Gamma(\rho_k+1)}{\Gamma(\rho_k+2-\alpha)}t^{\rho_k-\alpha+1}\right]\right|_{t=0} \tag{5.31}$$

in which all the terms vanish except the one with $\rho_k - \alpha + 1 = 0$. For $\rho_k - \alpha > -1$ there exists a fractional derivative of the function, $D_t^{\alpha}\left[f(t)\right]$, from which we can deduce that $t^{\rho_k-\alpha+1}$ has the form t^{ε} with $\varepsilon > 0 \Rightarrow t^{\varepsilon}|_{t=0} = 0$ so that (5.31) reduces to

$$D_t^{\alpha-1}\left[f(t)\right]\Big|_{t=0} = C_{\overline{n}}\Gamma(\alpha) \tag{5.32}$$

which is (5.30). Finally, the Laplace transform of the fractional derivative, from (5.28), is

$$\mathcal{LT}\left\{D_t^{\alpha}\left[f(t)\right]; s\right\} = s^{\alpha}\widetilde{f}(s) - f^{(\alpha-1)}(0) \tag{5.33}$$

where we have the initial condition for the fractional derivative

$$f^{(\alpha-1)}(0) \equiv D_t^{\alpha-1}\left[f(t)\right]\Big|_{t=0}. \tag{5.34}$$

Familiar Example Consider, for example, the case where $\alpha = 1/2$, so that

$$D_t^{1/2}\left[E_{1/2}^t\right] = e^t. \tag{5.35}$$

The Laplace transform of (5.35) is given by, applying (5.33) to the left-hand side of the equation,

$$\mathcal{LT}\left\{D_t^{1/2}\left[E_{1/2}^t\right];s\right\} = s^{1/2}\frac{s^{1/2}}{s-1} - D_t^{-1/2}\left[E_{1/2}^t\right]\Big|_{t=0}$$

$$= \frac{s}{s-1} - e^t\Big|_{t=0} = \frac{1}{s-1}. \tag{5.36}$$

We see that (5.36) is the same as we would obtain by taking the Laplace transform of the right-hand side of (5.35).

Laplace Transform of αth Derivative Using the same procedure, we have more generally that

$$\mathcal{LT}\left\{D_t^{\alpha}\left[f\left(t\right)\right];s\right\} = \overline{\sum_{k=0}}C_k\frac{\Gamma\left(\rho_k+1\right)}{s^{\rho_k+1-\alpha}}$$

$$= s^{\alpha}\left[\sum_{k=0}C_k\frac{\Gamma\left(\rho_k+1\right)}{s^{\rho_k+1}} - \sum_{\overline{n}=0}^{[\alpha]}C_{\overline{n}}\frac{\Gamma\left(\rho_{\overline{n}}+1\right)}{s^{\rho_{\overline{n}}+1}}\right] \tag{5.37}$$

because the last terms are not present in the summation with an overbar. However, $\rho_{\overline{n}} - \alpha + 1 = -\overline{n}$, because $t^{\rho_{\overline{n}}}$ is a $C\left(\alpha\right)$ constant and $\overline{n} = 0, 1, \cdots, [\alpha]$. Now writing the Laplace transform of the fractional derivative of the function

$$\mathcal{LT}\left\{D_t^{\alpha}\left[f\left(t\right)\right];s\right\} = s^{\alpha}\widetilde{f}\left(s\right) - s^{\alpha}\sum_{\overline{n}=0}^{[\alpha]}C_{\overline{n}}\frac{\Gamma\left(\alpha-\overline{n}\right)}{s^{\alpha-\overline{n}}}$$

$$= s^{\alpha}\widetilde{f}\left(s\right) - \sum_{\overline{n}=0}^{[\alpha]}C_{\overline{n}}\Gamma\left(\alpha-\overline{n}\right)s^{\overline{n}} \tag{5.38}$$

but using the value for the constant

$$C_{\overline{n}} = \frac{D_t^{\alpha-1-\overline{n}}\left[f\left(t\right)\right]}{\Gamma\left(\alpha-\overline{n}\right)}\Bigg|_{t=0}$$

reduces (5.38) to the form

$$\mathcal{LT}\left\{D_t^{\alpha}\left[f\left(t\right)\right];s\right\} = s^{\alpha}\widetilde{f}\left(s\right) - \sum_{\overline{n}=0}^{[\alpha]}s^{\overline{n}}f^{\left(\alpha-1-\overline{n}\right)}\left(0\right), \tag{5.39}$$

where again we see the dependence on the initial value of the fractional derivatives arising from the function constants.

Explicit Fractional Derivatives Consider, for example, the fractional derivative of the generalized exponential

$$\begin{aligned}
\mathcal{LT}\left\{D_t^{3/2}\left[E_{1/2}^t\right]\right\} &= s^{3/2}\frac{s^{1/2}}{s-1} - D_t^{3/2-1}\left[E_{1/2}^t\right]\Big|_{t=0} - s\,D^{3/2-2}\left[E_{1/2}^t\right]\Big|_{t=0} \\
&= \frac{s^2}{s-1} - D^{1/2}\left[E_{1/2}^t\right]\Big|_{t=0} - s\,D^{-1/2}\left[E_{1/2}^t\right]\Big|_{t=0} \\
&= \frac{s^2}{s-1} - 1 - s = \frac{1}{s-1}.
\end{aligned} \tag{5.40}$$

Here again we obtain the same result as (5.36), indicating that any half-integer derivative of the generalized exponential $E_{1/2}^t$ will yield e^t.

A More General Example In our examples we observe that the Laplace transforms of the generalized exponentials do not change because e^t does not contain $C(\alpha)$ constants so that Theorem 4 in its original form is applicable. Thus,

$$\mathcal{LT}\left\{D_t^\alpha\left[e^t\right]; s\right\} = s^\alpha \mathcal{LT}\left\{e^t; s\right\} = \frac{s^\alpha}{s-1} \tag{5.41}$$

or in the new form with the initial fractional derivative of the function

$$\mathcal{LT}\left\{D_t^\alpha\left[e^t\right]; s\right\} = s^\alpha\frac{1}{s-1} - D_t^{\alpha-1}\left[e^t\right]\Big|_{t=0} = \frac{s^\alpha}{s-1} - E_{\alpha-1}^t\Big|_{t=0}$$

but we know that for $0 < \alpha < 1$,

$$E_{\alpha-1}^t\Big|_{t=0} = 0$$

and the two expressions are identical. In the case $\alpha > 1$ we also know that $D_t^\alpha\left[e^t\right]$ does not exist for $\alpha \notin \mathcal{N}$, as we discuss in the next section.

5.2 Generalized Exponentials

Generalized Laplace Transform We now try to generalize the equation for the nth order derivative of a Laplace transform given by (5.5) by replacing the exponential function e^{-st} with the generalized exponential function E_α^{-st} to obtain a formula like (5.6)[1]. But there is a fundamental difference between Fourier and Laplace transforms. In the case of the Fourier transforms convergence is determined by the form of the function being transformed. On the other hand, the convergence of the Laplace transform of a function is determined by the exponential factor e^{-st}, in addition to the form of the function. Thus, because the generalized exponential E_α^{-st} does not have an exponential decay as $t \to \infty$, but rather has an inverse power-law decay, it is not clear that the Laplace transform can be suitably generalized.

---

[1] The results in this lecture are taken from Bologna [2] and suitably revised.

Transform of Generalized Exponentials   Let us consider a few examples of Laplace transforms involving recurrent functions. We restrict our considerations to $\alpha < 1$, and from Theorem 4:

$$LT\left\{D_t^\alpha\left[t^\mu\right];s\right\} = \frac{\Gamma\left(\mu+1\right)}{s^{\mu+1-\alpha}}, \text{ for } 0 < \alpha < 1; \mu > 1, \qquad (5.42)$$

so that using the series expression for the generalized exponential we obtain

$$LT\left\{E_\alpha^{\lambda t};s\right\} = LT\left\{D_t^\alpha\left[\frac{e^{\lambda t}}{\lambda^\alpha}\right];s\right\} = \left(\frac{s}{\lambda}\right)^\alpha \frac{1}{s-\lambda}. \qquad (5.43)$$

Here we have used the definition of the generalized exponential having a parameter multiplying the independent variable.

Transform of Generalized Trigonometric Functions   Using (5.43) we can construct the Laplace transforms of the generalized trigonometric functions. Consider the generalized sine

$$
\begin{aligned}
LT\left\{\sin_\alpha \omega t; s\right\} &= \frac{1}{2i}\left[\left(\frac{s}{i\omega}\right)^\alpha \frac{1}{s-i\omega} - \left(\frac{s}{-i\omega}\right)^\alpha \frac{1}{s+i\omega}\right] \\
&= \left(\frac{s}{\omega}\right)^\alpha \frac{1}{s^2+\omega^2}\left[-s\sin\left(\alpha\pi/2\right) + \omega\cos\left(\alpha\pi/2\right)\right]
\end{aligned}
$$
$$\qquad (5.44)$$

and the generalized cosine

$$
\begin{aligned}
LT\left\{\cos_\alpha \omega t; s\right\} &= \frac{1}{2}\left[\left(\frac{s}{i\omega}\right)^\alpha \frac{1}{s-i\omega} + \left(\frac{s}{-i\omega}\right)^\alpha \frac{1}{s+i\omega}\right] \\
&= \left(\frac{s}{\omega}\right)^\alpha \frac{1}{s^2+\omega^2}\left[s\cos\left(\alpha\pi/2\right) + \omega\sin\left(\alpha\pi/2\right)\right].
\end{aligned}
$$
$$\qquad (5.45)$$

Of course, we can also construct the Laplace transforms of the fractional derivatives of the ordinary trigonometric functions such as

$$LT\left\{D_t^\alpha\left[\sin\omega t\right];s\right\} = s^\alpha LT\left\{\sin\omega t; s\right\} = s^\alpha \frac{\omega}{s^2+\omega^2} \qquad (5.46)$$

and

$$LT\left\{D_t^\alpha\left[\cos\omega t\right];s\right\} = s^\alpha LT\left\{\cos\omega t; s\right\} = \frac{s^{\alpha+1}}{s^2+\omega^2}. \qquad (5.47)$$

Note that (5.44) reduces to (5.46) for $\alpha = 0$ and similarly (5.45) reduces to (5.47) for $\alpha = 0$. The $\alpha = 1$ case requires a bit more thought because of the initial value terms in (5.22). In this latter case (5.46) is replaced with

$$\mathcal{LT}\left\{D_t\left[\sin\omega t\right];s\right\} = s\mathcal{LT}\left\{\sin\omega t;s\right\} = \frac{s\omega}{s^2+\omega^2} = \omega\mathcal{LT}\left\{\cos\omega t;s\right\}$$

so that the Laplace transform of the derivative of the sine function is just ω times the Laplace transform of the cosine function as it should be. On the other hand (5.47) is replaced with

$$\begin{aligned}
\mathcal{LT}\left\{D_t\left[\cos\omega t\right];s\right\} &= s\mathcal{LT}\left\{\cos\omega t;s\right\} - 1 \\
&= \frac{s^2}{s^2+\omega^2} - 1 = -\omega\mathcal{LT}\left\{\sin\omega t;s\right\},
\end{aligned}$$

where the minus one comes from the initial value, and the resulting expression gives the Laplace transform of the derivative of the cosine function as $-\omega$ times the Laplace transform of the sine function as it should.

Transform of Generalized Hyperbolic Functions For the generalized hyperbolic functions we set $*E_\alpha^{-\lambda t} \equiv E_\alpha^{-\lambda t}$ so that from (5.43) and Theorem 4 we obtain again for the case $\alpha < 1$

$$\mathcal{LT}\left\{E_\alpha^{-\lambda t};s\right\} = \mathcal{LT}\left\{D_t^\alpha\left[\frac{e^{-\lambda t}}{\lambda^\alpha}\right];s\right\} = \left(\frac{s}{\lambda}\right)^\alpha\frac{1}{s+\lambda} \qquad (5.48)$$

so that the Laplace transform of the generalized hyperbolic sine function is

$$\begin{aligned}
\mathcal{LT}\left\{\sinh_\alpha\omega t;s\right\} &= \frac{1}{2}\mathcal{LT}\left\{E_\alpha^{\omega t} - E_\alpha^{-\omega t};s\right\} \\
&= \frac{1}{2}\left(\frac{s}{\omega}\right)^\alpha\left[\frac{1}{s-\omega} - \frac{1}{s+\omega}\right] \\
&= \left(\frac{s}{\omega}\right)^\alpha\frac{\omega}{s^2-\omega^2} \qquad (5.49)
\end{aligned}$$

and that of the fractional derivative of the hyperbolic sine function is

$$\begin{aligned}
\mathcal{LT}\left\{D_t^\alpha\left[\sinh\omega t\right];s\right\} &= s^\alpha\mathcal{LT}\left\{\sinh\omega t;s\right\} \\
&= \frac{s^\alpha}{2}\left[\frac{1}{s-\omega} - \frac{1}{s+\omega}\right] \\
&= \frac{s^\alpha\omega}{s^2-\omega^2}. \qquad (5.50)
\end{aligned}$$

Note that (5.49) and (5.50) differ from each other by a scale factor ω^α. In the same way we construct the Laplace transform of the generalized hyperbolic cosine function to be

$$\mathcal{LT}\{\cosh_\alpha \omega t; s\} = \frac{1}{2}\mathcal{LT}\{E_\alpha^{\omega t} + E_\alpha^{-\omega t}; s\}$$

$$= \frac{1}{2}\left(\frac{s}{\omega}\right)^\alpha \cdot \left[\frac{1}{s-\omega} + \frac{1}{s+\omega}\right]$$

$$= \left(\frac{s}{\omega}\right)^\alpha \frac{s}{s^2 - \omega^2} \tag{5.51}$$

and that of the fractional derivative of the hyperbolic cosine function is

$$\mathcal{LT}\{D_t^\alpha[\cosh \omega t]; s\} = s^\alpha \mathcal{LT}\{\cosh \omega t; s\}$$

$$= \frac{s^\alpha}{2}\left[\frac{1}{s-\omega} + \frac{1}{s+\omega}\right]$$

$$= \frac{s^{\alpha+1}}{s^2 - \omega^2}. \tag{5.52}$$

Note that (5.51) and (5.52) differ from each other by the same scale factor ω^α that was found above.

No More Examples   We could, of course, continue in this way and generate the Laplace transforms of a great many quantities that arise in the fractional calculus, but this would not provide any additional insight into the kinds of physical processes for which these expressions would be useful. Besides, these transforms are catalogued elsewhere [8].

5.2.1   Solutions to Differential Equations

Typical Noninteger Differential Equation   In our discussion of linear fields, as well as in our considerations of the Hamiltonian model for stochastic processes, we focused on the linear harmonic oscillator, the workhorse of theoretical physics. We began our discussion on solving differential equations using Laplace transforms with this system. The solution was clearly dependent on the initial displacement of the oscillator and the initial velocity. Let us examine this familiar procedure for solving differential equations when applied to a fractional differential equation. Of course, to do this we implement the generalized form we have just discussed.

Case with Unequal Roots   Consider the fractional differential equation

$$D_t X(t) + aD_t^{1/2} X(t) + bX(t) = 0 \tag{5.53}$$

which using Theorem 4 can be written as

$$s \overset{\approx}{X}(s) - X(0) + as^{1/2} \overset{\approx}{X}(s) + b \overset{\approx}{X}(s) = 0. \tag{5.54}$$

The Laplace transform of the solution to Equation (5.53) can be written, by rearranging terms in (5.54), as

$$\tilde{X}(s) = \frac{X(0)}{s + as^{1/2} + b} \qquad (5.55)$$

which exists as long as the initial condition is finite. The question is now one of finding the inverse Laplace transform of (5.55). Following Miller and Ross ([8], page 135) we expand the denominator in (5.55) in partial fractions

$$\frac{1}{s + as^{1/2} + b} = \frac{1}{r_+ - r_-} \left[\frac{1}{s^{1/2} - r_+} - \frac{1}{s^{1/2} - r_-} \right], \qquad (5.56)$$

where r_+ and r_- are the distinct roots of the denominator. Thus, the problem is reduced to finding the inverse Laplace transform of $\left(s^{1/2} - r_+\right)^{-1}$. From the algebraic identity

$$\frac{1}{s^{1/2} - r_+} = \frac{s^{1/2}}{s - r_+^2} + \frac{r_+}{s - r_+^2}$$

we can use (5.48) to obtain the inverse Laplace transform in terms of the generalized exponential

$$\mathcal{LT}^{-1} \left\{ \frac{1}{s^{1/2} - r_+}; t \right\} = r_+ E_{1/2}^{r_+^2 t} + r_+ E_0^{r_+^2 t}$$

with a similar expression for the other root. Thus, we obtain the solution to the fractional differential equation (5.53) to be

$$
\begin{aligned}
X(t) &= \mathcal{LT}^{-1} \left\{ \frac{X(0)}{s + as^{1/2} + b}; t \right\} \\
&= \frac{X(0)}{r_+ - r_-} \left[r_+ E_0^{r_+^2 t} - r_- E_0^{r_-^2 t} + r_+ E_{1/2}^{r_+^2 t} - r_- E_{1/2}^{r_-^2 t} \right] \qquad (5.57)
\end{aligned}
$$

which in terms of the Miller and Ross generalized exponential (5.57) is [their (3.7)]

$$
\begin{aligned}
X(t) &= \frac{X(0)}{r_+ - r_-} \left[r_+ E_t \left(0, r_+^2\right) - r_- E_t \left(0, r_-^2\right) \right. \\
&\quad \left. + E_t \left(-1/2, r_+^2\right) - E_t \left(-1/2, r_-^2\right) \right]. \qquad (5.58)
\end{aligned}
$$

Case with Equal Roots What about the case when the two roots are not distinct, but are degenerate? In that case we again follow Miller and Ross [8] and write instead of (5.55)

$$\tilde{X}(s) = \frac{X(0)}{\left(s^{1/2} - r\right)^2}. \qquad (5.59)$$

Here again we can factor the denominator as

$$\frac{1}{\left(s^{1/2} - r\right)^2} = \left(\frac{s^{1/2}}{s - r^2} + \frac{r}{s - r^2}\right)^2$$

so that we need to calculate the following inverse Laplace transform

$$X\left(t\right) = X\left(0\right)\left[\mathcal{LT}^{-1}\left\{\frac{s + r^2 + 2s^{1/2}r}{\left(s - r^2\right)^2}; t\right\}\right]. \tag{5.60}$$

We can obtain the inverse Laplace transform of the product of two functions using the convolution relation for Laplace transforms

$$\mathcal{LT}^{-1}\left\{\tilde{f}\left(s\right)\tilde{g}\left(s\right); t\right\} = \int_0^t f\left(t'\right)g\left(t - t'\right)dt' \tag{5.61}$$

so that if $\tilde{f}\left(s\right) = s^\alpha$ with $\alpha \neq$ integer and $\tilde{g}\left(s\right) = \left(s - \lambda\right)^{-n}$ with n a positive integer we have

$$\mathcal{LT}^{-1}\left\{\frac{s^\alpha}{\left(s - \lambda\right)^n}; t\right\} = \int_0^t e^{\lambda(t - t')}\frac{\left(t - t'\right)^{n-1}}{\Gamma\left(n\right)}\frac{t'^{-\alpha - 1}}{\Gamma\left(-\alpha\right)}dt'.$$

Using the binomial expansion under the integral yields

$$\mathcal{LT}^{-1}\left\{\frac{s^\alpha}{\left(s - \lambda\right)^n}; t\right\} = \sum_{k=0}^{n-1}\binom{n - 1}{k}\frac{\left(-1\right)^k t^{n-1+k}}{\Gamma\left(n\right)\Gamma\left(-\alpha\right)}\int_0^t e^{\lambda(t - t')}t'^{k-\alpha - 1}dt',$$

where the integral can be evaluated to yield the series corresponding to the generalized exponential

$$\mathcal{LT}^{-1}\left\{\frac{s^\alpha}{\left(s - \lambda\right)^n}; t\right\} = \sum_{k=0}^{n-1}\left(-1\right)^k\binom{n - 1}{k}\frac{t^{n-1+k}\lambda^{\alpha - k}\Gamma\left(k - \alpha\right)}{\Gamma\left(n\right)\Gamma\left(-\alpha\right)}E_{\alpha - k}^{\lambda t}. \tag{5.62}$$

Thus, using (5.62) we can write for $n = 2$,

$$\mathcal{LT}^{-1}\left\{\frac{s^\alpha}{\left(s - \lambda\right)^2}; t\right\} = \lambda^\alpha t E_\alpha^{\lambda t} + \alpha\lambda^{\alpha - 1}E_{\alpha - 1}^{\lambda t} \tag{5.63}$$

so that we can integrate the individual terms in (5.60) to obtain for the solution to the fractional differential equation with equal roots

$$X\left(t\right) = X\left(0\right)\left[\left(1 + 2r^2 t\right)E_0^{r^2 t} + 2r^2 t E_{1/2}^{r^2 t} + E_{-1/2}^{r^2 t}\right] \tag{5.64}$$

with the same solution in the notation of Miller and Ross [8].

Proceed by Analogy We note in passing that the solution to fractional differential equations of this kind can in general be written in terms of the generalized exponential and products of the generalized exponentials and powers of t just as shown in the series (5.62). Miller and Ross give an excellent discussion of how to use this technique for solving fractional differential equations with rational indices and we refer the reader there for additional examples.

5.2.2 Incomplete Symmetry

A Familiar Integral We now turn our attention to the symmetry of a number of similar but distinct integrals. Consider the integral

$$I(\omega) = \int_0^\infty \frac{\cos \omega t}{t^2 + a^2} dt. \tag{5.65}$$

It is a simple matter to evaluate (5.65) using contour integration and/or the shift property of the integral and the Laplace transform. For example,

$$\mathcal{LT}\{\cos \omega t \; ; s\} = \frac{s}{s^2 + t^2}, \tag{5.66}$$

where we have taken the Laplace transform with respect to the frequency parameter ω. Inserting (5.66) into (5.65) we get

$$\mathcal{LT}\{I(\omega) \; ; s\} = \int_0^\infty \frac{s}{s^2 + t^2} \frac{1}{t^2 + a^2} dt = \frac{\pi}{2a} \frac{1}{s + a}, \tag{5.67}$$

where the integral is evaluated using the calculus of residues. Taking the inverse Laplace transform of (5.67) with respect to the s parameter we obtain

$$I(\omega) = \frac{\pi}{2a} e^{-\omega a}. \tag{5.68}$$

Another Familiar Integral If we try to evaluate the integral

$$\bar{I}(\omega) = \int_0^\infty \frac{\sin \omega t}{t^2 + a^2} dt \tag{5.69}$$

in the same way we did (5.65) we encounter some difficulties and find that there is no symmetry between the integrals. However, we know that the sine and cosine functions lead to many similar results and we suspect that this should also be the case for the above two integrals. To emphasize the symmetry between (5.65) and (5.69) we first break the symmetry of (5.65) as follows

$$I_\alpha(\omega) = \int_0^\infty \frac{t^\alpha \cos \omega t}{t^2 + a^2} dt \quad \text{with } |\alpha| < 1. \tag{5.70}$$

Now again taking the Laplace transform with respect to the ω parameter and using the calculus of residues we have

$$LT\left\{I_\alpha\left(\omega\right);s\right\} = \int_0^\infty \frac{s}{s^2+t^2}\frac{t^\alpha}{t^2+a^2}dt$$

$$= \frac{\pi a^{\alpha-1}}{2\cos\left(\alpha\pi/2\right)}\left[\frac{s}{s^2-a^2}-\left(\frac{s}{a}\right)^\alpha\frac{a}{s^2-a^2}\right]$$

which taking the inverse Laplace transforms using (5.48) yields

$$I_\alpha\left(\omega\right) = \frac{\pi a^{\alpha-1}}{2\cos\left(\alpha\pi/2\right)}\left[\cosh\omega a-\sinh_\alpha\omega a\right]. \tag{5.71}$$

Now we break the symmetry of the sine integral and write

$$\bar{I}_\alpha\left(\omega\right) = \int_0^\infty \frac{t^\alpha\sin\omega t}{t^2+a^2}dt \quad \text{with } |\alpha|<1 \tag{5.72}$$

and in the same way obtain

$$\bar{I}_\alpha\left(\omega\right) = \frac{\pi a^{\alpha-1}}{2\sin\left(\alpha\pi/2\right)}\left[\sinh_\alpha\omega a-\sinh\omega a\right]. \tag{5.73}$$

From $\bar{I}_\alpha\left(\omega\right)$ and $I_\alpha\left(\omega\right)$ we can see that the solutions to (5.71) and (5.73) are very similar. The problem arises in the limit $\alpha\to 0$ because we have the integrals converging to two different limits. For (5.71) we have

$$\lim_{\alpha\to 0}I_\alpha\left(\omega\right) = I\left(\omega\right) = \frac{\pi}{2a}e^{-\omega a}$$

which is just (5.68). But for relation (5.73) we have in the $\alpha\to 0$ limit

$$\bar{I}\left(\omega\right) = \lim_{\alpha\to 0}I_\alpha\left(\omega\right) = \lim_{\alpha\to 0}\frac{\pi a^{\alpha-1}}{2\sin\left(\alpha\pi/2\right)}\left[\sinh_\alpha\omega a-\sinh\omega a\right] \tag{5.74}$$

which is an indeterminate form; so using L'Hospital's rule we have

$$\bar{I}\left(\omega\right) = \lim_{\alpha\to 0}\frac{1}{a}\frac{\partial}{\partial\alpha}\left[\sinh_\alpha\omega a\right]. \tag{5.75}$$

The connection of (5.75) with the parametric fractional derivative

$$\frac{\partial}{\partial\alpha}D_t^\alpha\left[e^t\right]$$

should be clear. We can take the Laplace transform of this quantity to obtain

$$LT\left\{\frac{\partial}{\partial\alpha}D_t^\alpha\left[e^t\right];s\right\} = LT\left\{\frac{\partial}{\partial\alpha}E_\alpha^t;s\right\} = \frac{\partial}{\partial\alpha}\left[LT\left\{E_\alpha^t;s\right\}\right]$$

$$= \frac{\partial}{\partial\alpha}\left[\frac{s^\alpha}{s-1}\right] = \frac{s^\alpha\ln s}{s-1} \tag{5.76}$$

and therefore (5.75) reduces to

$$\tilde{I}(\omega) = \frac{\ln s}{s^2 - 1}. \tag{5.77}$$

In conclusion we can say that the generalization of integrals (5.65) and (5.69) gives similar results and the asymmetry of the $\alpha = 0$ case is an accidental asymmetry.

Product of Functions Consider as our last example:

$$I(t) = \frac{1}{\sqrt{\pi t}} \int_0^\infty dx e^{-\frac{x^2}{4t}} \cos x \tag{5.78}$$

which using the Efrös theorem [1] becomes

$$I(t) = e^{-t}. \tag{5.79}$$

Theorem 5 (Efrös): *If $\hat{G}(s)$ and $q(s)$ are analytic functions, and*

$$\hat{G}(s) e^{-q(s)\tau} = \mathcal{LT}\{g(t,\tau);s\} \tag{5.80}$$

then

$$\hat{G}(s)\tilde{f}(q(s)) = \mathcal{LT}\left\{\int_0^\infty f(\tau)g(t,\tau)\,d\tau;s\right\}. \tag{5.81}$$

To demonstrate the correctness of this theorem consider writing out the Laplace transform on the right-hand side of (5.81),

$$\mathcal{LT}\left\{\int_0^\infty f(\tau)g(t,\tau)\,d\tau;s\right\} = \int_0^\infty e^{-st}dt \int_0^\infty f(\tau)g(t,\tau)\,d\tau$$

$$= \int_0^\infty f(\tau)\,d\tau \int_0^\infty g(t,\tau)e^{-st}dt \tag{5.82}$$

provided we can interchange the order of integrations. The last integral in (5.82) is the Laplace transform of $g(t,\tau)$; hence by definition (5.80),

$$\mathcal{LT}\left\{\int_0^\infty f(\tau)g(t,\tau)\,d\tau;s\right\} = \hat{G}(s)\int_0^\infty e^{-q(s)\tau}f(\tau)\,d\tau$$

$$= \hat{G}(s)\tilde{f}(q(s)) \tag{5.83}$$

which was to be shown.

Proof of the Theorem To establish (5.79) using Efrös theorem we define the function

$$g(t,\tau) \equiv \frac{1}{\sqrt{\pi t}} e^{-\frac{\tau^2}{4t}}$$ (5.84)

whose Laplace transform is given by

$$\mathcal{LT}\left\{\frac{1}{\sqrt{\pi t}} e^{-\frac{\tau^2}{4t}} ; s\right\} = \frac{e^{-t\sqrt{s}}}{\sqrt{s}} = \overset{\smile}{G}(s) e^{-tq(s)}$$ (5.85)

so that

$$\overset{\smile}{G}(s) = \frac{1}{\sqrt{s}} \text{ and } q(s) = \sqrt{s}.$$

This allows us to write (5.83) as

$$\mathcal{LT}\left\{\int_0^\infty f(\tau) \frac{1}{\sqrt{\pi t}} e^{-\frac{\tau^2}{4t}} d\tau ; s\right\} = \frac{1}{\sqrt{s}} \tilde{f}(\sqrt{s})$$ (5.86)

so that identifying the function $f(\tau)$ with the cosine in (5.78) and using

$$\tilde{f}(s) = \mathcal{LT}\{\cos \tau ; s\} = \frac{s}{s^2 + 1}$$ (5.87)

to obtain

$$\tilde{f}(\sqrt{s}) = \frac{\sqrt{s}}{s+1}$$

we have

$$\mathcal{LT}\left\{\int_0^\infty f(\tau) \frac{1}{\sqrt{\pi t}} e^{-\frac{\tau^2}{4t}} d\tau ; s\right\} = \frac{1}{s+1}.$$

Therefore, taking the inverse Laplace transform of this expression, we obtain

$$\mathcal{LT}^{-1}\left\{\frac{1}{s+1} ; t\right\} = e^{-t}$$

so the $I(t)$ is given by (5.79) as was to be shown.

Applying the Theorem Again it seems that there is no symmetry when we replace the cosine function with the sine function in (5.78):

$$\overline{I}(t) = \frac{1}{\sqrt{\pi t}} \int_0^\infty dx e^{-\frac{x^2}{4t}} \sin x$$ (5.88)

but using the Efrös theorem we obtain for the Laplace transform of (5.88):

$$\mathcal{LT}\{\overline{I}(t) ; s\} = \frac{1}{\sqrt{s}} \frac{1}{s+1}.$$ (5.89)

In terms of the generalized exponentials we have by inverting the Laplace transform in (5.89)

$$\overline{I}(t) =^* E_{-1/2}^{-t} = E_{-1/2}^{-t} \tag{5.90}$$

which means that $I(t)$ and $\overline{I}(t)$ are particular cases of the general solution E_α^{-t} and it is in this sense that $I(t)$ and $\overline{I}(t)$ are similar. More precisely, if we consider the vector

$$\mathbf{u}(t) = \left(\begin{array}{c} I_\alpha(t) \\ \overline{I}_\alpha(t) \end{array} \right) = \left(\begin{array}{c} \frac{1}{\sqrt{\pi t}} \int_0^\infty dx e^{-\frac{x^2}{4t}} \cos_\alpha x \\ \frac{1}{\sqrt{\pi t}} \int_0^\infty dx e^{-\frac{x^2}{4t}} \sin_\alpha x \end{array} \right) \tag{5.91}$$

then by the Efrös theorem we obtain the Laplace transform of the vector

$$
\begin{aligned}
\mathcal{LT}\{\mathbf{u}(t); s\} &= \frac{s^{\alpha-1/2}}{s+1} \left(\begin{array}{c} \sqrt{s}\cos(\alpha\pi/2) + \sin(\alpha\pi/2) \\ -\sqrt{s}\sin(\alpha\pi/2) + \cos(\alpha\pi/2) \end{array} \right) \\
&= \frac{s^{\alpha/2}}{s+1} \mathcal{R} \left(\begin{array}{c} 1 \\ \frac{1}{\sqrt{s}} \end{array} \right),
\end{aligned} \tag{5.92}
$$

where the rotation matrix $\mathcal{R}$ is given by

$$\mathcal{R} = \left(\begin{array}{cc} \cos(\alpha\pi/2) & \sin(\alpha\pi/2) \\ -\sin(\alpha\pi/2) & \cos(\alpha\pi/2) \end{array} \right).$$

Using the equations for the Laplace transform of the generalized exponential function we can take the inverse of (5.92) and obtain

$$\mathbf{u}(t) = \mathcal{R}^{-1} \left(\begin{array}{c} E_{\alpha-1}^{-t} \\ E_{\alpha/2}^{-t} \end{array} \right). \tag{5.93}$$

Equation (5.93) indicates that the integrals are linear combinations of generalized exponentials along with sine and cosine functions evaluated at the parameter value $\alpha = 0$.

5.2.3 Integral Properties of the Generalized Exponential

Heterogeneity and Anisotropy In our previous lectures we have tried to bring to life the physics underlying the mathematical formalism and to sketch out the important features of the fractional calculus, often sacrificing rigor for vigor. However, in such areas as dynamics it is necessary to develop a certain amount of intuition regarding the mathematics in order to understand the phenomenon of interest. When time is intermittent and space is distorted, the phenomena often require a more encompassing view of what constitutes a function, or what is an acceptable change in that function. Therefore we further explore the properties of the generalized exponentials and trigonometric functions introduced earlier, but here we restrict our attention to a few of their properties that are of value to us later.

Decomposition of Generalized Exponential   Let us first consider the decomposition of the generalized exponential function E_α^{t+a}. We take the Laplace transform of this quantity

$$\mathcal{LT}\left\{E_\alpha^{t+a}; s\right\} = \int_0^\infty E_\alpha^{t+a} e^{-st} dt = e^{sa} \int_a^\infty E_\alpha^x e^{-sx} dx \qquad (5.94)$$

so that using the indefinite integral

$$\int^x E_\alpha^{at} e^{bt} dt = \frac{e^{bx}}{a+b}\left(E_\alpha^{ax} - \left(\frac{b}{a}\right)^\alpha {}^* E_\alpha^{-bx}\right) + c$$

we can express (5.94) as

$$\mathcal{LT}\left\{E_\alpha^{t+a}; s\right\} = e^{sa}\frac{e^{-sx}}{1-s}\left[E_\alpha^x - (-s)^\alpha {}^* E_\alpha^{-sx}\right]_a^\infty$$

and recalling that $(-1)^\alpha {}^* E_\alpha^{-sx} = E_\alpha^{sx}$ we obtain

$$\mathcal{LT}\left\{E_\alpha^{t+a}; s\right\} = \frac{s^\alpha e^{sa}}{s-1} - \frac{1}{1-s}\left[E_\alpha^a - s^\alpha E_\alpha^{sa}\right]$$

since $E_\alpha^{sx} \to e^{sx}$ as $x \to \infty$. Thus, we obtain

$$\mathcal{LT}\left\{E_\alpha^{t+a}; s\right\} = \frac{s^\alpha}{s-1}\left[e^{sa} - E_\alpha^{sa}\right] + \frac{E_\alpha^a}{s-1}. \qquad (5.95)$$

Convolution of Generalized Exponentials   Now we examine the convolution properties of generalized exponentials. We consider the quantity

$$\frac{s^{\alpha+\beta}}{s^2 - 1} \quad \text{with } \alpha + \beta < 1.$$

We can also write this quantity in the factored form

$$\frac{s^{\alpha+\beta}}{s^2 - 1} = \frac{s^\alpha}{s-1}\frac{s^\beta}{s+1}.$$

If we take the inverse Laplace transform of the left-hand side of this equation we obtain, using (5.49),

$$\mathcal{LT}^{-1}\left\{\frac{s^{\alpha+\beta}}{s^2 - 1}; t\right\} = \sinh_{\alpha+\beta}(t), \qquad (5.96)$$

but we also have, using the convolution theorem for Laplace transforms,

$$\mathcal{LT}^{-1}\left\{\frac{s^{\alpha+\beta}}{s^2 - 1}; t\right\} = \int_0^t \mathcal{LT}^{-1}\left\{\frac{s^\alpha}{s-1}; t-\xi\right\} \mathcal{LT}^{-1}\left\{\frac{s^\beta}{s+1}; \xi\right\} d\xi$$

$$= \int_0^t E_\alpha^{t-\xi} {}^* E_\beta^\xi d\xi = \sinh_{\alpha+\beta}(t). \qquad (5.97)$$

In particular for $\alpha = -\beta$ the convolution integral (5.97) reduces to

$$\int_0^t E_\alpha^{t-\xi} * E_{-\alpha}^\xi d\xi = \sinh t. \tag{5.98}$$

In the same way we can take, for example,

$$\frac{s^{\alpha+\beta}}{(s-1)^2} = s^{\alpha+\beta} \frac{1}{(s-1)^2}$$

whose inverse Laplace transform is given by

$$\mathcal{LT}^{-1}\left\{\frac{s^{\alpha+\beta}}{(s-1)^2}; t\right\} = \int_0^t \mathcal{LT}^{-1}\left\{\frac{s^\alpha}{s-1}; t-\xi\right\} \mathcal{LT}^{-1}\left\{\frac{s^\beta}{s-1}; \xi\right\} d\xi$$

$$= \int_0^t E_\alpha^{t-\xi} E_\beta^\xi d\xi = t E_{\alpha+\beta}^t + (\alpha+\beta) E_{\alpha+\beta-1}^t. \tag{5.99}$$

We can also obtain this result using

$$\mathcal{LT}^{-1}\left\{\frac{s^{\alpha+\beta}}{(s-1)^2}; t\right\} = D_t^{\alpha+\beta}\left[\mathcal{LT}^{-1}\left\{\frac{1}{s^2-1}; t\right\}\right] = D_t^{\alpha+\beta}\left[t e^t\right]$$

$$= t E_{\alpha+\beta}^t + (\alpha+\beta) E_{\alpha+\beta-1}^t. \tag{5.100}$$

Furthermore, setting $\alpha = -\beta$ in (5.99) we obtain the analytic result for the convolution of the generalized exponential with itself

$$\int_0^t E_\alpha^{t-\xi} E_{-\alpha}^\xi d\xi = t e^t. \tag{5.101}$$

More examples It is also possible to consider other possibilities using the factors

$$\frac{s^{\alpha+\beta}}{s^2+1} \quad \text{and} \quad \frac{s^{\alpha+\beta}}{(s+1)^2}$$

and the convolutions involving $*E_\alpha^{-t}$ or E_α^{it}. We do not pursue these formal results further here, but it would be useful for students to work out some of these integrals for themselves.

5.3   Fractional Green's Function

Continuous Green's Functions Green's functions are often used in the context of spatial fields to obtain the contributions from all the pieces of an extended source. For example, if $G(\mathbf{x}, t; \mathbf{x}_0)$ is the field at the observation point $\mathbf{x}$ at time

t induced by an initial unit point source at the point $\mathbf{x}_0$, then if $\rho(\mathbf{x})$ is the distribution of point sources, the convolution $G * \rho$ is the total contribution to the field at the observation point at a given time from all the initial source points[2]. This Green's function is the solution to the homogeneous dynamical equation describing the evolution of the field, for example, the diffusion equation or wave equation we discussed earlier. Here we do not concern ourselves with spatial effects, but for the moment address the less ambitious task of constructing the Green's function for systems of dynamical equations.

Green's Function Solution to Linear Equations   Given the linear equation

$$\mathcal{L}u = f \tag{5.102}$$

where $\mathcal{L}$ is an, as yet unspecified, linear operator, we may solve (5.102) in a number of different ways. We might attempt to find eigenvalues and eigenvectors of $\mathcal{L}$ such that

$$\mathcal{L}\phi_j = \lambda_j \phi_j \tag{5.103}$$

and expand the homogeneous term on the right-hand side of (5.102) in terms of this set of eigenfunctions $\{\phi_j\}$. We have occasion to use the eigenfunction expansion in later lectures. If the set of eigenfunctions is complete we can also construct the inverse operator $\mathcal{L}^{-1}$ such that

$$\mathcal{L}\mathcal{L}^{-1} = \mathcal{I} = \mathcal{L}^{-1}\mathcal{L}, \tag{5.104}$$

where $\mathcal{I}$ is the unit operator and we may write the formal solution to (5.102) as

$$u = \mathcal{L}^{-1}f. \tag{5.105}$$

There are a number of books devoted wholly, or in part, to the explicit construction of the inverse operator $\mathcal{L}^{-1}$ [3, 9, 11]. When $\mathcal{L}$ is a differential operator, the inverse operator is the integral

$$\mathcal{L}^{-1}u(x) = \int_\Omega G(x, x')\, u(x')\, dx' \tag{5.106}$$

whose kernel $G(x, x')$ is a Green's function. The integration is over the domain Ω defined by the problem of interest. Symbolically, we can write

$$u(x) = \mathcal{L}\mathcal{L}^{-1}u(x) = \mathcal{L}\int_\Omega G(x, x')\, u(x')\, dx', \tag{5.107}$$

where it is obvious, in the light of (5.104), that

---

[2] An excellent book by Podlubny [10] was published covering some of the same mathematical material after those lectures had been given for the first time. We subsequently modified some of the discussion on fractional Green's functions to incorporate Podlubny's insights.

$$\mathcal{L}G\left(x, x'\right) = \delta\left(x - x'\right) \tag{5.108}$$

reducing (5.106) to a tautology.

Only an Inverse Thus, it is only necessary for the linear operator $\mathcal{L}$ to have an inverse in order for us to make use of the Green's function technique. Furthermore, since the fractional derivative operators have corresponding fractional integrals we can formally construct the solutions to linear fractional-differential equations using the appropriately defined fractional Green's functions.

Fractional Differential Equation Consider, for example, the three-term linear fractional-differential equation with constant coefficients:

$$aD_t^\alpha\left[u\left(t\right)\right] + bD_t^\beta\left[u\left(t\right)\right] + cu\left(t\right) = f\left(t\right), \tag{5.109}$$

where a, b, and c are constants, α and β are both noninteger, and $f\left(t\right)$ is an arbitrary analytic function. Taking the Laplace transform of (5.109) we obtain

$$as^\alpha \tilde{u}\left(s\right) + bs^\beta \tilde{u}\left(s\right) + c\tilde{u}\left(s\right) = C + \tilde{f}\left(s\right), \tag{5.110}$$

where the constant is determined by the initial values emerging from taking the Laplace transform of (5.109),

$$C = \left\{aD_t^{\alpha-1}\left[u\left(t\right)\right] + bD_t^{\beta-1}\left[u\left(t\right)\right]\right\}_{t=0} \tag{5.111}$$

so that

$$\tilde{u}\left(s\right) = \frac{C}{as^\alpha + bs^\beta + c} + \frac{\tilde{f}\left(s\right)}{as^\alpha + bs^\beta + c}. \tag{5.112}$$

The inverse Laplace transform of (5.112) yields

$$u\left(t\right) = CG\left(a, b, c; t\right) + \int_0^t G\left(a, b, c; t - t'\right) f\left(t'\right) dt', \tag{5.113}$$

where $G\left(a, b, c; t\right)$ is the stationary Green's function in time.

Analytic Form for the Green's Function To obtain the analytic form for the Green's function consider its Laplace transform

$$\tilde{G}\left(a, b, c; s\right) = \frac{1}{as^\alpha + bs^\beta + c} = \frac{s^{-\beta}}{a\left(s^{\alpha-\beta} + \frac{b}{a}\right)} \frac{1}{1 + \frac{c}{a\left(s^{\alpha-\beta} + \frac{b}{a}\right)}}$$

$$= \frac{1}{c}\sum_{j=0}^\infty \left(-1\right)^j \left(\frac{c}{a}\right)^{j+1} \frac{s^{-\beta(j+1)}}{\left(s^{\alpha-\beta} + \frac{b}{a}\right)^{j+1}}, \tag{5.114}$$

where the summation comes from expanding the second factor in the middle equation in a geometric series. The inverse Laplace transform of this series is obtained by making use of the following Laplace transform of the weighted generalized Mittag-Leffler function

$$LT\left\{t^{\beta-1}E_{\alpha,\beta}\left(-\lambda t^{\alpha}\right);s\right\} = \sum_{j=0}^{\infty}\frac{(-\lambda)^{j}}{\Gamma\left(j\alpha+\beta\right)}LT\left\{t^{j\alpha+\beta-1};s\right\}$$

$$= \frac{1}{s^{\beta}}\sum_{j=0}^{\infty}\left(\frac{-\lambda}{s^{\alpha}}\right)^{j} = \frac{s^{\alpha-\beta}}{s^{\alpha}+\lambda}. \qquad (5.115)$$

Equation (5.115) can be generalized by taking the kth order parametric derivative which using the expression on the right-hand side is simply

$$LT\left\{t^{\beta-1}\frac{d^{k}}{d\lambda^{k}}E_{\alpha,\beta}\left(-\lambda t^{\alpha}\right);s\right\} = \Gamma\left(k+1\right)\frac{s^{\alpha-\beta}}{\left(s^{\alpha}+\lambda\right)^{k+1}}. \qquad (5.116)$$

We now define the kth derivative of the generalized Mittag-Leffler function as

$$E_{\alpha,\beta}^{(k)}\left(x\right) \equiv \frac{d^{k}}{dx^{k}}E_{\alpha,\beta}\left(x\right) = \sum_{j=0}^{\infty}\frac{\Gamma\left(j+k+1\right)}{\Gamma\left(\left[j+k\right]\alpha+\beta\right)}x^{j} \qquad (5.117)$$

and scaling the parameter λ by t^{α} in (5.116) we obtain

$$LT\left\{t^{k\alpha+\beta-1}E_{\alpha,\beta}^{(k)}\left(-\lambda t^{\alpha}\right);s\right\} = \Gamma\left(k+1\right)\frac{s^{\alpha-\beta}}{\left(s^{\alpha}+\lambda\right)^{k+1}} \qquad (5.118)$$

so that taking the inverse Laplace transform of (5.114) after some algebra yields the Green's function

$$G\left(a,b,c;t\right) = \frac{1}{a}\sum_{j=0}^{\infty}\left(-\frac{c}{a}\right)^{j}t^{\alpha(j+1)-1}E_{\alpha-\beta,\alpha+j\beta}^{(j)}\left(-\frac{b}{a}t^{\alpha-\beta}\right). \qquad (5.119)$$

Simpler Forms for the Green's Function   The Green's function given by (5.119) can be used to obtain the propagators for simpler problems as well. For example, if $a=0$ in (5.109) we obtain,

$$\tilde{G}\left(a=0,b,c;s\right) = \frac{1}{bs^{\beta}+c} = \frac{1}{b}\frac{1}{s^{\beta}+\frac{c}{b}} \qquad (5.120)$$

whose inverse Laplace transform is the two-term Green's function

$$G\left(a=0,b,c;t\right) = \frac{t^{\beta-1}}{b}E_{\beta,\beta}\left(-\frac{c}{b}t^{\beta}\right). \qquad (5.121)$$

Furthermore, if we set $c=0$ then using the definition of the Mittag-Leffler function we have for the one-term Green's function

$$G\left(a = 0, b, c = 0; t\right) = \frac{t^{\beta-1}}{b\Gamma\left(\beta\right)}. \tag{5.122}$$

Different Forms   It is straightforward to extend these results to differential equations with fractional derivatives having integer parts by modifying the definition of the constant C in (5.111). Assume that $(n-1) < \alpha < n$ and $(m-1) < \beta < m$ then we define the constants

$$a_k \equiv D_t^{\alpha-k}\left[u\left(t\right)\right]\big|_{t=0} \ , \ k = 0, 1, \cdots, [\alpha] \tag{5.123}$$

$$b_k \equiv D_t^{\beta-k}\left[u\left(t\right)\right]\big|_{t=0} \ , \ k = 0, 1, \cdots, [\beta] \tag{5.124}$$

so that the complete solution for the three-term fractional differential equation with constant coefficients is given by

$$u\left(t\right) = \sum_{k=0}^{[\alpha]} a_k G^{(k)}\left(a, b, c; t\right) + \sum_{k=0}^{[\beta]} b_k G^{(k)}\left(a, b, c; t\right) + \int_0^t G\left(a, b, c; t - t'\right) f\left(t'\right) dt',$$

where using the inverse Laplace transform of

$$\frac{s^{k-1}}{as^{\alpha} + bs^{\beta} + c}$$

we obtain the Green's functions

$$G^{(k)}\left(a, b, c; t\right) = \frac{1}{a} \sum_{j=0}^{\infty} \left(-\frac{c}{a}\right)^j t^{\alpha(j+1)-k} E_{\alpha-\beta, \alpha+j\beta-k+1}^{(j)}\left(-\frac{b}{a} t^{\alpha-\beta}\right). \tag{5.125}$$

Here again if we set $a = 0$ we obtain for the solution to the two-term fractional differential equation

$$u\left(t\right) = \sum_{k=0}^{[\beta]} \left\{ a_k t^{\beta-k} E_{\beta, \beta-k+1}\left(-\frac{c}{b} t^{\beta}\right) + \int_0^t E_{\beta, \beta-k+1}\left(-\frac{c}{b}\left(t - t'\right)^{\beta}\right) \right.$$
$$\left. \times \left(t - t'\right)^{\beta-1} f\left(t'\right) dt' \right\} \tag{5.126}$$

with inhomogeneous initial conditions.

5.4   Commentary

Solution to Dynamical Equations   We have seen that the Laplace transform and its inverse do indeed enable one to solve classes of differential equations that would otherwise be outside our capability. However, since there are many more equations that we cannot solve than there are those that we can solve, this in itself would not be of much interest to the physical scientist. It is the fact that these equations constitute mathematical models of physical phenomena and that their solutions imply interesting physical properties that make them attractive. Furthermore, other models of the phenomena of interest have resisted solution in the past.

Viscoelasticity and Fractional Derivatives   Thus, the application of fractional-differential equations to model the relaxation of viscoelastic materials has provided a new insight into the physical structure of such materials. The theory of rheology suggested by Scott Blair et al. [12] and developed in some detail by Glöckle and Nonnenmacher [4] through [7] provides an excellent fit to the relaxation of a wide range of materials over time scales that span 10 orders of magnitude. It is the only first-principles description of this phenomenon; assuming that the reader is now willing to accept a fractional-differential equation as a fundamental description of a dynamical physical process.

Green's Functions   The Green's function often takes the form of an infinite series, such as we found for certain fractional-differential equations. In the present case we could define an analytic function, the generalized Mittag-Leffler function, as an infinite series. Of course most special functions , such as the sine, cosine, Bessel, and other functions of theoretical physics can also be expressed as infinite series. The difference from the Green's function is that typically the series converges rather slowly, thus making it difficult to obtain general insight into the overall behavior of the phenomenon. On the other hand, Green's functions can often be expressed in terms of integral representations involving closed functions.

Boundary Value Problems   The Green's function described in the last few lectures does not solve a boundary value problem, but rather solves initial value problems. The Green's function $G(t, t_0)$ denotes the field at a point in space, say, at a time t induced by a unit point source at the initial time t_0. Thus, if the inhomogeneous term $f(t)$ is interpreted as the distribution of point sources in time, then the convolution in time $G * f$ is the total contribution to the field, at this point in space, at time t. The Green's function is therefore a solution which is homogeneous everywhere except at a single point.

Bibliography

[1] M. Ya Antimirov, A. A. Kolyshkin and R. Vaillancourt, *Applied Integral Transform Vol. 2*, CMR Monograph Series, Centre de Recherches Mathematiques, Universitè de Montreal

[2] M. Bologna, *Derivata ad Indice Reale*, Ets Editrice, Italy (1990).

[3] J. T. Cushing, *Applied Analytic Mathematics for Physical Scientists*, John Wiley, New York (1975).

[4] W. G. Glöckle and T. F. Nonnenmacher, *Macromolecules* **24**, 6426 (1991).

[5] W. G. Glöckle and T. F. Nonnenmacher, A fractional Calculus approach to self-similar protein dynamics, *Biophys. J.* **68**, 46-53 (1995).

[6] W. G. Glöckle and T. F. Nonnenmacher, Fox function representation of non-Debye relaxation processes, *J. Stat. Phys.* **71**, 741 (1993).

[7] W. G. Glöckle and T. F. Nonnenmacher, Fractional relaxation and the time-temperature superposition principle, *Rheol. Acta* **33**, 337 (1994).

[8] K. S. Miller and B. Ross, *An Introduction to the Fractional Calculus and Fractional Differential Equations*, John Wiley, New York (1993).

[9] P. M. Morse and H. Feshbach, *Methods of Theoretical Physics*, McGraw-Hill , New York (1953).

[10] I. Podlubny, *Fractional Differential Equations*, Academic, New York (1999).

[11] G. F. Roach, *Green's Functions*, second edition, Cambridge University Press, Cambridge, MA (1982).

[12] G. W. Scott Blair, B. C. Veinoglou and J. E. Caffyn, Limitations of the Newtonian time scale in relation to non-equilibrium rheological states and a theory of quasi-properties, *Proc. Roy. Soc. Ser. A* **187**, 69 (1947).

Chapter 6

Fractional Randomness

Different Kinds of Differential Equations The equations describing the evolution of complex physical phenomena can be put into a number of categories. This separation depends on whether the changes in the physical observables are relatively slow, regular, and describable by simple analytic functions, or if the changes are rapid, irregular, and not predictable, and therefore describable by fractal functions. Historically this led to the two categories of dynamics: deterministic equations of motion and stochastic equations of motion. However, since the early 1960s it has become increasingly clear that these two categories are not mutually exclusive, so other ways to draw distinctions among phenomena have become more popular.

Randomness Is Subtle Here, and in earlier lectures, we have introduced one of the most subtle concepts entering into our discussion of complex phenomena, that is, the existence and role of randomness. Herein the idea of randomness is related to the difference between an *act* and an *event*. Turner [39] distinguished between the two by noting that an event has a symmetry in time, in that there is no difference between knowing an event can happen and knowing that an event did happen, so no additional information is gained by the occurrence of an event. On the other hand, an act has an asymmetry in time, in that what is known about a process is fundamentally different before and after an act. An event may be predicted by the situation preceding it; an act may not. However, even though we cannot predict the outcome of an act, in retrospect we may say that the act is understandable given the preexisting situation. Turner also argued that this distinction allows for the notion of freedom to be reintroduced into a deterministic universe, and for a clear separation to be made between what a thing is (ontology) and how a thing is known (epistemology). Note that we use the term act to include such non-conscious processes as the self-organization made by the formation of stable vortices in turbulent fluid flow, the patterns on butterfly wings, and oscillating chemical reactions in the natural sciences, as well as the conscious development of myths, religions, and organizations in the social sciences.

What Causes Uncertainty? From one perspective the unpredictability of free acts has to do with the large number of elements in the system, so many in fact, that the behavior of the system ceases to be predictable. On the other hand, we now know that having only a few dynamical variables in a system does not ensure predictability or knowability. It has been demonstrated that the irregular time series observed in such disciplines as biology, chemical kinetics, economics, logic, physics, and physiology are, at least in part, due to chaos. Technically the term chaos may be defined to be a sensitive dependence of the solutions to a set of nonlinear, deterministic, dynamical equations, on initial conditions [29]. Practically, chaos means that the solutions to such equations look erratic and may pass all the traditional tests for randomness even though they are deterministic. Therefore, if we think of random time series as complex, then the output of a chaotic generator is complex. However, we know that something as simple as a one-dimensional quadratic map (the logistic equation) can generate a chaotic sequence [24]. Thus, using the traditional definition of complexity, it would appear that chaos implies the generation of complexity from simplicity. This is part of the Poincaré legacy of paradox. Another part of that legacy is the fact that chaos is a generic property of nonlinear dynamical systems, which is to say, chaos is ubiquitous; all systems change over time, and because all non-ideal systems are nonlinear, all systems manifest chaotic behavior to a greater or lesser extent.

Chaos Versus Noise A nonlinear system with only a few dynamical variables can generate random patterns and therefore has chaotic solutions. So we encounter the same restrictions on our ability to know and understand a system when there are only a few dynamical elements as when there are a great many dynamical elements, but for very different reasons. Let us refer to the latter random process as noise, the unpredictable influence of the environment on the system. Here the environment is assumed to have an infinite number of elements, all of which we do not know, but they are coupled to the system of interest and perturb it in a random, that is, unknown, way. By way of contrast, chaos is a consequence of the nonlinear deterministic interactions in an isolated dynamical system, resulting in erratic behavior of limited predictability. Chaos is an implicit property of a complex system, whereas noise is a property of the environment in contact with the system of interest. Chaos can therefore be controlled and predicted over short time intervals, whereas noise can neither be predicted nor controlled, except perhaps through the way it interacts with the system.

What Is Complexity? This distinction between chaos and noise highlights one of the difficulties in formulating an unambiguous measure of complexity. Since noise cannot be predicted or controlled it might be viewed as being complex, thus, systems with many degrees of freedom that manifest randomness might be considered complex. On the other hand, a system with only a few dynamical elements, when it is chaotic, might be considered to be simple. Thus, two systems with apparently the same erratic behavior can, because of the different causes of that behavior, have conflicting definitions of complexity. In this way the idea of complexity is ill-posed and a new approach to its definition is required since noise and chaos are often confused with each other.

Systems Theory In the early papers on systems theory it was argued that the increasing complexity of an evolving system can reach a threshold where the system is so complicated that it is impossible to follow the dynamics of the individual elements. At this point new properties often emerge and the new organization undergoes a completely different type of dynamics. The details of the interactions among the individual elements are substantially less important than is the structure, the geometrical pattern, of the new aggregate. This is the self-aggregating behavior observed in many biological, physical, and social phenomena. Increasing further the number of elements, or alternatively the number of relations, often leads to a complete disorganization and the stochastic approach becomes a good description of the system behavior. If randomness (noise) is now considered as something simple, as it is intuitively, one has to seek a measure of complexity that decreases in magnitude in the limit of the system having an infinite number of elements. So a viable measure of complexity must first increase and then decrease with continually increasing numbers of system elements and/or an increasing number of the relations among system elements.

Nonlinear Dynamical Systems The equations of motion that are generated by a system's Hamiltonian are called conservative, since they conserve the total energy of the system. Other systems are not easily described by a Hamiltonian, in part because they are not conservative, and dissipate energy over time. In both the conservative and nonconservative cases the equations of motion are deterministic, which is to say, given the initial conditions for the system the final state is unambiguously determined by the equations of motion. Of course, we now know that if the dynamical equations are nonlinear, an arbitrarily small change in the initial state of the system may lead to an unpredictably large change in the final state of the dynamical system. This sensitive dependence on the initial state of the system, chaos, was discovered, but not named, in solutions to deterministic, nonlinear, dynamical equations in the last century by the mathematician and astronomer H. Poincaré [33] in his study of the conservative motion of the planets in their orbits. Chaos is a consequence of nonintegrable Hamiltonian dynamics and its significance was not understood by the scientific community until the last half of the last century. The works of Kolmogorov [16], Arnold [2] and Moser [28] clarified what is now called KAM theory. The phenomenon of chaos only drew the attention of the general scientific community with the rediscovery of the work of the meteorologist Ed Lorenz on dissipative hydrodynamic systems in the early 1960s [19]. This rediscovery was made almost simultaneously by a number of investigators in the late 1970s. In the case of nonintegrable Hamiltonian KAM systems the trajectories for the individual particles break up due to instabilities, whereas in the dissipative, nonlinear, dynamical case studied by Lorenz, the trajectories evolve on a strange attractor. Thus, the single term *chaos* may mean quite different things depending on whether the system is conservative or dissipative.

Physical Models of Randomness   We do not have the space here to discuss nonlinear dynamics, in even a preliminary way, so we refer the reader to a number of excellent texts on the mathematics of chaos [17] and the applications of nonlinear dynamics to all manner of phenomena [36, 47]. Instead, we develop models of stochastic processes with inverse power-law memory. First, we use ordinary random walks driven by correlated fluctuations to discuss the difference between ordinary and anomalous diffusion. This latter formalism was developed by Montroll and Weiss [25] and is called a continuous-time random walk (CTRW). Second, we generate a random process using fractional-difference random walks and show how the fractional-differences impose an inverse power-law memory on a system's response. Finally, this latter model is generalized to the continuum case and is shown to give rise to fractional Brownian motion. The existence of inverse power-law spectra in complex phenomena is the signature of fractal statistical processes. The index of the inverse power-law spectrum is shown to be related to the fractal dimension of the time series being described, as we discuss using colored noise as an exemplar.

Probabilities and Their Evolution   Of course, when we discuss time series and their statistics, we must also pay attention to the corresponding probability densities and their evolution over time. We review the derivation of the phase space equation of evolution of the probability density, called the Fokker-Planck equation in physics and the Kolmogorov equation in mathematics. In its simplest form this is the diffusion equation with the mass density replaced with a probability density, and it describes normal diffusion. This latter equation describes the evolution of the Gauss distribution with a variance that increases linearly with time. In addition we also consider the less well-known derivation of the phase space equation for the Lévy distribution, and find that this equation has a fractional derivative in the phase space variable and Lévy processes are nondifferentiable in general.

6.1   Ordinary Random Walks

Errors in Measurement   To explain the regression of a system to equilibrium and therefore to give time a direction, physicists allow for a certain uncertainty in the measured values of dynamical variables. The uncertainty means that scientists typically only require that we measure a given quantity to within some predefined interval. Such an interval in the measuring process is referred to as coarse-graining. The coarse-grained measurements are traditionally done by discarding the fiction of a closed system and recognizing that every system has an environment with which it interacts. One obtains a description that is statistical in nature by explicitly eliminating the environmental variables from the description of the system dynamics. This is what was done using the Hamiltonian model in earlier lectures. The absolute predictability, which was apparently present in the deterministic nature of Newton's equations, is abandoned for a more tractable description of the system having many fewer variables. The new equations are more tractable, but the price is predictability. We cannot know the environment in a finite experiment. The environment is not under our control; only the system is accessible through experimental constraints. The environment's influence on the experimental system is unpredictable and unknowable except in an average sense, when the experiment is repeated again and again. It is this repeatability of the experiment that allows us to map out all the different ways the environment influences the system, through the construction of an ensemble distribution function. This function captures all the available information, features common to the majority of the experiments in the ensemble.

Random Walks The best physical model is the simplest one that can explain all the available experimental data with the fewest number of assumptions. One such model is the random walk, which in its simplest form provides a physical picture of ordinary diffusion. The real strength of this model, however, lies in the ease with which it can be generalized to describe more complex phenomena, such as random processes with long-term memory.

One-dimensional Random Walk A Bernoulli sequence, a fair coin toss, is often used to construct a simple random walk, which is the physical scientist's model of a stochastic (random) process that has normal or Gaussian statistics and no memory. We formally define a time-dependent random variable $X(t)$ using a Bernoulli time sequence $U_j = \pm 1$

$$X(t) = \sum_{j=1}^{N} U_j, \tag{6.1}$$

where we assume that each step is taken in a time interval Δt and the total time to take N steps is $t = N\Delta t$, when $t >> \Delta t$. Here we have a picture of a walker who takes a step forward or backward depending on the outcome of a coin toss. After N such steps we can determine how far the walker has traveled from the starting point by adding a plus one for each head and a minus one for each tail. In this way the sum of positive and negative ones determines how far the walker is from the origin, that is, from the walker's initial location on the one-dimensional lattice. This is the meaning of the dynamical variable $X(t)$ in (6.1), and therefore it is a random variable, that is, a random function of time (step number).

Stepping Out We need not restrict our comments to the simple random walk depicted in (6.1), but we can generalize the process on the right-hand side to steps of size $\{s_j\}$, rather than steps of unit length, and the probability of taking a step of size s_j is given by $p(s_j)$. In this latter case the simple random walk above is replaced with

$$X(t) = \sum_{j=1}^{N} s_j. \tag{6.2}$$

For large N, if the first two moments of the step lengths, $\langle s \rangle$ and $\langle s^2 \rangle$ of the transition probability $p(s)$, are finite, then the dynamical variable is a diffusion process. By a diffusion process we mean that the statistics of the random walk variable are Gaussian and the mean-square displacement increases linearly with time, as we show below. We do not derive the statistical distribution here, but instead argue from the CLT that the sum of a large number of statistically independent random variables with finite central moments has a Gaussian limiting distribution; see Montroll and West [26] or Montroll and Shlesinger [27] for details.

On Average We Do Not Move   Indicating the average over the distribution of step lengths by a bracket we obtain for the average displacement of the walker after N steps

$$
\begin{aligned}
\langle X(t) \rangle &= \sum_{j=1}^{N} \int_{-\infty}^{\infty} s_j p(s_j)\, ds_j \\
&= N \langle s \rangle.
\end{aligned}
\tag{6.3}
$$

The average step length is

$$\langle s \rangle = \int_{-\infty}^{\infty} s_j p(s_j)\, ds_j \tag{6.4}$$

independent of the index of the step, that is, independent of where in the random walk process it occurs. Thus, if the probability $p(s)$ is symmetrical, the probability of taking a step of a given length to the right is the same as taking a step of the same length to the left, so the mean step length vanishes, $\langle s \rangle = 0$.

On the Average Our Uncertainty Increases   The variance of the displacement can be calculated as the difference of squares,

$$\left\langle X(t)^2 \right\rangle - \langle X(t) \rangle^2 = \sum_{k=1}^{N} \sum_{j=1}^{N} \langle [s_j - \langle s \rangle][s_k - \langle s \rangle] \rangle \tag{6.5}$$

so that for statistically independent steps we have $\langle s_j s_k \rangle = \langle s^2 \rangle \delta_{j,k}$ and (6.5) reduces to

$$\left\langle X(t)^2 \right\rangle - \langle X(t) \rangle^2 = N\left[\langle s^2 \rangle - \langle s \rangle^2 \right]. \tag{6.6}$$

If we again replace N by the total time $t = N\Delta t$ and introduce the diffusion coefficient

$$D = \frac{\langle s^2 \rangle - \langle s \rangle^2}{2\Delta t} \tag{6.7}$$

into (6.6) we obtain for a symmetric distribution of step lengths in the continuous time limit

$$\left\langle X(t)^2 \right\rangle = 2Dt. \tag{6.8}$$

Note that we have not included the limiting procedure that is part of the usual definition of the diffusion coefficient involving the limit of the lattice spacing Δs and the time increment Δt. In general the diffusion coefficient is defined by

$$D = \lim_{\Delta s, \Delta t \to 0} \frac{(\Delta s)^2}{2\Delta t}. \tag{6.9}$$

It is understood, however, that the limits in (6.9) are to be taken in a real physical system. Aside from these technical details we see from (6.8) that a normal diffusion process is one in which the mean-square displacement increases linearly in time.

Stochastic Differential Equation We now take the continuum limit of the sum (6.2) and write the random walk process in the form of a stochastic differential equation

$$\frac{dX(t)}{dt} = \xi(t) \tag{6.10}$$

where you will recall our discussion of how we are to interpret such stochastic differential equations. The meaning of (6.10) is actually

$$X(t) = X(0) + \int_0^t \xi(\tau)\, d\tau, \tag{6.11}$$

where the sum in (6.2) is replaced by the integral over the random function ξ. We can only properly interpret (6.11) by specifying the statistical properties of the ξ-fluctuations. For example, if ξ is a stationary, symmetric, two-state random process taking the values $\pm W$, its moments are given by

$$\langle \xi(t) \rangle = 0, \tag{6.12}$$

$$\langle \xi(t+\tau)\xi(t) \rangle = \langle \xi^2 \rangle \Phi_\xi(\tau). \tag{6.13}$$

The statistical response of the random walk variable $X(t)$ remains Gaussian due to the central limit theorem. However, the response of the random walker variable over time is quite different depending on the nature of the correlations in the fluctuations, as described by the correlation function $\Phi_\xi(\tau)$.

Ordinary Diffusion For algebraic simplicity we set $X(0) = 0$ and obtain for the second moment of the random walk process (6.11) by using (6.13)

$$\left\langle X(t)^2 \right\rangle = \left\langle \xi^2 \right\rangle \int_0^t dt_1 \int_0^t dt_2 \Phi_\xi (t_1 - t_2). \tag{6.14}$$

In the case where the two-state process is flipping a coin the correlation function is a Dirac delta function $\Phi_\xi (t_1 - t_2) = 2D\delta (t_1 - t_2)/\left\langle \xi^2 \right\rangle$; that is, there is no memory in the fluctuations. In the absence of memory (6.14) can be directly integrated twice and reduces to (6.8). The process is that of ordinary diffusion.

Anomalous Diffusion On the other hand, if the fluctuations contain long-term correlations we model this memory as the inverse power-law correlation function, cf. (1.73)

$$\Phi_\xi (\tau) = \frac{\overline{T}^\beta}{\left[\overline{T}^2 + \tau^2 \right]^{\beta/2}}. \tag{6.15}$$

We can define a microscopic time scale by means of the correlation time, cf. (1.75),

$$\tau_c \equiv \int_0^\infty \Phi_\xi (\tau) \, d\tau. \tag{6.16}$$

This time scale, τ_c, exists and is a nonzero finite constant for $\beta > 1$ and diverges to infinity for $1 > \beta > 0$. Recall the discussion in Chapter 1. A finite correlation time allows us to separate the microscopic time scale for the process from the macroscopic time scales for the process, resulting in normal diffusion. On the other hand, a correlation time that diverges means that there is no separation between the microscopic and macroscopic time scales, since the microscopic time scale can be arbitrarily long and we therefore have anomalous diffusion. The latter situation can be seen by inserting (6.15) into (6.14) and integrating to obtain the asymptotic result

$$\left\langle X(t)^2 \right\rangle \approx Kt^{2H}, \tag{6.17}$$

where $K = \left\langle \xi^2 \right\rangle A(1-\beta)^{-1} (1-\beta)^{-2}$, and the power-law index is given by

$$H = 1 - \beta/2. \tag{6.18}$$

Thus, (6.17) is the signature of the phenomenon of anomalous diffusion when $H \neq 1/2$, and $H = 1/2$ for normal diffusion.

Various Kinds of Random Walks For $H > 1/2$ the random walker, after taking a step has a greater probability of the next step being in the same direction rather than in changing directions. The walker therefore tends to continue in the direction he has been walking; the higher the value of H the greater is this tendency to persist in one's direction of motion. This type of random walk has been called *persistent*. For $H < 1/2$, however, the random walker, after taking a step has a lesser probability of the next step being in the same direction rather than in changing directions. The walker actively chooses to avoid continuing in a given direction, and changes direction more frequently than in Brownian motion. This latter type of random walk has been called *anti-persistent*. These random walk models have successfully described the observed statistical properties of DNA sequences [1, 31, 41], the interbeat intervals of the human heart [32, 44], and the variation in the stride interval of normal walking [12, 45] to name a few biological applications. Note that (6.15) with $0 < \beta \le 1$ implies a persistent process, since according to (6.18), H is always greater than $1/2$.

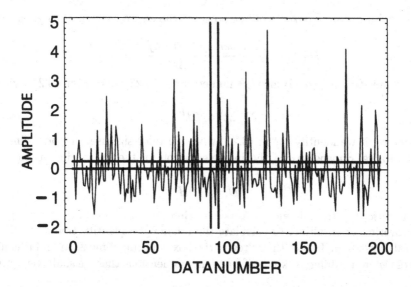

Figure 6.1 : *Here we have sketched a random time series with a box of amplitude unity and two units of width.*

Measure of Fractional Dimension   Before leaving this topic let us consider how we determine the fractal dimension of an erratic time series such as the random walk depicted in Figure 6.1. Here we evaluate the number of boxes of side Δt required to cover the curve, just as we did for the geometric forms discussed earlier. First we divide the time axis into intervals of size Δt and measure the average length of the curve for two points separated in time by such an interval. This will enable us to determine the number of boxes necessary to cover the curve in this interval

$$number\ of\ boxes\ of\ side\ \Delta t = \frac{area\ of\ graph}{area\ of\ box} = \frac{\langle |X(t + \Delta t) - X(t)| \rangle \Delta t}{\Delta t^2}.$$

$$(6.19)$$

For a Gaussian random process, with a power-law correlation function for increments of the process, the average length scales as a power of the separation in time given by Beran [5]

$$\langle |X(t + \Delta t) - X(t)| \rangle \propto \Delta t^H. \tag{6.20}$$

Of course this can be made quite general if we consider the sampling of a continuous time series given by $X(t_1), X(t_2), \cdots, X(t_N)$ such that the increments are stationary $X(t_2) - X(t_1), X(t_3) - X(t_2), \cdots, X(t_N) - X(t_{N-1})$ so that like (6.20) the mean-square value of the increments are [5],

$$\left\langle |X(t_{j+1}) - X(t_j)|^2 \right\rangle \propto (t_{j+1} - t_j)^{2H}. \tag{6.21}$$

The total number of boxes required to cover the entire curve is then given by

$$N(\Delta t) = \frac{number\ of\ boxes\ of\ side\ \Delta t}{\Delta t} \tag{6.22}$$

so that substituting (6.19) into the numerator of (6.22), and using (6.21), yields

$$N(\Delta t) \propto \Delta t^{H-2}. \tag{6.23}$$

Therefore, if we identify Δt in (6.23) with the ruler size used in the definition of the fractal dimension, we obtain the equation for the fractal dimension

$$D = 2 - H \tag{6.24}$$

for a stationary fractal random process. Here we have used box counting for the covering of an object to calculate the noninteger quantity that we call the fractal dimension. In fact this is really the box-counting dimension and it is just one of the many different kinds of fractional dimensions that we shall encounter.

6.2 Continuous-Time Random Walks

Lattice Random Walk   The most significant generalization of the random walk model was made by Montroll and Weiss [25] in the 1960s, where they included random intervals of time between successive steps in the walking process. This generalization is referred to as the continuous-time random walk (CTRW). The CTRW has been used to model a number of complicated statistical physical phenomena, from the microscopic structure of individual lattice sites to the stickiness of stability islands in chaotic dynamical systems [14]. In our discussion of random walks we concentrated on the dynamical variable. Let us now consider the evolution of the probability defined on a lattice:

$$P_n(\mathbf{x}) \equiv \text{probability that a walker arrives at } \mathbf{x} \text{ after the } n\text{th step,} \tag{6.25}$$

where $\mathbf{x}$ is a vector on an E-dimensional space lattice, locating the walker. Our presentation is limited to the case in which all lattice points are equivalent. At the moment a step is taken from a given lattice point $\mathbf{x}'$, to another one at $\mathbf{x}$, the probability of the transition $\mathbf{x}' \to \mathbf{x}$ is postulated to depend only on the spatial separation $\mathbf{x} - \mathbf{x}'$ and not on $\mathbf{x}$ and $\mathbf{x}'$ individually. Furthermore, the transition probability $p(\mathbf{x} - \mathbf{x}')$ is chosen to be independent of time.

Steps and Pauses The transition probability $p(\mathbf{x})$ is one of the quantities that characterizes the lattice walk. The second basic function for the characterization of the lattice walk is the pausing time probability density function $\psi(t)$ of the duration of pauses between transitions. We postulate this distribution to be independent of the site occupied by a walker during the pause before the next transition. We consider an extended walk to be composed of a sequence of alternating steps and pauses, with the step once executed, being instantaneous.

Master Equation The probabilities $P_n(\mathbf{x})$ satisfy the simple recurrence formula in the simple random walk

$$P_n(\mathbf{x}) = \sum_{\mathbf{x}'} p(\mathbf{x} - \mathbf{x}') P_n(\mathbf{x}') \tag{6.26}$$

and, since walkers are conserved during our random walk we have

$$\sum_{\mathbf{x}'} P_n(\mathbf{x}') = 1, \text{ and } \sum_{\mathbf{x}'} p(\mathbf{x}') = 1, \tag{6.27}$$

where the summation extends over all the lattice points. Now let us define $p(\mathbf{x}, t)\, dt$ to be the probability that the time between two successive steps is between t and $t + dt$, and the transition that follows is $\mathbf{x}$ lattice points. The transition probability now satisfies the new normalization conditions

$$\sum_{\mathbf{x}} p(\mathbf{x}, t) = \psi(t) \tag{6.28}$$

and

$$\int_0^\infty dt \sum_{\mathbf{x}} p(\mathbf{x}, t) = \int_0^\infty dt \psi(t) = 1. \tag{6.29}$$

Here $\psi(t)$ is called the waiting time distribution function and $dt\psi(t)$ is the probability that a walker will pause for a time t between steps.

Independent Pause Length and Step Size Montroll and Weiss [25] made the specific assumption that the length of a pause and the size of a step are mutually independent, so that the time-dependent transition probability can be factored

$$p(\mathbf{x}, t) = p(\mathbf{x}) \psi(t), \tag{6.30}$$

in which case the random walk is said to be factorable. The Fourier-Laplace transform of $p(\mathbf{x}, t)$ yields

$$p^* (\mathbf{k}, s) = \hat{p} (\mathbf{k}) \tilde{\psi} (s) \qquad (6.31)$$

where $\hat{p} (\mathbf{k})$ is the Fourier transform of the transition probability called the lattice structure function in solid state physics, and $\tilde{\psi} (s)$ is the Laplace transform of the waiting time distribution function.

Waiting Time Distribution Function Let $\psi_n (t)$ be the probability density for the time at which the nth transition occurs. The $\psi_n (t)$ satisfy the recurrence formulae

$$\psi_1 (t) \equiv \psi (t)$$

$$\psi_2 (t) \equiv \int_0^t \psi (t - t') \psi_1 (t') \, dt'$$

$$\cdot$$

$$\cdot$$

$$\psi_n (t) \equiv \int_0^t \psi (t - t') \psi_{n-1} (t') \, dt'. \qquad (6.32)$$

In the last equation, the integration variable represents the time at which the $(n-1)$st transition occurs, so that if the nth transition is to occur at time t the pausing time between the $(n-1)$st and the nth is $t - t'$. Thus, if the function $Q (\mathbf{x}, t)$ is defined as the probability that a walker is at the lattice site $\mathbf{x}$ at a time t *immediately* after a step has been taken, and if

$$\Psi (t) \equiv \text{probability density walker remains fixed in the interval } (0, t)$$

$$= 1 - \int_0^t \psi (t') \, dt' = \int_t^\infty \psi (t') \, dt', \qquad (6.33)$$

the relation between $P (\mathbf{x}, t)$, the probability density of a walker being at point $\mathbf{x}$ at time t, and $Q (\mathbf{x}, t)$ is

$$P (\mathbf{x}, t) = \delta_{\mathbf{x}, 0} \Psi_0 (t) + \int_0^t \Psi (t - t') Q (\mathbf{x}, t') \, dt'; \qquad (6.34)$$

that is, if a walker arrives at $\mathbf{x}$ at time t' and remains there for a time $(t - t')$, or the walker has not moved from the origin of the lattice, $P (\mathbf{x}, t)$ is the average over all arrival times with $0 < t' < t$. Note that the waiting time at the origin might be different so that $\Psi_0 (t)$ can be determined from (6.28). In turn the transition probability on the right-hand side of (6.34) satisfies its own recurrence relation

$$Q (\mathbf{x}, t) = p_0 (\mathbf{x}, t) + \sum_{\mathbf{x}'} \int_0^t p (\mathbf{x} - \mathbf{x}', t - t') Q (\mathbf{x}', t') \, dt'. \qquad (6.35)$$

Fourier-Laplace Transforms The linear convolution forms of (6.34) and (6.35) suggest solving these equations using the Fourier-Laplace transforms. We define the doubly transformed quantity, that is, the discrete Fourier transform in space and Laplace transform in time,

$$Q^* (\mathbf{k}, s) \equiv \sum_{\mathbf{x}} e^{i \mathbf{k} \cdot \mathbf{x}} \int_0^\infty dt e^{-st} Q (\mathbf{x}, t) \tag{6.36}$$

and denote other similarly transformed quantities in the same way. Thus, using the properties of convolutions we obtain from (6.35),

$$Q^* (\mathbf{k}, s) \equiv \frac{p_0^* (\mathbf{k}, s)}{1 - p^* (\mathbf{k}, s)} \tag{6.37}$$

and for the double transform of the probability density from (6.34),

$$P^* (\mathbf{k}, s) \equiv \tilde{\Psi}_0 (s) + \frac{\tilde{\Psi} (s) \, p_0^* (\mathbf{k}, s)}{1 - p^* (\mathbf{k}, s)}. \tag{6.38}$$

In the case where the transition from the origin does not play a special role $p_0^* (\mathbf{k}, s) = p^* (\mathbf{k}, s)$, $\tilde{\Psi}_0 (s) = \tilde{\Psi} (s)$, and (6.38) simplifies to

$$P^* (\mathbf{k}, s) \equiv \frac{\tilde{\Psi} (s)}{1 - p^* (\mathbf{k}, s)}. \tag{6.39}$$

This equation can be further simplified when the transition probability factors as in (6.30).

Relation Between Observation and Transition Times Weiss and Rubin [42] give an argument relating $\tilde{\Psi}_0 (s)$ and $\tilde{\Psi} (s)$ that requires the random walker to be observed at a random time. To demonstrate the lack of correlation between the time of observation and the time of the transition just prior to the observation, we need to assume that the mean time between jumps is finite

$$\langle t \rangle = \int_0^\infty t \psi (t) \, dt < \infty. \tag{6.40}$$

Then, since complete ignorance is assumed, the probability that the observation is made in a specific interval $(t, t+dt)$ is proportional to dt. The proportionality constant determines that

$$\psi_0 (t) = \frac{\Psi (t)}{\langle t \rangle} \tag{6.41}$$

so that $\psi_0 (t)$ integrates to unity. In this way the Laplace transform of the waiting time distribution is given by

$$\tilde{\psi}_0 (s) = \frac{1 - \tilde{\psi} (s)}{\langle t \rangle s}. \tag{6.42}$$

Thus, as pointed out by Weiss and Rubin the two distributions are the same only in the case of exponential pausing $\psi(t) = \exp[-t/\tau_p]/\tau_p$ where the time constant is $\tau_p = \langle t \rangle$. This identity may be expressed by saying that the negative-exponential waiting time density has no memory.

Moment Generating Functions   Let us consider the simplest situation of a factorable random walk and where the origin of time coincides with the beginning of the waiting time, in which case the double transform of the probability density is

$$P^*(\mathbf{k}, s) \equiv \frac{1 - \widetilde{\psi}(s)}{s} \frac{1}{1 - \widehat{p}(\mathbf{k})\widetilde{\psi}(s)}. \tag{6.43}$$

The doubly transformed probability density can be used as the moment generating function. Recall that the Fourier transform of the probability density is the characteristic function. For example, the first moment of the process, in one dimension, can be written as

$$\mathcal{LT}\{\langle l; t \rangle; s\} = -i\frac{\partial}{\partial k}P^*(k, s)\,|_{k=0}. \tag{6.44}$$

Now, using the definition of the structure function, we can write for the first moment

$$-i\frac{\partial}{\partial k}\widehat{p}(k)\,|_{k=0} = \sum_l e^{ikl}lp(l)\,|_{k=0} \equiv \mu_1 \tag{6.45}$$

and therefore inserting (6.43) into (6.44) and taking the appropriate derivative, we obtain

$$\mathcal{LT}\{\langle l; t \rangle; s\} = \frac{\mu_1\widetilde{\psi}(s)}{s\left(1 - \widetilde{\psi}(s)\right)}. \tag{6.46}$$

In the same way the second moment of the process, in one dimension, can be written

$$\mathcal{LT}\{\langle l^2; t \rangle; s\} = -\frac{\partial^2}{\partial k^2}P^*(k, s)\,|_{k=0} \tag{6.47}$$

so that

$$\frac{\partial^2}{\partial k^2}\widehat{p}(k)\,|_{k=0} = -\sum_l e^{ikl}l^2p(l)\,|_{k=0} \equiv -\mu_2 \tag{6.48}$$

and therefore inserting (6.43) into (6.47) and taking the appropriate derivatives, we obtain

$$\mathcal{LT}\{\langle l^2; t \rangle; s\} = \frac{\mu_2\widetilde{\psi}(s)}{s\left(1 - \widetilde{\psi}(s)\right)} + \frac{2}{s}\left[\frac{\mu_1\widetilde{\psi}(s)}{1 - \widetilde{\psi}(s)}\right]^2. \tag{6.49}$$

We now expand the exponential in the definition of the Laplace transform of the waiting time density to obtain

$$\tilde{\psi}(s) = \int_0^\infty e^{-st} \psi(t)\, dt = 1 - \langle t \rangle s + \frac{1}{2} \langle t^2 \rangle s^2 + \cdots, \qquad (6.50)$$

that when inserted into (6.46) and (6.49), keeping only the lowest-order terms in s, and inverse Laplace transforming yields the average displacement

$$\langle l; t \rangle \sim \frac{\mu_1}{\langle t \rangle} t. \qquad (6.51)$$

In the same way the variance of the displacement is determined to be

$$\langle l^2; t \rangle - \langle l; t \rangle^2 \sim \frac{\left[(\mu_2 - \mu_1^2) \langle t \rangle^2 + \left(\langle t^2 \rangle - \langle t \rangle^2 \right) \right]}{\langle t \rangle^3} t. \qquad (6.52)$$

Note that we are here assuming that the second moment of the waiting time density is finite. There are two contributions to the variance of the displacement of the random walker; one due to the variance in the step size and other due to the variance in the waiting time distribution. The assumption of a finite second moment results in both the mean and variance of the random walk increasing linearly with time.

Moments with Long-Time Memory Now we consider a waiting time distribution function where moments can diverge. For this purpose we examine an inverse power-law waiting time distribution function

$$\psi(t) \propto \frac{1}{t^{\beta+1}}, \, 0 < \beta < 1. \qquad (6.53)$$

Using this distribution function, we find that the mean-square displacement of the random walk depends on the size of the power-law index. Following Weiss and Rubin we can use an Abelian theorem [8] for Laplace transforms to obtain

$$\tilde{\psi}(s) \sim 1 - A\Gamma(1 - \beta) s^\beta \qquad (6.54)$$

for $|s| \to 0$. Note the difference between (6.54) and the Laplace transform of the Taylor expansion given by (6.50); that β is not an integer. Introducing $\tilde{\psi}(s)$ from (6.54) into (6.46) and (6.49) we obtain for the first moment

$$\mathcal{LT} \{ \langle l; t \rangle ; s \} \sim \frac{\mu_1}{A\Gamma(1 - \beta)} \frac{1}{s^{\beta+1}} \qquad (6.55)$$

and for the second moment

$$\mathcal{LT} \{ \langle l^2; t \rangle ; s \} \sim \frac{\mu_1}{A^2 \Gamma^2 (1 - \beta)} \frac{1}{s^{2\beta+1}} + \frac{\mu_1^2 + \sigma^2}{A\Gamma(1 - \beta)} \frac{1}{s^{\beta+1}}. \qquad (6.56)$$

Using a Tauberian theorem [8] for Laplace transforms together with the observation that the first and second moments are monotonic in time yields the asymptotic results

$$\langle l;t \rangle \sim \frac{\mu_1}{A\Gamma(1-\beta)} \frac{t^\beta}{\Gamma(1+\beta)} \tag{6.57}$$

and

$$\langle l^2;t \rangle - \langle l;t \rangle^2 \sim \frac{\mu_1^2 t^{2\beta}}{A^2\Gamma^2(1-\beta)} \left[\frac{2}{\Gamma(1+2\beta)} - \frac{1}{\Gamma^2(1+\beta)} \right] \tag{6.58}$$

where the second term on the right-hand side of (6.56) is negligible compared with the first for $\beta < 1$. However, the second term dominates when the first moment vanishes, $\mu_1 = 0$. From (6.57) and (6.58) we see that the variance is proportional to the square of the mean, whereas in (6.51) and (6.52) the two are proportional to each other. This kind of general relation, between the variance and mean,

$$Variance = aMean^b, \tag{6.59}$$

where a and b are constants, was first found by Taylor [37] in an ecology context.

Probability Density When the Steps Are Large and the Pauses Long
The probability density for these moments can be obtained by taking the inverse Fourier-Laplace transforms of (6.43) with the appropriate $\hat{p}(\mathbf{k})$ and $\tilde{\psi}(s)$ inserted. Consider the general case in which the lowest-order expansions of the two functions are

$$\hat{p}(\mathbf{k}) \sim 1 - A|\mathbf{k}|^\alpha \tag{6.60}$$

$$\tilde{\psi}(s) \sim 1 - Bs^\beta \tag{6.61}$$

so that (6.43) becomes, to lowest-order in both transform variables,

$$P^*_{asy}(\mathbf{k},s) \sim \frac{s^{\beta-1}}{s^\beta + D'|\mathbf{k}|^\alpha} \tag{6.62}$$

yielding the double transform of the asymptotic distribution, with $D' = A/B$. Following Uchaikin [40] we express (6.62) by means of the integral

$$P^*_{asy}(\mathbf{k},s) = s^{\beta-1} \int_0^\infty \exp\left[-\xi\left(s^\beta + D'|\mathbf{k}|^\alpha\right)\right] d\xi, \tag{6.63}$$

which on inverting the Laplace transform yields

$$\hat{P}_{asy}(\mathbf{k},t) = \int_0^\infty \exp\left[-\xi D'|\mathbf{k}|^\alpha\right] d\xi \mathcal{L}T^{-1}\left\{s^{\beta-1}e^{-\xi s^\beta};t\right\}. \tag{6.64}$$

Integrating the inverse Laplace transform integral by parts yields

$$\hat{P}_{asy}(\mathbf{k},t) = \int_0^\infty \exp\left[-\xi D' |\mathbf{k}|^\alpha\right] d\xi \frac{t}{\beta\xi} \mathcal{L}T^{-1}\left\{e^{-\xi s^\beta} ; t\right\}$$

so that with the change of variables $\eta = \xi^{1/\beta} s$ we obtain

$$\hat{P}_{asy}(\mathbf{k},t) = \frac{t}{\beta} \int_0^\infty \exp\left[-\xi D' |\mathbf{k}|^\alpha\right] \frac{d\xi}{\xi^{1+1/\beta}} \mathcal{L}T^{-1}\left\{e^{-\eta^\beta} ; \xi^{-1/\beta}t\right\}. \quad (6.65)$$

Using the Laplace transform we define the one-sided Lévy distribution

$$\psi^{(\beta,1)}(s) = \mathcal{L}T\left\{\psi^{(\beta,1)}(t) ; s\right\} = e^{-s^\beta}. \quad (6.66)$$

We give a more extensive discussion of the properties of Lévy distributions after introducing fractional diffusion equations. For the moment we proceed in the inversion of (6.62) so that inserting (6.66) into (6.65) gives us

$$\hat{P}_{asy}(\mathbf{k},t) = \frac{t}{\beta} \int_0^\infty \exp\left[-\xi D' |\mathbf{k}|^\alpha\right] \frac{d\xi}{\xi^{1+1/\beta}} \psi^{(\beta,1)}\left(\xi^{-1/\beta}t\right)$$

or a simple change of variables simplifies this equation to

$$\hat{P}_{asy}(\mathbf{k},t) = \int_0^\infty \exp\left[-D' |\mathbf{k}|^\alpha (t/\tau)^\beta\right] \psi^{(\beta,1)}(\tau) d\tau. \quad (6.67)$$

Finally, we can obtain the space-time probability density using the E-dimensional Lévy distribution expressed in terms of Bessel functions [26, 27]

$$P_{asy}(\mathbf{x},t) = \int_{-\infty}^\infty e^{i\mathbf{k}\cdot\mathbf{x}} \hat{P}_{asy}(\mathbf{k},t) \frac{d^E k}{(2\pi)^E} \quad (6.68)$$

$$\psi_E^{(\alpha,\beta)}(x) = \int_{-\infty}^\infty \frac{d^E k}{(2\pi)^E} e^{i\mathbf{k}\cdot\mathbf{x}} \exp\left[-D' |\mathbf{k}|^\alpha (t/\tau)^\beta\right]$$

$$= \int_0^\infty \frac{k^{E-1} dk}{(2\pi)^E} J_{E/2-1}(kx)(kx)^{1-E/2} e^{-k^\alpha} \quad (6.69)$$

so that we have the asymptotic probability density

$$P_{asy}(\mathbf{x},t) = \left(D't^\beta\right)^{-E/\alpha} P_E^{(\alpha,\beta)}\left(\left[D't^\beta\right]^{-1/\alpha} x\right), \quad (6.70)$$

where the two-parameter probability density is defined by

$$P_E^{(\alpha,\beta)}(r) = \int_0^\infty \psi_E^{(\alpha,\beta)}\left(r\tau^{\beta/\alpha}\right) \psi^{(\beta,1)}(\tau) \tau^{E\beta/\alpha} d\tau. \quad (6.71)$$

It is clear from (6.70) that the diffusion of the ensemble of particles is determined by the ratio β/α, the index for time divided by the index for space. The

process has superdiffusive behavior if $\beta > \alpha/2$ and is subdiffusive if $\beta < \alpha/2$
. Uchaikin [40] calls the function given by (6.71) the anomalous diffusion distribution (ADD). When $\beta = \alpha/2$ the width of the distribution grows linearly in time as does an ordinary diffusion process; however, the distribution is not Gaussian and depends on the index α.

6.3   Fractional Random Walks

Dynamically Generated Memory   We modeled simple stochastic processes using both discrete and continuous equations in the preceding section. We had to *assume* long-term memory in the random fluctuations in order to generate a system response that contained long-time memory. We now turn our attention to constructing a dynamical model that imposes the long-time memory onto the process by means of the dynamical equations, rather than through the assumed properties of the fluctuations. One way of introducing long-time memory into a process is through nonlinear dynamical maps. Certain dynamical maps have intermittent chaotic fluctuations that generate inverse power-law spectra and algebraic correlation functions. The power-law behavior of the spectrum is a manifestation of long-time memory. We have studied this dynamical approach elsewhere [1], here we take a different tack. We follow an approach that is more directly compatible with the fractional calculus. [1]

6.3.1   Inverse Power-Law Spectra

Long-Time Memory   We find that in order to model long-term memory in nondifferentiable phenomena we need to generalize the concept of differencing to include fractional values. In the same spirit as the random walk model, this approach to modeling long-time memory provides us with a conceptually straightforward mathematical representation of rather complex processes.

Fractional Differences in Walking   Let us define the discrete shift operator B such that its operation on a discrete data set Y_j shifts the index by one unit

$$BY_j = Y_{j-1} \tag{6.72}$$

thereby shifting the data value to one unit of time earlier. This is a particularization of the time shift operator introduced in (2.35) with $\tau = 1$. A simple random walk can be written in terms of the shift operator as

$$(1 - B)Y_j = \xi_j, \tag{6.73}$$

where ξ_j is the discrete random process driving the system. Note that (6.73) is just a symbolic way of writing the series for the random walk (6.2), where

---

[1] The information in this section is taken primarily from West [46] and is suitably modified for the purposes of the lecture.

each step size has the value $s_j = \xi_j$. We generalize this simple random walk by considering the fractional difference equation

$$(1 - B)^\alpha Y_j = \xi_j, \tag{6.74}$$

where α is not an integer.

What Does the Fractional Difference Equation Mean? Now we must find the proper interpretation of (6.74) and to do this we follow, in part, the discussion of the operator $(1 - B)^\alpha$ given by West [46], based on the work of Hosking [13]. The operator can be defined in a natural way for integer α using the binomial theorem

$$(1 - B)^\alpha = \sum_{k=0}^{\alpha} \binom{\alpha}{k} (-B)^k, \tag{6.75}$$

where, of course, this procedure is only defined when operating on an appropriate function. This expansion may also be carried out for noninteger α by expressing the binomial coefficient in terms of gamma functions and extending the upper limit of the sum to infinity. The solution to the fractional difference equation (6.74) can therefore be written as [13]

$$Y_j = (1 - B)^{-\alpha} \xi_j \tag{6.76}$$

which in terms of the binomial expansion for $|\alpha| < 1$ becomes

$$Y_j = \sum_{k=0}^{\infty} \binom{-\alpha}{k} (-1)^k B^k \xi_j = \sum_{k=0}^{\infty} \binom{-\alpha}{k} (-1)^k \xi_{j-k}, \tag{6.77}$$

where using some identities among gamma functions we obtain for the binomial coefficient

$$\binom{-\alpha}{k} = (-1)^k \frac{\Gamma(\alpha + k)}{\Gamma(k+1)\Gamma(\alpha)}. \tag{6.78}$$

Note that the series in (6.77) extend to infinity due to the fact that α is not an integer. If α were an integer a gamma function would diverge and cut the series off at the value $k = \alpha$.

Properties of the Solution The solution to the fractional-difference stochastic equation (6.74) given by (6.77) clearly couples the present response of the system Y_j to fluctuations that occurred infinitely far back in time through ξ_{j-k} as $k \to \infty$. The size of the influence of these infinitely remote fluctuations are determined by the magnitude of the binomial coefficients, since these coefficients are essentially the coupling strengths of the fluctuations to the system. We can estimate the strength of the system-environment coupling using Stirling's approximation for gamma functions

$$\frac{\Gamma(k+\alpha)}{\Gamma(k+\beta)} \propto k^{\alpha-\beta}, \; k >> \alpha, \beta, \tag{6.79}$$

so that the coupling strength in (6.78) becomes

$$(-1)^k \begin{pmatrix} -\alpha \\ k \end{pmatrix} \propto \frac{k^{\alpha-1}}{\Gamma(\alpha)} \tag{6.80}$$

for $k \to \infty$ since $k >> |\alpha|$. Thus, the strength of the contribution to (6.77) decreases with increasing time lag as an inverse power law as long as $|\alpha| < 1$. We can, in fact, show that α always lies in the interval $-1/2 \leq \alpha \leq 1/2$ since any power of the form $(1-B)^q$ in the original fractional-difference equation can be written as $q = n + \alpha$, where n is an integer and α is in the desired range. Therefore, we can always transform the variable of interest from Y_j to $X_j = (1-B)^n Y_j$ and the discussion proceeds in terms of the new variable with the appropriate fractional difference.

Analytic Spectrum   The spectrum of the time series (6.77) is obtained by taking its discrete Fourier transform and averaging the square of the modulus of the Fourier amplitude over an ensemble of realizations of the random fluctuations driving the fractional-difference equation. Recalling that the discrete convolution of two functions is the product of their Fourier coefficients, we can write the Fourier amplitudes from (6.77) as

$$\hat{Y}_\omega = \hat{\Theta}_\omega \hat{\xi}_\omega \tag{6.81}$$

where we have denoted the coefficients in the series (6.77) as Θ_k and the corresponding Fourier coefficient as $\hat{\Theta}_\omega$. We define the spectrum for the random walk process by

$$S(\omega) = \left\langle \left| \hat{Y}_\omega \right|^2 \right\rangle_\xi, \tag{6.82}$$

where the brackets denote an average over an ensemble of realizations of the ξ-fluctuations driving the system. Substituting (6.81) into (6.82) and assuming that the random fluctuations have a white noise spectrum of unit strength, that is, $\langle \xi_j \xi_k \rangle_\xi = \delta_{j,k}$, we obtain

$$S(\omega) = \left| \hat{\Theta}_\omega \right|^2. \tag{6.83}$$

The Fourier coefficient $\hat{\Theta}_\omega$ is calculated by taking the discrete Fourier transform of the coefficient Θ_k, so that

$$\hat{\Theta}_\omega = \sum_{k=0}^{\infty} \frac{\Gamma(\alpha+k)}{\Gamma(k+1)\Gamma(\alpha)} e^{-ik\omega}$$

$$= \left(1 - e^{-ik\omega}\right)^{-\alpha}, \qquad (6.84)$$

which when substituted into (6.83) and rearranging terms yields

$$S(\omega) = \frac{1}{[2\sin(\omega/2)]^{2\alpha}} \qquad (6.85)$$

as the spectrum of the fractional-difference process driven by unit-strength white noise. Therefore, we obtain the inverse power-law spectrum [5, 13, 46]

$$S(\omega) \approx \frac{1}{\omega^{2\alpha}}, \text{ as } \omega \to 0, \qquad (6.86)$$

where we have used $\sin x \sim x$ as $x \to 0$.

Inverse Power-Law Spectrum We see from the infinite series representation of the fractional-difference process that the statistics of the random walk are the same as those of the fluctuations, since (6.77) is linear. Thus, since the statistics of the ξ-fluctuations are assumed to be Gaussian, so too are the statistics of the observed process. However, whereas the ξ-spectrum is flat, characteristic of white noise, the Y-spectrum is an inverse power law, characteristic of fractal stochastic processes. From these analytic results we conclude that the process defined by the fractional-difference stochastic equation is analogous to fractional Brownian motion (fBm). The analogy is complete if we set $\alpha = H - 1/2$ so that the spectrum (6.86) reads

$$S(\omega) \approx \frac{1}{\omega^{2H-1}}, \text{ as } \omega \to 0. \qquad (6.87)$$

Random Walk Interpretation In the language of random walks the inverse power law (6.87) for $1 \geq H > 1/2$, or equivalently for $0 < \alpha \leq 1/2$, implies persistence. In the same way for $1/2 \geq H > 0$, or equivalently for $-1/2 \leq \alpha < 0$, the spectrum increases as a power law in frequency and the process is antipersistent. In 1981 Hosking [13] recognized that fractional-difference processes exhibit long-term persistence and antipersistent behavior. Thus, the long-time memory that was assumed in the previous section is here a consequence of the fractional dynamics describing the evolution of the process.

6.3.2 Fractional Brownian Motion

Two Different Models Two approaches have been used to model physical phenomena with long-time memory. The first approach uses discrete fractional-difference stochastic equations, where we determined in the last section that fractional differences induce long-time memory onto uncorrelated stochastic processes. The second method uses low-dimensional nonlinear, deterministic, dynamical equations having intermittent chaotic solutions, as we mentioned earlier. Both of these approaches explain erratic behavior in time series, but neither has been directly connected to the probability density in these lectures.

In fact the discussion of fractional differences has been called the discrete analogue of fBm, that is, a process with long-time memory and Gaussian statistics [46]. We now examine the continuous case and construct the equations of motion for a filtered delta-correlated process with Gaussian statistics, which is in fact given by a fractional integral.

Fractional Differential Stochastic Equation Let us consider the continuum version of the fractional-difference stochastic equation (6.74):

$$D_t^\beta [Y(t)] = \xi(t). \tag{6.88}$$

The proper interpretation of the fractional stochastic equation, we know from earlier discussions, is actually an integral equation of the form

$$Y_\beta(t) \equiv D_t^{-\beta} [\xi(t)],$$

which can be written explicitly as

$$Y_\beta(t) = \frac{1}{\Gamma(\beta)} \int_0^t \frac{\xi(\tau)\,d\tau}{(t-\tau)^{1-\beta}}. \tag{6.89}$$

Using the power-law index $H = \beta - 1/2$ we write (6.89) as

$$Y_H(t) = \frac{1}{\Gamma(H+1/2)} \int_0^t (t-\tau)^{H-1/2} \xi(\tau)\,d\tau \tag{6.90}$$

which is one choice for the continuum analogue of the fractional-difference stochastic process. Note that our choice of (6.90) differs from the one made by Mandelbrot and van Ness [20] to describe fBm.

Characteristic Function We can determine the statistical distribution for the dynamic variable $Y_H(t)$ by using the characteristic function defined as the Fourier transform of the probability density

$$\phi(k,t) = \int_{-\infty}^{\infty} e^{iky} P(y,t)\,dy. \tag{6.91}$$

Once the characteristic function has been independently determined the probability density may be obtained by means of the inverse Fourier transform

$$P(y,t) = \int_{-\infty}^{\infty} e^{-iky} \phi(k,t) \frac{dk}{2\pi}. \tag{6.92}$$

Therefore, using (6.91) we can calculate the characteristic function for the filtered process (6.90),

$$\phi_H(k,t) = \left\langle e^{ikY_H} \right\rangle, \tag{6.93}$$

where the brackets denote an average over an ensemble of realizations of the filtered function. Using the Gaussian property of the ξ-fluctuations in (6.90) we can write (6.93) as

$$
\begin{aligned}
\phi_H(k,t) = \ & \exp\left[\frac{-k^2}{2\Gamma(H+1/2)^2}\int_0^t dt_1\,(t-t_1)^{H-1/2}\right.\\
& \left.\times\int_0^t dt_2\,(t-t_2)^{H-1/2}\,\langle\xi(t_1)\xi(t_2)\rangle\right].
\end{aligned}
\tag{6.94}
$$

The fact that a Wiener process is delta correlated in time

$$
\langle\xi(t_1)\xi(t_2)\rangle = \langle\xi^2\rangle\,\delta(t_1-t_2)
$$

allows us to carry out the time integrations in (6.94) and obtain

$$
\phi_H(k,t) = \exp\left[-k^2\sigma_H^2(t)/2\right],
\tag{6.95}
$$

where the variance of the random process is given by

$$
\sigma_H^2(t) = \frac{\langle\xi^2\rangle}{\Gamma(H+1/2)^2}\int_0^t (t-t')^{2H-1}\,dt' = \sigma_H^2 t^{2H}
\tag{6.96}
$$

and

$$
\sigma_H^2 \equiv \frac{\langle\xi^2\rangle}{2H\Gamma(H+1/2)^2}.
\tag{6.97}
$$

The inverse Fourier transform of the characteristic function (6.95) then yields the Gaussian probability density

$$
P(y,t) = \frac{1}{2\pi}\int_{-\infty}^{\infty} e^{-iky}\exp\left[-k^2\sigma_H^2(t)/2\right]dk = \frac{\exp\left[-y^2/2\sigma_H^2(t)\right]}{\sqrt{2\pi\sigma_H^2(t)}}.
\tag{6.98}
$$

Anomalous Diffusion Process Thus, we have determined that the statistics of the solution to a fractional-differential stochastic equation driven by a Wiener process are Gaussian with a variance that increases as a power law in time. For $H > 1/2$ these fluctuations diffuse faster than a normal diffusion process. For $H < 1/2$ the fluctuations diffuse slower than normal diffusion processes. The fractional integral therefore transforms a Wiener process into an anomalous diffusion process.

Mandelbrot and van Ness We had noted that the choice of (6.90) differs from the choice made by Mandelbrot and van Ness [20] for a filtered Wiener process, theirs being

$$Y_H'(t) = \frac{1}{\Gamma(H+1/2)} \int_{-\infty}^{t} (t-\tau)^{H-1/2} \xi(\tau) \, d\tau, \qquad (6.99)$$

where we see that the lower limit of the integral extends to $-\infty$. In the Mandelbrot and van Ness case the characteristic function (6.93), which was so well-behaved using the definition we adopted, now vanishes identically. This vanishing of the characteristic function implies that the dynamical variable $Y_H'(t)$ is not stationary. However, if we scale time with the parameter λ to obtain

$$Y_H'(\lambda t) = \frac{1}{\Gamma(H+1/2)} \int_{-\infty}^{\lambda t} (\lambda t-\tau)^{H-1/2} \xi(\tau) \, d\tau \qquad (6.100)$$

and recalling the scaling of a Wiener process

$$\xi(\tau) = \xi(\lambda \tau') = \lambda^{-1/2} \xi(\tau')$$

we obtain from (6.100)

$$Y_H'(\lambda t) = \lambda^H Y_H'(t), \qquad (6.101)$$

resulting in this filtered Wiener process also being self-similar.

Stationary Statistical Process A stationary process is constructed from the nonstationary process $Y_H'(t)$ by using increments to obtain

$$\Delta Y_H(t) \equiv Y_H'(t) - Y_H'(0) = \int_{-\infty}^{t} K(t-\tau) \xi(\tau) \, d\tau, \qquad (6.102)$$

where the kernel is that given by Mandelbrot and van Ness [20]

$$
\begin{aligned}
K(t-\tau) &= (t-\tau)^{H-1/2} \, ; & 0 \le \tau \le t \\
&= (t-\tau)^{H-1/2} - (-\tau)^{H-1/2} \, ; & \tau \le 0.
\end{aligned}
\qquad (6.103)
$$

Notice that taking the difference in (6.102) removes the explicit influence of the time series in the interval $(-\infty, 0)$ from $Y_H'(t)$ so that $\Delta Y_H(t)$ is more like $Y_H(t)$ that only filters the Wiener process from zero to t, than it is like $Y_H'(t)$. Since the kernel scales in the same way as the process in (6.99) the increments are also self-similar.

Again Inverse Power Law Let us now determine the spectrum of the increments. To do this we reexpress (6.102) using a shifted variable

$$\Delta Y_H(t) = \int_{0}^{\infty} K(t') \xi(t-t') \, dt'. \qquad (6.104)$$

The Fourier transform of the convolution of two functions is the product of the Fourier amplitudes so that we have from (6.104),

$$\Delta \hat{Y}_H (\omega) = \hat{K} (\omega) \, \hat{\xi} (\omega).$$

Thus, assuming the strength of the ξ-fluctuations have unit value, the spectrum of the system response is given by

$$S'_H (\omega) = \left| \hat{K} (\omega) \right|^2, \qquad (6.105)$$

but the kernel has the same overall weighting in time as used in (6.90) so that

$$S'_H (\omega) \propto \frac{1}{\omega^{2H-1}}. \qquad (6.106)$$

Thus, we see that the fractional-difference equation has the same statistics (Gaussian) and the same spectrum (inverse power law) as *fBm*. A different procedure involving the explicit calculation of the correlation of increments is given in Beran [5] and yields the same spectrum.

6.4 Fractal Stochastic Time Series

Two Kinds of Fluctuations The erratic nature of experimental time series has been attributed to two quite different sources. [2] The one with the longest pedigree is *noise*, an externally induced random modification of the dynamical process of interest. The random walks discussed above might be put into this category since they are phenomenological in nature. The recent competitor to noise is *chaos*, an intrinsic dynamical property of the underlying process. Noise is a property of the environment, whereas chaos is a consequence of the deterministic nonlinear interactive nature of the system variables. Thus, an analysis of the fluctuations in the case of noise does not provide information about the system, whereas for chaos the fluctuations themselves contain information about the intrinsic dynamical properties of the phenomenon. The processing of the time series would be quite different depending on which of the two mechanisms is determined to be responsible for the observed fluctuations.

Inverse Power Laws and Fractal Dimensions When the spectrum of the time series has an inverse power law, the fluctuations are said to be *colored*, and the traditional measures such as the correlation dimension [11] cannot discriminate between chaos and colored noise. The discrimination problem has been solved in a number of ways [24] and does not concern us here. What is of concern is the fact that what intermittent chaos and colored noise have in common is a possibility of a noninteger dimension, and that the algorithms developed to calculate the dimension of erratic time series are often unreliable and require a certain amount of human judgment as pointed out by Theiler et al. [38].

---

[2] The major part of this section is taken from Zhang and West [48] and suitably modified for the present discussion.

Graphs and Tails It is useful to examine the properties of colored noise using a particular analytic representation of a random time series having a prescribed inverse power-law spectrum. We investigate the dimension of the graph of the time series $X(t)$. The graph of $X(t)$ is the set of points $\{t, X(t)\}$. The other interesting dimension is the dimension of the trail. The trail of the time series $\{X_j(t)\}$, $j = 1, 2, \cdots, n$, is the set $\{X_1(t), X_2(t), \cdots, X_n(t)\}$ in state space. The latter concept is often used in dynamical systems. The two types of dimension, that of the graph and that of the trail, are usually different and are often confused with each other in technical discussions. Here we concentrate on the dimension of the graph, since what is usually available is one long experimental time series, rather than an ensemble of realizations of a given time series.

6.4.1   Colored Noise

Fourier Series Representation The colored noise we study is written in the general form of a Fourier series:

$$
\begin{aligned}
X(t) &= \lim_{N \to \infty} X_N(t) \\
&= \lim_{N \to \infty} \sum_{k=1}^{N} A(\omega_k) \cos[\omega_k t + \phi_k] \quad (6.107)
\end{aligned}
$$

where N is a positive integer, the set of phases $\{\phi_k\}$ are random variables with a uniform distribution on the interval $(0, 2\pi)$, ω_k is the dispersion relation for the frequency, and the set of amplitudes $\{A(\omega_k)\}$ gives the strength of each periodic component contributing to the time series. For colored noise we assume that the amplitude is $A(\omega_k) = \omega_k^{-\alpha}$ where we set the coefficient of the frequency factor to unity, since an overall constant multiplier does not change the dynamical behavior of the series. We further restrict our discussion to a linear dispersion relation such that $\omega_k = k\omega_0$ where ω_0 is a fundamental frequency (unit). Hence, the time series we study is

$$
X(t) = \lim_{N \to \infty} \sum_{k=1}^{N} (k\omega_0)^{-\alpha} \cos[k\omega_0 t + \phi_k] \quad (6.108)
$$

and from the definition, $X(t)$ and $X_N(t)$ have the same discrete power spectrum:

$$
S(\omega_k) = \frac{1}{\omega_k^{2\alpha}}, \quad (6.109)
$$

which is the Fourier transform of the correlation function $\langle X(t+\tau) X(t) \rangle_\phi$. The bracket with the phase subscript denotes an average over an ensemble of realizations of the uniformly distributed random phases. For $\alpha = 0$ (6.109) is associated with white noise, for which the series (6.108) would be an explicit realization.

Time Series Intervals An interesting quantity is the difference in the time series

$$\Delta(t,\tau) = X(t+\tau) - X(t)$$
$$= -2\sum_{k=1}^{\infty}(k\omega_0)^{-\alpha}\sin[k\omega_0 t + \phi_k + k\omega_0\tau/2]\ \sin[k\omega_0\tau/2], \quad (6.110)$$

from which we can determine if the intervals of the process are statistically stationary. The mean of the difference function vanishes

$$\langle\Delta(t,\tau)\rangle_\phi = -2\sum_{k=1}^{\infty}(k\omega_0)^{-\alpha}\sin[k\omega_0\tau/2]$$
$$\times\int_0^{2\pi}\frac{d\phi_k}{2\pi}\sin[k\omega_0 t + \phi_k + k\omega_0\tau/2]$$
$$= 0 \qquad (6.111)$$

because a periodic function integrated over one period vanishes. In this way we see that the mean of the time series is independent of the origin of time.

Variance Dependence on Time The variance of the difference function is given by

$$\sigma^2[\Delta(\tau)] = \left\langle\Delta(t,\tau)^2\right\rangle_\phi = 2\sum_{k=1}^{\infty}(k\omega_0)^{-2\alpha}\sin^2[k\omega_0\tau/2]. \qquad (6.112)$$

What we wish to determine is how the series in (6.112) behaves for small τ. Zhang and West [48] have used the analytic properties of the series to obtain the variance of the difference variable

$$\sigma^2[\Delta(\tau)] = \begin{cases} O\left(\tau^{2\alpha-1}\right) & \text{for } 1 < 2\alpha < 3 \\ O(\tau) & \text{for } 3 \le 2\alpha \end{cases} \qquad (6.113)$$

as $\tau \to 0$. The difference variable is therefore stationary in time.

Characteristic Function The statistics of the difference function can be determined using the characteristic function

$$\phi(K,\tau) = \left\langle e^{iK\Delta(t,\tau)}\right\rangle_\phi \qquad (6.114)$$

and the probability density is given by the inverse Fourier transform of the characteristic function. Inserting the definition of the difference variable into (6.114) yields

$$\phi(K,\tau) = \prod_{k=1}^{\infty} J_0\left(2K\,(k\omega_0)^{-\alpha}\sin\left[k\omega_0\tau/2\right]\right). \tag{6.115}$$

The small argument approximation for the zeroth order Bessel function yields

$$J_0(Kz) \approx \exp\left[-K^2 z^2/4\right] \text{ for } Kz << 1$$

which is appropriate for small τ, so that the characteristic function (6.115) becomes

$$\phi(K,\tau) \approx \exp\left[-2K^2\sigma^2(\tau)\right], \tag{6.116}$$

where the variance is given by

$$\sigma^2(\tau) \equiv 2\sum_{k=1}^{\infty}(k\omega_0)^{-2\alpha}\sin^2\left[k\omega_0\tau/2\right] \tag{6.117}$$

which is the same as (6.112). The probability density for the time series is therefore given by

$$P(x,\tau) = \frac{1}{2\pi}\int_{-\infty}^{\infty} e^{iKx}\exp\left[-2K^2\sigma^2(\tau)\right]dK = \frac{\exp\left[-\frac{x^2}{2\sigma^2(\tau)}\right]}{\sqrt{2\pi\sigma^2(\tau)}} \tag{6.118}$$

so that the statistical distribution of a sufficiently long, colored noise, time series $\Delta(t,\tau)$ is Gaussian, or more importantly, fractional Brownian motion.

Behavior of the Time Series Difference   The restrictions on the variance of the probability density are given by (6.113) so we may write for the time series

$$\Delta(t,\tau) \overset{p}{=} \begin{cases} O\left(\tau^{(2\alpha-1)/2}\right) & \text{for } 1 < 2\alpha < 3 \\ O(\tau) & \text{for } 3 \le 2\alpha \end{cases} \tag{6.119}$$

where " $\overset{p}{=}$ " denotes the equality with a probability arbitrarily close to unity. The condition (6.119) implies that the Lipschitz-Hölder exponent, H, for $X(t)$

$$|X(t+\tau) - X(t)| \le \tau^H \tag{6.120}$$

is given by

$$H \overset{p}{=} \begin{cases} \frac{2\alpha-1}{2} & \text{for } 1 < 2\alpha < 3 \\ 1 & \text{for } 3 \le 2\alpha. \end{cases} \tag{6.121}$$

The proof of (6.121) is given in Falconer [10], but is seen to be reasonable just by comparing (6.119) and (6.120).

6.4.2 Box Dimension

Mathematical Definition of Dimension Since the series we study are continuous in time over a finite interval, an appropriate dimension to investigate is the dimension of the graph of the time series in the (t, x)-coordinate frame. We select that the box dimension of a set is defined as [10],

$$D_B = \lim_{\varepsilon \to 0} \frac{\ln [N(\varepsilon)]}{\ln \left(\frac{1}{\varepsilon}\right)}, \qquad (6.122)$$

where ε is the diameter of a box in the Euclidean space in which the set lies and $N(\varepsilon)$ is the (minimum) number of such boxes required to cover the set. As we discussed earlier this definition of dimension emerges from the idea of a measurement of scale size ε. Let us now use (6.122) to determine the dimension of $X(t)$.

Stack Boxes to Measure Dimension We use a regular square box cover for $X(t)$. Let the sides of the boxes be length ε, where ε is an increment of time. We restrict our considerations to time series of length $\overline{T}$, where $\overline{T} = 2\pi/w_0$ is actually the fundamental period of the function $X(t)$. We need $M = \overline{T}/\varepsilon$ columns of boxes to cover the time series. In each column, from (6.119) we have:

$$\frac{\Delta(\varepsilon)}{\varepsilon} \overset{p}{=} \begin{cases} O\left(\varepsilon^{(2\alpha-1)/2-1}\right) & \text{for } 1 < 2\alpha < 3 \\ O(1) & \text{for } 3 \le 2\alpha \end{cases} \qquad (6.123)$$

boxes. Hence, the total number of boxes needed to cover the graph of $X(t)$ is obtained by dividing (6.123) by ε to obtain

$$N(\varepsilon) \overset{p}{=} \begin{cases} O\left(\varepsilon^{(2\alpha-1)/2-2}\right) & \text{for } 1 < 2\alpha < 3 \\ O(\varepsilon^{-1}) & \text{for } 3 \le 2\alpha. \end{cases} \qquad (6.124)$$

Substituting this expression for the number of boxes needed to cover the time series into the definition of the box dimension, (6.122), yields

$$D_B \overset{p}{=} \begin{cases} \frac{5-2\alpha}{2} & \text{for } 1 < 2\alpha < 3 \\ 1 & \text{for } 3 \le 2\alpha. \end{cases} \qquad (6.125)$$

In other words, when the Lipschitz-Hölder exponent is known, we have, from (6.121) and (6.125),

$$D_B = 2 - H \qquad (6.126)$$

for the box dimension of the graph in a finite interval [10]. The box counting dimension is just one of the many different kinds of fractional dimensions that we encounter. Recall that (6.122) is also used to define the similarity dimension. We continue to refer to these and other such noninteger dimensions generically as the fractal dimension, except where that might lead to some confusion.

6.4.3   Power-Law Correlations

When Things Are Not Independent   Other measures can also be used to determine the nature of the statistics of fluctuating time series. Let us sample a continuous time series a finite number of times and use the samples as our data. We denote the N data elements of this time series by $\{X_j\}$ $j = 1, 2, \cdots, N$. Note that the index here refers to time and not to a member of an ensemble such as in (6.107). The correlation coefficient is defined as the ratio of the covariance to the variance of the data set

$$r_k = \frac{Cov_k}{VarX},$$ (6.127)

where the variance in X is

$$VarX = \frac{1}{N} \sum_{j=1}^{N} \left(X_j - \overline{X}\right)^2$$ (6.128)

and the covariance for datapoints separated in time by k units is

$$Cov_k = \frac{1}{N - k} \sum_{j=1}^{N-k} \left(X_j - \overline{X}\right)\left(X_{j+k} - \overline{X}\right).$$ (6.129)

These are the standard definitions applied to the raw data and the correlation coefficient (6.127) has a value restricted to the interval $(-1, 1)$ for a given lag time k. Note that $N >> k$ in (6.129) and in the physics literature $N - k$ is replaced with N.

When Is the Correlation Function Inverse Power Law?   When the statistics of the data set $\{X_j\}$ are described by a probability density having an inverse power-law form there is significant probability that extreme values of the random variable are nonnegligible. This is to be contrasted with Gaussian statistics where the probability of extreme values of the variate are exponentially small. In fact it may be the existence of these extreme values that determines the behavior in complex phenomena. In a similar way the correlation function has an inverse power-law form when there are long-time correlations in the data set. The scaling of the time series ties events together in a statistically orderly sequence in time (space) giving rise to an inverse power-law spectrum, just as it ties the relative frequencies of the occurrence of events together in the distribution function.

Correlation for a Fractal Data Set   We are all familiar with the decrease in a correlation function as the time interval between the events being correlated increases. In considering the structure and function of complex phenomena it is reasonable to expect that neighboring temporal regions are more alike with respect to a property than are time domains more widely separated, independently of the kind of event being considered. We know from previous discussions that a fractal time series has correlations extending over longer time intervals than would ordinarily be expected than, say, those generated by simple random walks. Van Beek et al. [4] show that the correlations between nearest neighbors, in a spatially heterogeneous system, were defined by the fractal dimension describing the degree of heterogeneity in a size-independent fashion. As discussed by Bassingthwaighte et al. [3] this heterogeneity is expressible directly in terms of the correlation coefficient between datapoints lagged by k data elements

$$r_k = \frac{1}{2}\left[(k+1)^{4-2D} - 2k^{4-2D} + (k-1)^{4-2D}\right] \qquad (6.130)$$

which is valid for $k \geq 1$.

Nearest Neighbors Determine the Fractal Dimension   We can see that in the case of nearest neighbor correlations where $k = 1$ (6.130) reduces to

$$r_1 = 2^{3-2D} - 1 \qquad (6.131)$$

independently of the actual size of the unit time interval. Thus, the single time lag autocorrelation function is given in terms of the fractal dimension D of the time series. This is one of those simple expressions that is worth remembering because it encapsulates many of the properties of the time series that are of importance. If there is no correlation in the time series $r_1 = 0$, the local irregularities are completely random and the fractal dimension is $D = 1.5$. If, on the other hand, the nearest neighbors are perfectly correlated $r_1 = 1$, the irregularities are uniform at all times and the fractal dimension is $D = 1.0$. Most time series have fractal dimensions that fall somewhere between the two extremes of Brownian motion $(D = 1.5)$ and complete regularity $(D = 1.0)$.

Brownian Motion Time Series   In (6.130) we see that if the fractal dimension is $D = 1.5$ then the samples are uncorrelated independently of k; that is, $r_k = 0$ for all k. This is a truly remarkable result, for it states that no matter how we coarse-grain the data for the time series generated by Brownian motion the correlation remains zero. Values of the correlation dimension above 1.5 indicate negative correlation between neighbors. As the fractal dimension approaches 2.0, its maximum value for a time series, the correlation coefficient approaches -1.0. As pointed out by Bassingthwaighte et al. [3] this does not work exactly, but fits with the general perspective that smooth scalar signals have a fractal dimension near unity, and do not fill much of the plane. Consider the scalar function $X(t)$. To fill the (t, x)-plane as fully as possible with one value of the dependent variable for each value of the independent variable requires that the X-values at neighboring t-values be negatively correlated and thereby change as much as possible.

We Can See the Dimension   In Figure 6.2 we show the functional form for the correlation coefficient (6.130) for various values of the fractal dimension as a function of the delay index k. For values of the fractal dimension close to 1.0 the correlation coefficient is nearly one. As the value of the dimension approaches 1.5 it is apparent that r_k drops rapidly from the value of one and then slowly decays after a k of one or two. This means that the time series rapidly decorrelates from its nearest and next-nearest neighbors, but then remains correlated at some low level out to appreciable distances, which is to say long times. It is clear from the figure that the time series are still significantly correlated after 10 time steps for a dimension less than 1.4.

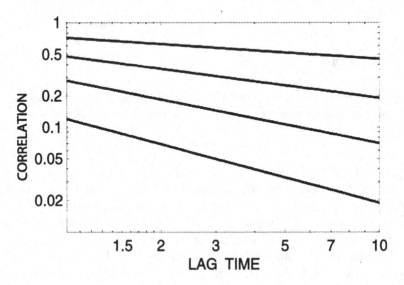

Figure 6.2: *The correlation coefficient given by (6.130) is graphed for four different values of the fractal dimension D = 1.1, 1.2, 1.3 and 1.4.*

Inverse Power-Law Correlation Function   We can see that for $1.5 \geq D \geq 1.0$ the correlation coefficient has a long tail. To explicitly show the functional form of this tail we Taylor expand (6.130)

$$r_k = \frac{k^{4-2D}}{2}\left[\left(1+\frac{1}{k}\right)^{4-2D} - 2 + \left(1-\frac{1}{k}\right)^{4-2D}\right]$$

so that combining terms to lowest order in inverse powers of the lag time gives

$$r_k \approx \frac{(4-2D)(3-2D)}{2k^{2D-2}}. \tag{6.132}$$

The correlation coefficient has an inverse power law as the lowest-order term for $D > 1$. It is clear that this correlation coefficient decays very slowly with k, much more slowly than would an exponential, for example. In fractal signals with the fractal dimension greater than approximately 1.3, the values of the correlation coefficient for $k > 2$ are so low, less than 0.1, that it is easy to understand why fractal signals would historically be confused with uncorrelated noise.

6.5 Evolution of Probability Densities

Brownian Motion In the physical sciences, instead of following the paths of individual particles, we often group the particle trajectories into collections called ensembles. This was first done by Maxwell and later by Boltzmann in their development of the kinetic theory of gases in the nineteenth century. Most students today learn of this through the study of diffusion. The phenomenon of diffusion was first observed by the Dutch physician, Jan Ingen-Housz in 1784, who, while in the Austrian court of Empress Maria Theresa, observed that finely powdered charcoal floating on an alcohol surface executed a highly erratic random motion. A similar observation was made by the Scottish botanist Robert Brown in 1828, who noted the erratic motion of pollen grains suspended in fluids. Einstein, in the tradition of Boltzmann, correctly explained this process in terms of the imbalance in the number of impacts of the lighter ambient fluid particles on the larger particle, whether charcoal, pollen or any other foreign particle. Thus, diffusion is also referred to as Brownian motion, a term used by Boltzmann as early as 1900. In earlier lectures we briefly discussed the stochastic equations describing the motion of a Brownian particle, that is, the Langevin equation. We now wish to discuss the evolution of the probability density associated with the Langevin equation.

6.5.1 The Fokker-Planck Equation

Chain Condition The change in time of a stationary stochastic processes using the conditional transition probability density $P(y, t \mid y', t')$ for the dynamical variable $X(t)$ to lie in the range $(y, y + dy)$ conditional on $X(t') = y'$ is given by the chain condition

$$P(y, t + \Delta t \mid x_0) = \int_\Omega P(y, t + \Delta t \mid y', t) P(y', t \mid x_0) \, dy', \qquad (6.133)$$

where Ω is the domain of the variate. This equation is often used as the starting equation for the analysis of Brownian motion. Here $P(y, t \mid x_0)$ is the probability that the process undergoes a transition from the initial value x_0 to a final value y at time t through a sequence of intermediate values; that is, the process goes from x_0 to y' in the time t, then from y' to y in the time interval Δt and the intermediate values of the state variable are integrated over. Equation (6.133) was introduced by Bachelier in 1900 in his PhD thesis on speculation in the French stock market. The nonphysical context of the application of this equation was probably the reason why this work went unnoticed for 50 years, even though the mathematical content of his paper was equivalent to that found in the Einstein papers on diffusion published five and more years later.

Infinite Order Equation of Evolution For small time intervals Δt, one assumes that the moments of the transition probability can be expanded in a

Taylor series in powers of Δt. To first-order in the time interval the moments can be written

$$\int (y - y')^n P(y, t + \Delta t \mid y', t) \, dy \equiv n! B_n(y', t) \Delta t \qquad (6.134)$$

for all $n \geq 0$. We next multiply (6.134) by an arbitrary function $F(y)$, integrate over y, and then introduce (6.133),

$$\int F(y) P(y, t + \Delta t \mid x_0) \, dy$$
$$= \int dy \int dy' F(y) P(y, t + \Delta t \mid y', t) P(y', t \mid x_0). \qquad (6.135)$$

Representing the unknown function by its Taylor series expansion about y'

$$F(y) = F(y') + \sum_{n=1}^{\infty} \frac{(y - y')^n}{n!} F^{(n)}(y') \qquad (6.136)$$

in the right-hand side of (6.135) and using the defining equation (6.134) yields

$$\int F(y) P(y, t + \Delta t \mid x_0) \, dy$$
$$= \int dy' F(y') P(y', t \mid x_0)$$
$$+ \Delta t \sum_{n=1}^{\infty} \int dy' P(y', t \mid x_0) F^{(n)}(y') B_n(y', t). \qquad (6.137)$$

In the limit $\Delta t \to 0$, by moving the first integral on the right-hand side to the left-hand side and dividing by Δt, (6.137) yields

$$\frac{\partial}{\partial t} \langle F(y) \rangle = \sum_{n=1}^{\infty} \left\langle B_n(y, t) \frac{\partial^n F(y)}{\partial y^n} \right\rangle, \qquad (6.138)$$

where the brackets signify the average over the ensemble of realizations of the process in phase space

$$\langle F(y) \rangle \equiv \int dy F(y) P(y, t \mid x_0). \qquad (6.139)$$

If we set the function to be a Dirac delta function in phase space, $F(y) = \delta(y - x)$ in (6.138), then

$$\langle F(y) \rangle \equiv P(x, t \mid x_0) = \langle \delta(y - x) \rangle. \qquad (6.140)$$

This choice of the function in (6.138), followed by an integration by parts on the right-hand side, leads to the evolution equation

$$\frac{\partial P(x,t \mid x_0)}{\partial t} = \sum_{n=1}^{\infty} (-1)^n \frac{\partial^n}{\partial x^n} [B_n(x,t) P(x,t \mid x_0)]. \tag{6.141}$$

The First Two Moments The moments $B_n(x,t)$ appearing in (6.141) and defined by (6.134) can be expressed as averages (moments) of the solution to the dynamical equation. If $X(t,x_0)$ is the solution to the equations of motion subject to the initial condition $X(t=0,x_0) = x_0$ then the moments can be written as

$$n! B_n(x,t) = \lim_{\Delta t \to 0} \left\langle \frac{1}{\Delta t} [X(t+\Delta t, x_0) - X(t,x_0)]^n \mid X(t,x_0) = x \right\rangle, \tag{6.142}$$

where the averages in (6.142) are conditional on the value of the dynamical variable being equal to the value of the phase space variable $X(t,x_0) = x$, and thus only includes a subset of the members of the ensemble. To evaluate the moments in (6.142) we need to evaluate the difference $\Delta X(t) = X(t+\Delta t, x_0) - X(t,x_0)$ and its powers to first-order in Δt. Without going into the details we merely mention here that the first two moments are given by [18]:

$$B_1(x,t) = \lim_{\Delta t \to 0} \frac{\langle \Delta X(t) \mid X(t,x_0) = x \rangle}{\Delta t}, \tag{6.143}$$

$$B_2(x,t) = \lim_{\Delta t \to 0} \frac{\left\langle \Delta X(t)^2 \mid X(t,x_0) = x \right\rangle}{2\Delta t} \tag{6.144}$$

and for a dynamical process driven by a delta-correlated, Gaussian, random force

$$B_n(x,t) = 0 \quad \text{for } n > 2 \tag{6.145}$$

since

$$\langle \Delta X(t)^n \mid X(t,x_0) = x \rangle \propto \Delta t^{\alpha_n} \quad \text{with } \alpha_n > 1 \text{ for } n > 2.$$

Thus, for this latter process, we can rewrite (6.141) as the Fokker-Planck equation

$$\frac{\partial}{\partial t} P(x,t \mid x_0) = -\frac{\partial}{\partial x} [B_1(x,t) P(x,t \mid x_0)] + \frac{\partial^2}{\partial x^2} [B_2(x,t) P(x,t \mid x_0)] \tag{6.146}$$

since the higher-order terms vanish as $\Delta t \to 0$.

Standard Diffusion   A number of examples for various choices of the average drift $B_1(x, t)$ and the state-dependent diffusion coefficient $B_2(x, t)$ have been listed in the review article of Montroll and West [26]. An especially interesting choice is that corresponding to simple Brownian motion in which $B_1(x, t) = 0$ and $B_2(x, t) = 2D$, a constant, reducing (6.146) to the diffusion equation of Einstein

$$\frac{\partial}{\partial t} P(x, t \mid x_0) = D \frac{\partial^2}{\partial x^2} P(x, t \mid x_0). \tag{6.147}$$

The properly normalized solution to this equation with the initial delta function condition

$$P(x, t = 0 \mid x_0) = \delta(x - x_0) \tag{6.148}$$

is the Gaussian distribution

$$P(x, t \mid x_0) = \frac{1}{\sqrt{4\pi D t}} exp\left[-\frac{(x - x_0)^2}{2D t}\right], \tag{6.149}$$

where we see that the second moment increases linearly with time

$$\left\langle X(t)^2 \right\rangle = 2D t. \tag{6.150}$$

This linear growth of the second moment with time is one of the defining characteristics of ordinary diffusion.

Anomalous Diffusion   We are also interested in processes, called anomalous diffusion, for which

$$\left\langle X(t)^2 \right\rangle = 2D t^{2H}, \tag{6.151}$$

where $0 < H \leq 1$. Anomalous diffusion processes *cannot* be described by the solution to a Fokker-Planck equation, a fact we emphasize below when we examine phenomena having such second moments. The literature is replete with examples of complex phenomena having this non-traditional diffusive behavior: transport in heterogeneous catalysis [7], transport processes in heterogeneous rocks [15], self-diffusion in micelle systems [30], behavior in dynamical systems [6], and even in economic time series [35]. This list is by no means complete. Anomalous diffusion can also be found in physiological phenomena [3] and in social systems [46]

6.5.2   The Lévy Evolution Equation

Back to the Chain Condition   In the above discussion we derived the Fokker-Planck equation from the chain condition for the probability density. This partial differential equation describes the evolution of the probability density in phase space over time. The smooth continuous nature of this equation for the probability density is a direct consequence of the moments of the underlying process changing continuously with the time interval Δt. If the moments of the process changed discontinuously with Δt, or if the higher moments did not vanish as powers of the time interval greater than one, then the phase space equation of evolution would be more complicated than the FPE. One feature of the FPE, that any other phase space equation must have, is that the solution must be irreversible in time. To examine these more general equations of evolution let us go back to the chain condition (6.133) and write

$$P\left(x_1, t_1 \mid x_2, t_2\right) = \int_{\Omega} P\left(x_1, t_1 \mid x_3, t_3\right) P\left(x_3, t_3 \mid x_2, t_2\right) dx_3, \qquad (6.152)$$

where Ω is the domain of support of the probability density. Equation (6.152) is the general description of the evolution of the probability density for an infinitely divisible stable process and whose solution is the most general form of a Markov probability density. When the range of the variate is unbounded $\Omega = (-\infty, \infty)$ and the process under consideration has translational invariance, so the probability density is independent of the origin of the coordinate system $P\left(x_1, t_1 \mid x_2, t_2\right) = P\left(x_1 - x_2, t_1 - t_2\right)$, (6.152) becomes

$$P\left(x_1 - x_2, t_1 - t_2\right) = \int_{-\infty}^{\infty} P\left(x_1 - x_3, t_1 - t_3\right) P\left(x_3 - x_2, t_3 - t_2\right) dx_3.$$
$$(6.153)$$

The stationary chain equation (6.153) is more simply expressed in terms of characteristic functions, the Fourier transform of the probability density, as the product

$$\phi\left(k, t_1 - t_2\right) = \phi\left(k, t_1 - t_3\right) \phi\left(k, t_3 - t_2\right) \qquad (6.154)$$

using the convolution property of Fourier transforms. Montroll and West [26] noticed that, since the probability density resulting from the characteristic function $\phi\left(k, t\right)$ satisfies the chain condition, it is an infinitely divisible distribution. The most general form of $\phi\left(k, t\right)$, for infinitely divisible distributions, was obtained by Paul Lévy in 1937.

Most General Characteristic Function Here we merely sketch how to obtain the general solution to (6.154). Take the logarithm of the equation to obtain

$$\log \phi\left(k, t_1 - t_2\right) = \log \phi\left(k, t_1 - t_3\right) + \log \phi\left(k, t_3 - t_2\right) \qquad (6.155)$$

from which it is clear that the characteristic function factors into a function of k, say $g\left(k\right)$, and a function of time. In order for the intermediate time t_3 to vanish from the solution the function of time must be linear. Thus, the form of the solution to (6.155) is

$$\phi(k, t) = e^{g(k)t}.$$ (6.156)

Since the probability density is normalizable at all times, the real part of $g(k)$ must be negative definite. In order for the characteristic function to retain the product form at all spatial scales, it must be infinitely divisible. If we scale the Fourier variable k be a constant factor b, then in order for the probability density to be infinitely divisible $g(k)$ must be homogeneous

$$g(bk) = b^{\alpha}g(k), \ b > 0.$$ (6.157)

The homogeneity requirement (6.157) implies that

$$g(k) = -b(\alpha)|k|^{\alpha},$$ (6.158)

where $b(\alpha)$ is a complex function dependent on the parameter α, with a positive definite real part. Thus, we have for the characteristic function

$$\phi(k, t) = e^{-b(\alpha)|k|^{\alpha}t}.$$ (6.159)

The symmetrical solution to the chain condition is obtained by setting $b(\alpha) = b$, a positive constant independent of α. The most general solution to the chain condition is obtained using

$$b(\alpha) = b\left[1 + iC\omega(k, \alpha)\frac{k}{|k|}\right],$$ (6.160)

where C is a real parameter, $\omega(k, \alpha)$ is a real function, and the imaginary part of the coefficient determines the skewness of the distribution.

Lévy Characteristic Function   The most general characteristic function is given by

$$\phi_L(k, t) = \exp\left[-bt|k|^{\alpha}\left(1 + iC\omega(k, \alpha)\frac{k}{|k|}\right)\right],$$ (6.161)

where α, b, and C are real constants. In order for the inverse Fourier transform of $\phi_L(k, t)$ to be a probability density, the parameters have the range of values $0 < \alpha \leq 2$, so that the probability density is positive definite $b > 0$, so that the probability density is normalizable and $-1 \leq C \leq 1$ determines the degree of skewness of the distribution. The function $\omega(k, \alpha)$ is defined by

$$\omega(k, \alpha) = \begin{cases} \tan(\alpha\pi/2) & \text{if } \alpha \neq 1 \\ \frac{2}{\pi}\ln|k| & \text{if } \alpha = 1 \end{cases},$$ (6.162)

but we do not pursue the derivation of this function here.

Equation of Evolution The equation of evolution for the probability density is obtained by taking the time derivative of the characteristic function

$$\frac{\partial \phi_L(k,t)}{\partial t} = -b(\alpha) |k|^\alpha \phi_L(k,t). \tag{6.163}$$

The inverse Fourier transform of (6.163) yields

$$\frac{\partial P_L(x,t)}{\partial t} = \mathcal{FT}^{-1}\{-b(\alpha) |k|^\alpha \phi_L(k,t) ; x\} \tag{6.164}$$

so that using the convolution property of the product of Fourier amplitudes we obtain [43]

$$\frac{\partial P_L(x,t)}{\partial t} = \frac{b(\alpha)\Gamma(\alpha+1)\sin(\alpha\pi/2)}{\pi} \int_{-\infty}^{\infty} \frac{P_L(x',t)}{|x-x'|^{\alpha+1}} dx'. \tag{6.165}$$

Thus, the equation of evolution of the Lévy probability density, rather than being a partial differential equation of the Fokker-Planck type, is an integro-differential equation of the fractional derivative type. The fractional derivative in (6.165) has the form of a Riesz potential as discussed, for example, in Samko et al. ([34], page 214). The discontinuous form of the Lévy statistics makes the description of the evolution of the underlying process by a fractional diffusion equation not only desirable, but necessary. We discuss the solutions to (6.165) in subsequent lectures.

6.6 Langevin Equation with Lévy Statistics

Lévy Random Variables Let us now examine the response of a linear system to Lévy fluctuations. These fluctuations are represented by a stationary differential Markov process $dL(t)$. For such processes, given the sequence of times $t_0, t_1, \cdots, t_n$, the differences $L(t_1)-L(t_0), L(t_2)-L(t_1), \cdots, L(t_n)-L(t_{n-1})$ are mutually independent random variables and the distribution of $L(t+\tau)-L(t)$ is independent of t. The statistics of the Lévy fluctuations are further specified in terms of the characteristic function (6.161). Thus, the probability distribution $P(L, t+t_0 | L_0, t_0) = P(L-L_0, t)$ is given by

$$P(L-L_0, t) = \int_{-\infty}^{\infty} \frac{dk}{2\pi} e^{-ik(L-L_0)} e^{-bt|k|^\alpha}, \tag{6.166}$$

where we have selected the symmetric Lévy process with $C=0$ in (6.161) to represent the fluctuations.

Lévy Random Force Driving a Linear System Consider the Langevin equation

$$dv(t) + \lambda v(t) dt = dL(t), \tag{6.167}$$

where $v(t)$ is the dynamical variable, say, the velocity, λ is the dissipation parameter, and the fluctuations $dL(t)$ are a differential Markov process whose statistical properties are specified by (6.166). As we know, if the random force had Gaussian statistics and was delta correlated in time, we would have an Ornstein-Uhlenbeck process. The variance of the system response would increase linearly in time for early times and be constant at late times. However, when the random force is Lévy stable the second moment of the system response is infinite.

Solution to the Langevin Equation The linear dynamical equation can be formally integrated to yield

$$v(t, v_0) = v_0 e^{-\lambda t} + \int_0^t e^{-\lambda(t-\tau)} dL(\tau), \qquad (6.168)$$

where v_0 is the initial value. West and Seshadri [43] used the phase space equations to determine the conditional probability density for this process. However, here we can use the characteristic function given by

$$\phi(k, t | v_0) = \int_{-\infty}^{\infty} e^{ikv} P(v, t | v_0) \, dv \qquad (6.169)$$

to determine the complete dynamical properties of the system response. Another way to express the characteristic function is in terms of the solution to the Langevin equation

$$\begin{aligned}
\phi(k, t | v_0) &= \left\langle e^{ikv(t)} \right\rangle \\
&= \exp\left[ike^{-\lambda t}\right] \left\langle \exp\left[ik \int_0^t e^{-\lambda(t-\tau)} dL(\tau)\right]\right\rangle. \quad (6.170)
\end{aligned}$$

Doob [9] has shown that for a differential Lévy process described by (6.161), for an arbitrary analytic function $g(\tau)$,

$$\left\langle \exp\left[i \int_0^t g(\tau) \, dL(\tau)\right]\right\rangle = \exp\left[-b \int_0^t |g(\tau)|^\alpha \, d\tau\right]. \qquad (6.171)$$

Thus, (6.170) can be evaluated to yield

$$\phi(k, t | v_0) = \exp\left[ike^{-\lambda t}\right] \exp\left[-\sigma_{\alpha\lambda}^2(t) |k|^\alpha\right], \qquad (6.172)$$

where

$$\sigma_{\mu\lambda}^2(t) \equiv \frac{b}{\lambda\alpha}\left(1 - e^{-\alpha\lambda t}\right) \qquad (6.173)$$

which agrees with the result obtained by Doob [9] and also West and Seshadri [43]. The conditional probability density is then given by

$$P(v,t\,|v_0) = \int_{-\infty}^{\infty} \frac{dk}{2\pi} \exp - \left[ik\left(v - v_0 e^{-\lambda t}\right) - \sigma_{\alpha\lambda}^2(t)\,|k|^{\alpha}\right], \qquad (6.174)$$

where the Fourier transform is taken with respect to the variable $v - v_0 e^{-\lambda t}$. Hence, the solution to the linear Langevin equation driven by a random force with Lévy statistics, itself has Lévy stable statistics in the variable $v - v_0 e^{-\lambda t}$ with exponent α and parameter $\sigma_{\alpha\lambda}^2(t)$ [43].

Asymptotic Solution   In the long time limit the characteristic function reduces to its asymptotic form

$$\phi(k, \infty\,|v_0) = \exp\left[-\frac{b}{\alpha\lambda}\,|k|^{\alpha}\right] \qquad (6.175)$$

independent of the initial state of the system. At long times the probability distribution $P(v,t)$ attains a steady-state form P_{ss},

$$P_{ss}(v) = \int_{-\infty}^{\infty} \frac{dk}{2\pi} \exp\left[-ikv - \frac{b}{\alpha\lambda}\,|k|^{\alpha}\right]. \qquad (6.176)$$

Thus, the linear dissipation in the Langevin equation leads to a steady-state in the presence of Lévy fluctuations. The variance of the $v(t)$ is, however, infinite for all $t > 0$ and, in particular, for the steady-state distribution (6.176).

6.7   Commentary

Brownian and Lévy Processes   In the last few lectures we have attempted to blend the old and the new. The traditional random walk models were first used to help understand the dynamics of ordinary diffusion. Subsequently random walks were generalized, using the concept of fractional differences, to include memory, thereby giving rise to second moments that do not increase linearly with time. The spectra of these processes are inverse power laws. We discussed some of the mathematical properties of series having such power-law correlation functions and established how to relate the index of the inverse power law to the fractal dimension of the underlying time series. In the continuum limit these latter phenomena were described by fractional stochastic equations. The solution to the linear form of such equations gave rise to fBm. After the stochastic dynamical equations we discussed the phase space equations of motion using the corresponding Fokker-Planck equation to describe the evolution of the probability density and the evolution of a Lévy process for which there is no phase space differential equation of evolution, but instead there is a fractional partial differential equation.

Scaling of Distributions   From the functional form of the probability density shown in (6.149) it is clear that the diffusion process smoothly fills the available space over a time interval determined by the variance of the displacement (6.150). The space-time Gaussian distribution satisfies the scaling law

$$P\left(\lambda^{1/2}x, \lambda t\right) = \lambda^{-1/2}P\left(x, t\right) \qquad (6.177)$$

so that the distribution for the random variable $\lambda^{1/2}X\left(\lambda t\right)$ is, up to the overall constant $\lambda^{-1/2}$, the same as that for the original dynamical variable $X\left(t\right)$. This scaling relation establishes that the random irregularities are generated at each scale in a statistically identical manner; that is, if the fluctuations are known in a given time interval they can be determined in a second larger time interval by scale transformation. Thus, the diffusion process is invariant in distribution under the transformation that changes the time interval by a factor λ and the space interval by a factor $\lambda^{1/2}$. Such a distribution that scales space and time by different factors is called self-affine. When the distribution is invariant under a transformation that scales space and time by the same factor it is called self-similar. The same considerations apply to curves, they can be either self-similar or self-affine depending on how they scale. Thus, the scaling property of the physical concentration of the diffusing quantity is determined by the scaling of the Gaussian distribution.

Scaling of Dynamical Variables   If the time series for a physical observable is self-affine then the macrostructure of the time series repeats itself in ever smaller intervals of time. This telescoping effect, present in fractal processes, is discussed more fully below. Here we mention it so as to indicate that all these scaling features have a common vocabulary and mode of description. We note that scaling can occur in two distinct ways. One is through the time series itself, so that if $X\left(\lambda t\right) = \lambda^{-1/2}X\left(t\right)$ the geometric structure of the time series is self-affine. On the other hand when this equality is only true in the sense of a distribution, then it is the probability density that is self-affine as is shown in (6.177). In the case of the probability density one would not achieve the perfect scaling on the dynamical process just discussed; the statistical behavior would be repeated on larger and larger time scales implying a scaling of the statistics rather than a scaling of the geometrical structure of the time series.

Clustering Behavior   Hierarchies in physics often have to do with homogeneity and clustering. Homogeneity concerns the smoothness or sameness of a process in space and/or time, whereas clustering addresses the heterogeneity or differences in a process across space and/or time. The clustering of a process refers to regions of activity interspersed between regions of inactivity, such that in time one has bursts of activity separated by periods of quiescence. A hierarchy provides a classification scheme, according to some property of the physical system, that allows for the identification of a pattern across space and/or time. Let us examine a spatial example of a system with such scaling.

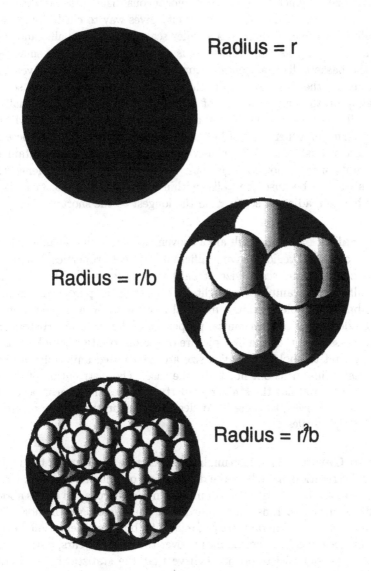

Radius = r

Radius = r/b

Radius = r/b

Figure 6.3: *A depiction of balls within balls within balls to denote the heterogeneous nature of the distribution of mass in a fractal hierarchical phenomenon.*

Balls Within Balls Within Balls Consider a physical object that from a large distance appears to be a huge sphere uniformly filled with dust. Upon closer inspection we see that this object is not uniformly filled with dust, but instead it is apparently filled with basketballs and each of these basketballs is uniformly filled with dust. Coming still closer to the object we see that each of the basketballs is filled with ping-pong balls, and it is the ping-pong balls that are uniformly filled with dust. In this picture the original view of the spherical object was one in which there was a homogeneous distribution of dust within. However, as we draw closer the homogeneity gives way to clustering, first on a large scale (basketballs), then on a smaller scale (ping-pong balls),and then on smaller scales, say, peas. The clustering is a consequence of those small regions outside the basketballs that contain empty space, and then those even smaller regions outside the ping-pong balls that contain empty space, and so on. If the breakup on this initial illusion of homogeneity is continued to smaller and smaller ball sizes, we can describe the original distribution of dust within the sphere in terms of a hierarchy of ball sizes that disrupts the smoothness of the distribution of dust; that is, the distribution of intervals that contain either dust or empty space is not homogeneous. Furthermore, the heterogeneity is not without structure because the balls, within balls, within balls, give rise to a coupling between adjacent scales from the longest to the shortest.

Spatial Scaling   Here the balls are a convenient fiction that enable us to grasp the notion of spatial regions having different physical properties, but in such a way that connects one scale (size of ball) to another scale (the next smaller size of ball). For example, we might be interested in porous media so that the distribution of holes within a material is what we want to focus on, since this would determine the mechanical properties of the material. Historically we would have assumed that the first picture of the distribution of holes was accurate, being that the holes of a given size are distributed uniformly throughout the material. This turns out not to be the case. The distribution of pore sizes in a porous material has the scaling property that ties the largest pore sizes to the smallest pore sizes, so there is no clear scale separation of the macroscopic and microscopic scales.

Stars also Cluster   The dust might also describe the stars in the night sky and their distribution, as well as the distribution of galaxies that look like stars to the naked eye. This distribution of matter in the heavens can tell us about the origins of the universe. Einstein assumed that the universe was homogeneous in his articulation of the general theory of relativity, but the observational evidence against this view has been accumulating over the past 80 years, until today the majority of astrophysicists do not believe that the statistical distribution of stars (galaxies) is homogeneous. In fact, like the distribution of pores in porous material, it is believed that the distribution of matter in the universe does not possess a fundamental scale and manifests spatial clustering.

Central Limit Theorem   Our discussion has spanned scales from the microscopic to the cosmological using a single concept that interrelates hierarchy of spatial scales, that being the notion of fractals or scaling. So what is the appropriate statistical distribution to describe physical phenomena that lack a fundamental scale? First of all if the distribution of interest lacks a fundamental scale then the notion of a narrow distribution of step sizes used in the random walk sketch of the CLT breaks down. A second moment cannot exist, for if it did exist, then the distribution would be attracted to the Gaussian. Furthermore, the size of the fluctuations in a measurement, that is, the random walk displacement, is no longer of order $\sqrt{N}$ but is more accurately given as $N^{1/\alpha}$, where since $0 < \alpha \leq 2$ we have $N^{1/\alpha} \geq \sqrt{N}$. Therefore when there is no fundamental scale in the statistical process the fluctuations in the data are greater than those anticipated by Gauss. This is the situation where the probability of the individual step lengths is rather broad.

Generalization of the Central Limit Theorem The Gaussian distribution suppresses large-scale fluctuations in such a way that the mean and variance alone are sufficient to characterize the statistical process. As the extreme values become increasingly important the Gaussian distribution breaks down and a more general limit distribution appears. When the scaling persists, the most general class of distributions that satisfy the CLT is the infinitely divisible distributions, whose very name describes the scaling behavior of the phenomena they represent. The properly normalized random walk variable $X(N)/N^{1/\alpha}$ is attracted to an α-stable Lévy distribution, just as $X(N)/\sqrt{N}$ is attracted to the Gaussian distribution in the usual CLT. Montroll and West [26], among others, showed that asymptotically the α-stable Lévy distribution decreases as an inverse power law, so that moments of order μ, when $\mu \geq \alpha$, diverge. The cause of these divergences is the overabundance of large scale fluctuations in the phenomenon, more than would be expected from Gauss' assumption that the means is the best characterization of the process. In fact the divergence suggests that the process is dominated by the outliers rather than the central values of the fluctuations. Here we see a major difference between phenomena described by a Gaussian distribution and those described by the more general Lévy stable process. In the former the interactions are short-range and the fluctuations are typically small, whereas in the latter the interactions can be long-range and the fluctuations can typically be very large.

Hyperbolic Random Variables Variables whose statistical distribution satisfies the inverse power law given by the asymptotic form of the Lévy distribution are also called hyperbolic random variables. Such hyperbolic distributions preserve self-affinity and have been shown to have realizations of trajectories with fractional dimension D. Thus, a Lévy stable process with index α has a trajectory (graph of x versus t) with a fractional dimension $D = 2 - 1/\alpha$. A process described by a function, the graph of which has a fractional dimension greater than unity, is said to be fractal, that is, a fractal process is one described by a function that exhibits a fractal dimension, as we saw earlier. A diffusion process with a Gaussian distribution would therefore, because $\alpha = 2$, have a fractal dimension $D = 1.5$, as is well known. But we have argued that truly complex statistical phenomena are not described by Gaussian statistics, but rather by Lévy statistics.

Bibliography

[1] P. Allegrini, M. Barbi, P. Grigolini and B. J. West, Dynamical Model for DNA sequences, *Phys. Rev. E* **52**, 5281-96 (1995).

[2] V. I. Arnold, *Russ. Math. Survey* **18**, 9 (1963).; *ibid.* **18**, 85 (1963).

[3] J. B. Bassingthwaighte, L. S. Liebovitch and B. J. West, *Fractal Physiology*, Oxford University Press, Oxford (1994).

[4] J. H. G. M. van Beek, S. A. Roger and J. B. Bassingthwaighre, Regional myocardial flow heterogeneity explained with fractal networks, *Am. J. Physiol.* **257**, H1670-80 (1989).

[5] J. Beran, *Statistics of Long-Memory Processes*, Monographs on Statistics and Applied Probability 61, Chapman & Hall, New York (1994).

[6] M. Bologna, P. Grigolini and J. Riccardi, Lévy diffusion as an effect of sporadic randomness, *Phys. Rev. E* **60**, 6435-6442 (1999).

[7] O. V. Bychuk and B. O'Shaughnessy, *Phys. Rev. Lett.* **74**, 1795 (1994); *J. Chem. Phys.* **101**, 772 (1994).

[8] G. Doetsch, *Theorie und Anwendungen der Laplace Transformation*, Dover, New York, (reprint), (1945).

[9] J. L. Doob, *Stochastic Processes*, John Wiley, New York (1953).

[10] K. Falconer, *Fractal Geometry*, John Wiley, New York (1990).

[11] P. Grassberger and I. Procaccia, *Physica D* **9**, 189 (1983).

[12] J. M. Hausdorff, C. -K. Peng, Z. Ladin, J. Y. Wei and A. L. Goldberger, Is walking a random walk - Evidence for long-range correlations in stride interval of human gait, *J. Appl. Physiol.* **78**, 349 (1995).

[13] J. T. M. Hosking, Fractional Differencing, *Biometrika* **68**, 165-178 (1981).

[14] J. Klafter, G. Zumofen and M. F. Shlesinger, Lévy description of anomalous diffusion in dynamical systems, in *Lévy Fights and Related Topics in Physics*, M. F. Shlesinger, G. M. Zaslavsky and U. Frisch, eds., Springer, Berlin (1995).

[15] J. Klafter, A. Blumen, G. Zumofen and M.F. Shlesinger, *Physica A* **168**, 637 (1990).

[16] A. N. Kolmogorov, *Dokl. Adad. Nauk SSSR* **98**, 527 (1954).

[17] A. J. Lichtenberg and M. A. Lieberman, *Regular and Stochastic Motion*, Springer-Verlag, New York (1983); L. E. Reichl, *The Transition to Chaos*, Springer-Verlag, New York (1992).

[18] K. Lindenberg, K. E. Shuler, V. Seshadre and B. J. West, in *Probabilistic Analysis and Related Topics*, vol. 3, A.T. Bharucha-Reid, ed., Academic, New York (1983).

[19] E. N. Lorenz, Deterministic nonperiodic flows, *J. Atmos. Sci.* **20**, 130 (1963).

[20] B. B. Mandelbrot and J. W. van Ness, Fractional Brownian motions, fractional noise and applications, *SIAM Rev.* **10**, 422 (1968).

[21] B. B. Mandelbrot, *The Fractal Geometry of Nature*, W.H. Freeman, San Francisco (1982).

[22] B. B. Mandelbrot, *Fractals, Form, Chance and Dimension*, W.H. Freeman, San Francisco (1977).

[23] R. M. May, Simple mathematical models with very complicated dynamics, *Nature* **261**, 459 (1976).

[24] G. Sugihara and R. M. May, *Nature* **344**, 734 (1990); A. A. Tsonis and J. B. Elsner, *Nature* **358**, 217 (1992); D. J. Wales, *Nature* **350**, 485 (1991).

[25] E. W. Montroll and G. H. Weiss, *J. Math. Phys.* **6**, 167 (1965).

[26] E. W. Montroll and B. J. West, On an enriched collection of stochastic processes, in *Fluctuation Phenomena*, 61-206, E. W. Montroll and J. L. Lebowitz, eds., second edition, North-Holland Personal Library, North-Holland, Amsterdam (1987); first edition (1979).

[27] E. W. Montroll and M. F. Shlesinger, On the wonderful world of random walks, in *Nonequilibrium Phenomena II: From Stochastics to Hydrodynamics*, 1-121, E. W. Montroll and J. . Lebowitz, eds., North-Holland, Amsterdam (1983).

[28] J. Moser, *Nachr. Akad. Wiss. Gottingen II, Math. Phys. K*d., 1 (1968).

[29] E. Ott, *Chaos in Dynamical Systems*, Cambridge University Press, New York (1993).

[30] A. Ott, J. P. Bouchaud, D. Langevin and W. Urbach, *Phys. Rev. Lett.* **65**, 2201 (1994).

[31] C. K. Peng, S. Buldyrev, A. L. Goldberger, S. Havlin, F. Sciortino, M. Simons, and H. E. Stanley, Long-range correlations in nucleotide sequences, *Nature* **356**, 168 (1992).

[32] C. K. Peng, J. Mietus, J. M. Hausdorff, , S. Havlin, H. G. Stanley and A. L. Goldberger, Long-range anticorrelations and non-Gaussian behavior of the heartbeat, *Phys. Rev. Lett.* **70**, 1343 (1993).

[33] H. Poincaré, *The Foundations of Science*, translated by J.M. Cattel, Science Press, New York (1929).

[34] S. G. Samko, A. A. Kilbas and O. I. Marichev, *Fractional Integrals and Derivatives*, Gordon and Breach, New York (1993).

[35] P. Santini, Lévy scaling in random walks with fluctuating variance, *Phys. Rev. E* **61**, 93 (2000).

[36] M. Schroeder, *Fractals, Chaos, Power Laws*, W.H. Freeman, New York (1991).

[37] L. R. Taylor, *Nature* **189**, 732 (1961).

[38] J. Theiler, B. Goldrilsion, A. Longtin, S. Eubank and J. D. Farmer, in *Nonlinear Modeling and Forecasting*, 163, M. Casdagli and S. Eubank, eds., Addison-Wesley, New York (1992).

[39] F. Turner, Forward to *Chaos, Complexity and Sociology*, R. A. Eve, S. Horsfall and M. E. Lee, eds., SAGE, Thousand Oaks (1997).

[40] V. V. Uchaikin, Montroll-Weiss problem, fractional diffusion equations and stable distributions, Int. J. Theor. Phys. **39** (8), 2087-2105 (2000).

[41] R. Voss, Evolution of long-range fractal correlations and 1/f-noise in DNA base sequences, *Phys. Rev. Lett.* **68**, 3805 (1992).

[42] G. H. Weiss and R. J. Rubin, Random walks: Theory and selected applications, *Advances in Chemical Physics*, vol. **52**, eds. I. Prigogine and S.A. Rice, John Wiley, New York (1983).

[43] B. J. West and V. Seshadri, Linear systems with Lévy fluctuations, *Physica A* **113**, 203-216 (1982).

[44] B. J. West, R. Zhang, A. W. Sanders, S. Miniyar, J. H. Zucherman and B. D. Levine, Fractal fluctuations in transcranial Doppler signals, *Phys. Rev. E* **59**, 1 (1999).

[45] B. J. West and L. Griffin, Allometric control, inverse power laws and human gait, *Chaos, Solitons & Fractals* **10**, 1519 (1999); Allometric control of human gait, *Fractals* **6**, 101 (1998).

[46] B. J. West, *Physiology, Promiscuity and Prophecy at the Millennium: A Tale of Tails*, Studies of Nonlinear Phenomena in the Life Sciences vol. **7**, World Scientific, Singapore (1999).

[47] B.J. West and T. Nonnenmacher, An ant in a gurge, *Phys. Lett. A* **278**, 255 (2000).

[48] W. Zhang and B. J. West, Analysis and numerical computation of the dimension of colored noise and deterministic time series with power-law spectra, *Fractals* **4**, 91 (1996).

Chapter 7

Fractional Rheology

Describing the Flow of Materials Rheology is concerned with the flow and deformation of material. Traditionally it is the study of the behavior of material bodies treated as continuous media rather than as aggregates of interacting particles. The macroscopic equations of motion can, in principle, be obtained by coarse-graining the microscopic force laws, much as the Navier-Stokes equations of classical hydrodynamics are obtained by averaging the microscopic momentum equations of the individual particles in a fluid. The macroscopic equations are not as simple for a solid as they are for a liquid in that the symmetry, compressibility, and temperature properties are quite different in the two cases. These differences and others are due to the fact that the interactions among the particles are strong and long-range in a solid and the interactions among the particles are weaker and shorter-range in a liquid. The theoretical difficulties in constructing the averages necessary to go from the microscopic to the macroscopic domains are quite interesting, but their pursuit would lead us too far afield. Therefore we restrict our discussion to the classical models of the 19th century and use phenomenological arguments to generalize the traditional rheological equations to the fractional calculus.

Between Liquids and Solids Rheology, as a discipline, has a 300-year history, so it would be arrogant of us to try to even scratch the surface of this body of work in these few pages. This science covers hydrodynamics, aerodynamics, and the flow of gases, and much of metallurgy is also included in its domain of study. In practice, however, rheology has come to be defined somewhat more narrowly and to be more concerned with the properties of materials that manifest behavior intermediate between those of solids and liquids. However, our purpose is not to review what is known in this area, but rather to discuss how the traditional theories, based on constitutive relations and differential equations, have failed to model the complex behaviors that emerge in the dynamics of complex materials, for example, polymers. For this limited purpose we develop a restricted history concerning springs and dash pots and end with a brief description of the standard model that has been used to describe the dynamics of a variety of materials. The generalization of the standard model that we follow was developed by Nonnenmacher and colleagues and introduces the concepts of fractional relaxation and fractional creep into rheology. This leads to a fractional theory of rheology, or the more restricted application of the fractional calculus to the viscoelasticity of materials.

7.1   History and Definitions

Strain Is Deformation   When a force is applied to a solid body, that body will change in shape as a result of the applied force. Sometimes the change persists after the force is removed and other times the original shape of the body returns. A diving board will bend under the weight of the athlete, but after the person dives, the board, after a few oscillations, returns to its original horizontal condition. A rubber band will stretch when pulled apart and snap back when released, unless, of course, it is pulled too far. Clay can be compressed when we get stuck in the mud with a car, and deformed through shear in the sculpting of a bust. All these various deforming forces can collectively be called stresses. The deformation undergone by the body under *stress* is called *strain*.

Elastic Versus Plastic   When an object undergoes deformation as the result of stress, it may or may not return to its original shape when the stress is removed. As we noted, the diving board returns to its original shape when the diver leaves the platform. The rubber band returns to its flaccid state when the pull on its ends is released. The property of a material to lose its deformation with the loss of stress is known as elasticity. This return to the original shape does not happen with clay, however. In the case of clay the substance is said to be plastic rather than elastic; wax also behaves in this way. However, an elastic material does not return to its original shape when its elastic limit has been exceeded. This means that a certain level of stress produces irreversible strains or deformations of the material that remain after the stress is released. The time-dependent behavior of the elastic materials described above is called *viscoelasticity*.

Not Well Understood   Material deformation is one of those physical phenomena that, like turbulence, most students of physics have heard about. Also like turbulence, they have perhaps formed opinions, but they have probably not studied the phenomenon to any great extent. What is generally known in this regard is based on the application of Newton's laws of motion to material bodies, rather than to particles alone, resulting in differential equations for use in the mechanics of deformable materials; see, for example, Chapter 4 of Findley et al. [2]. However, material bodies of the same mass and the same geometrical shape respond to the same external forces in different ways. This empirical fact is described by the constitutive equations and is attributable to the difference of the internal constitutions of the materials. As emphasized by Findley et al. [2], real materials behave in such complex ways that it is presently impossible to characterize this behavior with a single equation that is valid over the entire range of possible temperature and deformation. Thus, separate constitutive equations are used to describe various kinds of idealized material responses. The simplest example is that of a Hookean (linear) elastic solid in which the stress σ (force per unit area) is proportional to the strain ε (deformation per unit area). Note that we do not complicate the present discussion by considering the fact that stress and strain are both second-order tensors and the proportionality constant is a fourth-order tensor containing $3^4 = 81$ elastic moduli. We keep things at their simplest level, treating stress and strain as scalar functions of time.

Deformation Has Memory Another name for rheology, popular in Russia, is *hereditary solid mechanics* [17]. As this more descriptive name implies, the phenomena are time-dependent; that is, the discipline is concerned with materials that manifest delayed strain-rate effects over time in response to applied stresses. In general, most materials exhibit linear behavior under small stress levels and this is the region of concern to us here. However, the same materials often have nonlinear responses at high stress levels. We restrict our discussion of rheology to the linear domain because this yields a mathematically tractable theory while still describing a board class of rich phenomena. In particular, we restrict our discussion to the phenomenon of creep under constant stress and stress relaxation under constant strain. It is not that the nonlinear properties of materials do not interest us, for in fact they do, and we would focus on such behavior in a more complete review of rheology. However, we are here interested in the rich complexity of viscoelastic materials that occurs in the linear domain, in particular the influence of memory on the dynamics and the possibility of modeling these effects using the fractional calculus.

Stress Is Proportional to Strain The basic modeling elements of rheology are linear springs and linear viscous dash pots, with the additional condition that inertial effects are negligible. If $\mathcal{R}$ is the linear spring constant (Young's modulus) then Hooke's law states that the stress $\sigma(t)$ and strain $\varepsilon(t)$ are proportional as

$$\sigma = \mathcal{R}\varepsilon. \tag{7.1}$$

Lest we think that human nature has changed very much in the past 300 years, it should be noted that Hooke was very much concerned that he would not receive full credit for his discovery of this relation. Therefore, in 1676 he expressed (7.1) in the form of an anagram

$$ceiiinossssttuv$$

and challenged the scientific community to decipher it. He gave his rivals, among them Newton, two years to solve the puzzle and perhaps take credit for his discovery regarding springs. At the end of the two years, 1678, he gave the solution as *ut tensio, sic vis.*

The translation of the Latin is

as stretch, so force.

Here we read strain as the relative stretch of the spring and stress as the applied force. It seemed to Hooke and his peers that no matter what was done in science (Natural Philosophy) Newton received the credit and no doubt a certain amount of that was true. It may be a consequence of the above ploy that (7.1) is still known as Hooke's law today. In any case, in 1687 Newton proposed his famous law relating stress to velocity gradients in fluids, and the modern era of rheology began [19].

Relaxation of the Strain   The Hookean spring, (7.1), exhibits a reversible effect, so how do we incorporate the realism of dissipation into such phenomenological material equations? As we mentioned, Newton introduced the notion of linear viscosity to represent dissipation in fluids, however, it was not until 1845 that Stokes developed Newton's hypothesis and showed that the rate of shear of physical fluids is proportional to the applied stress. Now we address the question of how to model a material intermediate between an elastic solid and a viscous fluid. Maxwell, in 1877, was the first scientist to address the problem of relaxation in such materials. The force of viscous resistance to the motion of the piston in a dash pot is determined by the flow of viscous fluid through a gap between the piston and the wall of the cylinder. A dash pot has, since its use by Maxwell, been used to represent dissipation in materials, such that the stress is related to the time rate of change in the strain

$$\sigma = \eta \frac{d\varepsilon}{dt} = \eta \, \dot{\varepsilon}, \tag{7.2}$$

where the constant η is the coefficient of viscosity. Equation (7.2) shows that the strain rate $\dot{\varepsilon}$ is proportional to the stress implying that the rate of deformation of the dash pot is constant when subjected to a constant stress.

Maxwell's Model   Maxwell used the linear nature of the stress-strain relations to construct a two-element model of material response: a combination of elastic and viscous elements. This two-element model is depicted schematically in Figure 7.1. The Maxwell model is, expressed in our notation [2],

$$\dot{\varepsilon} = \frac{\dot{\sigma}}{R} + \frac{\sigma}{\eta} \tag{7.3}$$

which can be solved under a variety of initial conditions. In the case where the strain rate is constant, that is, there is a steady motion of the body which continually increases the displacement, the solution is

$$\sigma = \eta \,\dot{\varepsilon} + Ce^{-t/\tau} \tag{7.4}$$

showing that the strain tends to a constant value, depending on the rate of displacement, and C is a constant. The product $\mathcal{R}\tau$ is equal to the coefficient of viscosity η and τ is the relaxation time of the elastic force. In a fluid the relaxation time is on the order of fractions of a second, whereas in solids it can be from several hours to days. The Maxwell model is the simplest that can be made, but even he realized that it was an oversimplification and that there are classes of materials for which the relaxation time would not be independent of stress.

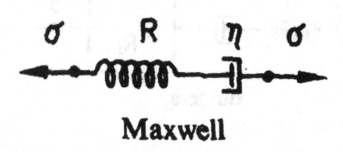

Maxwell

Figure 7.1: *Maxwell model with the two elements, a linear spring with constant $\mathcal{R}$ connected in series to a viscous dash pot with a viscosity coefficient η.*

Kelvin's Model Kelvin, on the other hand, used the linear nature of the strain-stress relations to obtain the two-element model, expressed in our notation [2],

$$\dot{\varepsilon} + \frac{\mathcal{R}}{\eta}\varepsilon = \frac{\sigma}{\eta}. \tag{7.5}$$

Here again, the two-element model is depicted schematically in Figure 7.2 as a part of a more general model. The Maxwell model is the mechanical analogue of electrical elements coupled in series, whereas the Kelvin model is the mechanical analogue of electrical elements coupled in parallel. In the latter case the force σ is balanced by the elastic force of the spring $\mathcal{R}\varepsilon$ and a force of viscous resistance to the motion of a piston in the dash pot $\eta \,\dot{\varepsilon}$, due to the flow of viscous fluid through a gap between the piston and the walls of the cylinder. This model is alternatively referred to as Voight's model, and a body whose behavior is described by (7.5) is sometimes called a Voight body [17].

What the Two Models Mean   The Maxwell model yields a strain that increases linearly with time for a constant applied stress. Reducing the applied stress to zero after a finite time yields a permanent strain that does not disappear. A constant applied strain yields an exponentially decaying stress. This is the stress relaxation phenomenon away from the initial stress of the Maxwell model under constant strain. On the other hand, in the Kelvin model a constant applied stress yields an increasing strain with a decreasing rate to a constant asymptotic value. This is the phenomenon of creep. If the applied stress is removed after a finite time, the strain decreases exponentially until the strain becomes zero asymptotically.

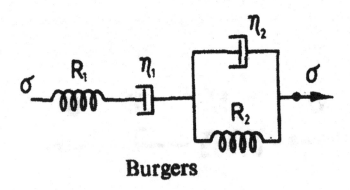

Figure 7.2: *The Kelvin model has two elements; a linear spring connected in parallel with a viscous dash pot. This is shown connected in series to a Maxwell model. The total construct shown here is referred to as the Burgers model or under the conditions specified in the text, it is called the standard model.*

7.1.1   Complex Moduli

Creep Compliance   As in any mature discipline there is a certain amount of jargon that one must master. For example, in terms of the *creep compliance*, a material property that is denoted by $J(t)$, strain is represented by

$$\varepsilon(t) = \sigma_0 J(t), \qquad (7.6)$$

where σ_0 is a constant initial stress. The creep compliance is the creep strain per unit of applied stress. Another quantity is the *relaxation modulus* denoted by $G(t)$, so stress is represented by

$$\sigma(t) = \varepsilon_0 G(t), \qquad (7.7)$$

where ε_0 is a constant initial strain. The relaxation modulus is the stress per unit of applied strain. Note that both the creep compliance and relaxation modulus change from material to material.

Experimental Measurements Creep and stress relaxation experiments provide data from approximately 10 seconds to 10 years (3×10^8 seconds) and dynamical experiments may provide data from 10^{-8} seconds to approximately 10^3 seconds. The response of viscoelastic materials at very short times is best done using oscillatory, rather than static, loading. The representation most suitable to oscillatory loading is the complex Fourier transform. For the case of a monochromatic applied force of frequency ω, the stress is denoted by the real part of

$$\sigma(t) = \sigma_0 e^{i\omega t}. \tag{7.8}$$

In general we can express the frequency dependence of the stress by taking its Fourier transform. Consider the Maxwell model of a viscoelastic material given by (7.3). Using the notation for Fourier transforms introduced earlier we write for the Fourier transform of the Maxwell model

$$\hat{\sigma}(\omega)\left[1 + i\omega\frac{\eta}{\mathcal{R}}\right] = i\omega\eta\varepsilon_0, \tag{7.9}$$

where we recall that the caret over the variable denotes its Fourier amplitude. Thus, from (7.9) we have for the complex stress relaxation modulus

$$\hat{G}(\omega) = \frac{\hat{\sigma}(\omega)}{\varepsilon_0} = \frac{i\omega\eta}{1 + i\omega\eta/\mathcal{R}} \tag{7.10}$$

or in terms of real and imaginary parts

$$\hat{G}(\omega) = \frac{\mathcal{R}\eta^2\omega^2}{\mathcal{R}^2 + \eta^2\omega^2} + i\frac{\mathcal{R}^2\eta\omega}{\mathcal{R}^2 + \eta^2\omega^2}. \tag{7.11}$$

An alternative representation of the relaxation modulus is in terms of amplitude and phase, given by

$$\left|\hat{G}(\omega)\right| = \frac{\mathcal{R}\eta\omega}{\sqrt{\mathcal{R}^2 + \eta^2\omega^2}} \tag{7.12}$$

and

$$\tan\delta = \frac{\mathcal{R}}{\omega\eta}, \tag{7.13}$$

respectively. For low frequencies the Maxwell material exhibits fluidlike behavior, $\left|\hat{G}(\omega)\right| \to 0$, and flow occurs. On the other hand, the Maxwell material behaves as does an elastic body at high frequency where $\tan\delta$, the measure of energy loss, goes to zero logarithmically with frequency.

A Different Frequency Dependence A similar analysis can be applied to the constitutive equations for a Kelvin model given by (7.5). The constitutive equation for the Kelvin material in terms of Fourier amplitudes is

$$\overset{\wedge}{\sigma}(\omega) = \varepsilon_0 \left[\mathcal{R} + i\omega\eta \right] \tag{7.14}$$

so that using the complex stress relaxation equation (7.10) yields

$$\hat{G}(\omega) = \mathcal{R} + i\omega\eta. \tag{7.15}$$

The amplitude and phase of the complex stress relaxation for the Kelvin constitutive equation are

$$\left| \hat{G}(\omega) \right| = \sqrt{\mathcal{R}^2 + \eta^2\omega^2} \tag{7.16}$$

and

$$\tan\delta = \frac{\eta\omega}{\mathcal{R}}, \tag{7.17}$$

respectively. For low frequencies the magnitude of the complex modulus for the Kelvin model is approximately equal to the spring constant $\mathcal{R}$, whereas at high frequencies the dynamic modulus increases rapidly, which is to say that the material becomes stiffer and stiffer. If the viscosity of the dash pot η increases and at the same time the spring constant is held fixed, the retardation time, defined by $\eta/\mathcal{R}$, increases as does the energy loss.

7.1.2   The Sandard Model

Combination of Maxwell and Kelvin Neither the Maxwell nor Kelvin models introduced above gives an accurate description of the time-dependence of most viscoelastic materials. The most realistic linear model is given by a series combination of the Maxwell and Kelvin models, called the Burgers model, and is depicted in Figure 7.2. If the parameters in the Maxwell model are $\mathcal{R}_1$ and η_1 and those of the Kelvin model are $\mathcal{R}_2$ and η_2 then the stress-strain constitutive relation for the Burgers model is

$$\sigma + \left[\eta_1 \left(\frac{1}{\mathcal{R}_1} + \frac{1}{\mathcal{R}_2} \right) + \frac{\eta_2}{\mathcal{R}_2} \right] \dot{\sigma} + \frac{\eta_1\eta_2}{\mathcal{R}_1\mathcal{R}_2} \ddot{\sigma}$$

$$= \quad \eta_1 \dot{\varepsilon} + \frac{\eta_1\eta_2}{\mathcal{R}_2} \ddot{\varepsilon} \tag{7.18}$$

obtained by following the analysis in Findley et al. [2]. Now applying the definition of Laplace transforms, rather than Fourier transforms, to (7.18) yields

$$\widetilde{\sigma}(s) + \left[\eta_1 \left(\frac{1}{\mathcal{R}_1} + \frac{1}{\mathcal{R}_2} \right) + \frac{\eta_2}{\mathcal{R}_2} \right] s\widetilde{\sigma}(s)$$

$$= \quad -\frac{\eta_1\eta_2}{\mathcal{R}_1\mathcal{R}_2} s^2\widetilde{\sigma}(s) + \eta_1 s\widetilde{\varepsilon}(s) + \frac{\eta_1\eta_2}{\mathcal{R}_2} s^2\widetilde{\varepsilon}(s), \tag{7.19}$$

which can be solved for a number of different initial conditions, including those corresponding to the phenomena of relaxation and creep. In (7.19) we have set all the initial values to zero.

Limit to Find Standard Model We can use (7.19) to construct the standard model which describes several different materials having the same type of constitutive equation. Thus, these models are mechanically equivalent and quantitatively the same with the proper choice of constants. The constitutive equation for the standard model is obtained by multiplying (7.19) by $\mathcal{R}_1 \mathcal{R}_2 / \eta_1 s \left(\mathcal{R}_1 + \mathcal{R}_2\right)$ and considering the limit $\eta_1 \to \infty$ to obtain after inverse Laplace transforming the resulting equation [2],

$$\sigma\left(t\right) + \frac{\eta_2}{\mathcal{R}_1 + \mathcal{R}_2} \dot{\sigma}\left(t\right) = \frac{\mathcal{R}_1 \mathcal{R}_2}{\mathcal{R}_1 + \mathcal{R}_2} \varepsilon\left(t\right) + \frac{\mathcal{R}_1 \eta_2}{\mathcal{R}_1 + \mathcal{R}_2} \dot{\varepsilon}\left(t\right). \tag{7.20}$$

Equation (7.20) can alternatively be directly obtained by putting a spring in series with a Voight body. This is equivalent to setting $\eta_1 = \infty$ so that the dash pot does not move in a Maxwell body. Choosing the parameters $\mathcal{R} = \mathcal{R}_1, \eta = \eta_2$, $\lambda = \left(\mathcal{R}_1 + \mathcal{R}_2\right) / \eta$, and $\mu = \mathcal{R}_2 / \eta$ the new constitutive equation becomes

$$\dot{\sigma} + \lambda \sigma = \mathcal{R}\left(\dot{\varepsilon} + \mu \varepsilon\right). \tag{7.21}$$

If either $\sigma\left(t\right)$ or $\varepsilon\left(t\right)$ is given, then (7.21) is a differential equation for the remaining quantity which is unknown. By integrating (7.21) we obtain an integral equation for the unknown quantity.

Formal Solutions Consider the antiderivative operator in time D_t^{-1} in terms of which we can rewrite (7.21),

$$\left(1 + \lambda D_t^{-1}\right) \sigma = \mathcal{R}\left(1 + \mu D_t^{-1}\right) \varepsilon$$

so that we have either the stress as an integral over the applied strain

$$\sigma = \frac{\mathcal{R}\left(1 + \mu D_t^{-1}\right)}{\left(1 + \lambda D_t^{-1}\right)} \varepsilon \tag{7.22}$$

or the strain as an integral over the applied stress

$$\varepsilon = \frac{\left(1 + \lambda D_t^{-1}\right)}{\mathcal{R}\left(1 + \mu D_t^{-1}\right)} \sigma. \tag{7.23}$$

We can now write (7.22) as

$$\sigma = \mathcal{R}\left[1 + \left(\mu - \lambda\right) \frac{D_t^{-1}}{1 + \lambda D_t^{-1}}\right] \varepsilon, \tag{7.24}$$

where we now wish to show that (7.24) is equivalent to an integral equation.

Integral Representation of the Formal Solution   Here we use resolvent operators to express (7.24) as an integral equation. We denote an integral kernel operator by $\mathcal{K}$ and the corresponding kernel function by K, so that given a Volterra-type integral equation of the second kind for the functions $u\,(t)$ and $v\,(t)$:

$$u = v + \lambda\, \mathcal{K} * u, \tag{7.25}$$

where the asterisk denotes a convolution. Thus, (7.25) represents the integral equation

$$u\,(t) = v\,(t) + \lambda \int_0^t K\,(t-s)\,u\,(s)\,ds. \tag{7.26}$$

The operator equation (7.25) can be iterated to obtain

$$
\begin{aligned}
u &= v + \lambda\, \mathcal{K} * v + (\lambda\, \mathcal{K})^2 * u \\
&= v + \lambda\, \mathcal{K} * v + (\lambda\, \mathcal{K})^2 * v + \cdots + (\lambda\, \mathcal{K})^m * u
\end{aligned}
\tag{7.27}
$$

and as $m \to \infty$ we obtain a Neumann series on the right hand side. Ignoring the mathematical subtleties associated with this series (see, for that discussion, Rabotnrov [17]), we introduce the resolvent operator $\mathcal{Q}$ by

$$u = \left(1 + \lambda\, \mathcal{K} + (\lambda\, \mathcal{K})^2 + \ldots\right) * v = (1 + \lambda\, \mathcal{Q}) * v. \tag{7.28}$$

Comparing (7.25) and (7.28) we formally obtain the operator identity

$$\frac{1}{1 - \lambda\mathcal{K}} = 1 + \lambda\, \mathcal{Q}, \tag{7.29}$$

or in terms of an integral equation for the resolvent equivalent to (7.26),

$$Q\,(t) = K\,(t) + \lambda \int_0^t K\,(t-s)\,Q\,(s)\,ds. \tag{7.30}$$

After some additional algebra we can also write (7.29) as the operator equation

$$\mathcal{Q}\,(\lambda) = \frac{\mathcal{K}}{1 - \lambda\mathcal{K}}, \tag{7.31}$$

where we have explicitly indicated the dependence of the resolvent $\mathcal{Q}$ on the parameter λ generated by $\mathcal{K}$. Thus, for a properly defined kernel we rewrite the constitutive equation (7.24) as

$$\sigma = \mathcal{R}\,[1 + (\mu - \lambda)\,\mathcal{Q}\,(\lambda)] * \varepsilon, \tag{7.32}$$

where the kernel operator $\mathcal{K}$ is given by the antiderivative D_t^{-1} and we want to find the corresponding resolvent operator.

Solution for Strain with Exponential Relaxation Let us assume that the K-kernel has the familiar form

$$K_\alpha(t) = \begin{cases} \frac{t^\alpha}{\Gamma(1-\alpha)}, & t > 0 \\ 0, & t < 0, -1 < \alpha \le 0. \end{cases} \tag{7.33}$$

In this case the resolvent kernel satisfying (7.30) is given by the infinite series [17],

$$\mathcal{E}_\alpha(\beta, t) = t^\alpha \sum_{k=0}^{\infty} \frac{\beta^k t^{k(1+\alpha)}}{\Gamma[(k+1)(\alpha+1)]} \tag{7.34}$$

which is often called a fractional exponential. Now using the fact that K_0 is equivalent to D_t^{-1} it is possible to write $Q_0(-\lambda)$ as the resolvent operator equivalent to $\mathcal{E}_0(-\lambda, t)$ so that

$$\sigma = \mathcal{R}[1 - (\lambda - \mu) Q_0(-\lambda)] * \varepsilon. \tag{7.35}$$

The corresponding integral equation solution for the strain in terms of the Mittag-Leffler function in the kernel is

$$\sigma(t) = \mathcal{R}\varepsilon(t) + \mathcal{R}(\lambda - \mu) \int_0^t \mathcal{E}_0(-\lambda, t - t') \varepsilon(t') dt'. \tag{7.36}$$

Note from (7.34) that the fractional exponential with $\alpha = 0$ is the ordinary exponential reducing (7.36) to

$$\sigma(t) = \mathcal{R}\varepsilon(t) + \mathcal{R}(\lambda - \mu) \int_0^t e^{-\lambda(t-t')} \varepsilon(t') dt' \tag{7.37}$$

indicating an exponential relaxation kernel.

Solution for Stress with an Exponential Relaxation In an entirely similar manner, the inversion of relation (7.35) is found to be

$$\varepsilon = \frac{1}{\mathcal{R}}[1 + (\lambda - \mu) Q_0(-\mu)] * \sigma, \tag{7.38}$$

so the integral equation solution for the stress is given by

$$\varepsilon(t) = \frac{1}{\mathcal{R}}\sigma(t) + \frac{(\lambda - \mu)}{\mathcal{R}} \int_0^t \mathcal{E}_0(-\mu, t - t') \sigma(t') dt', \tag{7.39}$$

which again has an exponential relaxation kernel. The quantities $1/\lambda$ and $1/\mu$ have the dimensions of time; the former is the characteristic relaxation time, and the latter is the characteristic creep time. We see that the decay rates in the two kernels are different, with the creep time in the stress equation being always greater than the relaxation time in the strain equation. It is clear that the above form of the solutions is that of a Green's function applied to an inhomogeneous term, just as we found previously.

What It Means The standard equation is widely used in viscoelastic materials. It exhibits a solidlike character with retarded elasticity (instantaneous elastic deformation and delayed elastic deformation); liquidlike character, viscous flow plus delayed elasticity; and instantaneous elastic response followed by viscous flow and delayed elasticity. This is the equation that was subsequently generalized to the fractional calculus by Nonnenmacher et al.

7.1.3 Fractional Memory

Fractional Memory Is Not New The generalization of the standard model to a fractional constitutive model followed here is phenomenological rather than fundamental. A more aesthetically appealing model would be one that started from the microscopic equations of motion and because of the lack of a time scale separation resulted in fractional-derivatives in the macroscopic variables. However, we have not yet succeeded in implementing this strategy. We noted that stress is proportional to strain for solids; that is, stress is proportional to the 0th derivative of the strain, as given by Hooke. For fluids stress is proportional to the 1st derivative of the strain, as given by Newton. It was this observation that motivated Scott Blair et al. [19] to suggest that a material with properties intermediate to that of a solid and a fluid, for example, a polymer, should be modeled by a fractional derivative of the strain, that is, a constitutive equation with a derivative between the 0th and 1st. The basic law of deformation that replaces Hooke's law is then

$$\sigma(t) = \mathcal{R} \, \tau^\alpha \, D_t^\alpha \left[\varepsilon(t) \right], \tag{7.40}$$

where $\mathcal{R}$, τ, and α are phenomenological constants that depend on the material. This is the generalization of Hooke's law to fractionally responsive materials. The Fourier transform of (7.40) yields

$$\hat{\sigma}(\omega) = \mathcal{R} \, \tau^\alpha \, (i\omega)^\alpha \, \hat{\varepsilon}(\omega) \tag{7.41}$$

from which the complex modulus is given by

$$\hat{G}(\omega) = \frac{\hat{\sigma}(\omega)}{\hat{\varepsilon}(\omega)} = \mathcal{R} \, (i\omega\tau)^\alpha . \tag{7.42}$$

We now demonstrate that, in fact, the complex modulus is *not* given by the inverse Fourier transform of (7.42).

Integral Representations of Creep and Relaxation The Boltzmann superposition principle states that the sum of the strain outputs resulting from each component of stress input is the same as the overall strain output resulting from the combined stress inputs. Assuming that the individual stress inputs are differential and then summing over them yields a Stieltjes integral in the continuum limit [2],

$$\varepsilon(t) = \int_0^t J(t - t') \frac{\partial \sigma(t')}{\partial t'} dt'. \tag{7.43}$$

This is an integral representation of creep and can be used to describe and predict the creep strains under a given stress history, that is, if the creep compliance function is known. An equivalent argument may be constructed for differential changes in stress. Thus, by interchanging stress and strain in (7.43) and replacing the creep compliance by the stress relaxation function we obtain

$$\sigma(t) = \int_0^t G(t - t') \frac{\partial \varepsilon(t')}{\partial t'} dt'. \tag{7.44}$$

Of course, these equations are defined for static initial values problems where $t \geq 0$. Things look different in the complex case.

Complex stress relaxation   The stress-strain relation in the case where there is a variable strain history is given by [2]

$$\sigma(t) = \int_{-\infty}^t G(t - t') \frac{\partial \varepsilon(t')}{\partial t'} dt', \tag{7.45}$$

where the lower limit on the integral is taken to be $-\infty$ because we assume an oscillatory strain has been operating for a sufficiently long time that the transitory vibrations have been damped away. A change of variables in (7.45) and some algebraic manipulations yield

$$\hat{G}(\omega) = i\omega \int_0^\infty G(t) e^{i\omega t} dt \tag{7.46}$$

so that using Euler's relation we obtain the two one-sided Fourier transforms, the real part of the complex modulus

$$\hat{G}_r(\omega) = \omega \int_0^\infty dt\, G(t)\, \sin \omega t, \tag{7.47}$$

and the imaginary part of the complex modulus

$$\hat{G}_i(\omega) = \omega \int_0^\infty dt\, G(t) \cos \omega t. \tag{7.48}$$

Inverting the Fourier transforms in (7.47) and (7.48) yields

$$\begin{aligned} G(t) &= \frac{2}{\pi} \int_0^\infty \frac{\hat{G}_r(\omega)}{\omega} d\omega \sin \omega t \\ &= \frac{2}{\pi} \int_0^\infty \frac{\hat{G}_i(\omega)}{\omega} d\omega \cos \omega t. \end{aligned} \tag{7.49}$$

Therefore, using either the real or imaginary parts of the complex modulus from (7.42) in (7.49) we obtain

$$
\begin{aligned}
G\left(t\right) &= \frac{2}{\pi}\int_{0}^{\infty}\mathcal{R}\,\tau^{\alpha}\,\omega^{\alpha-1}d\omega\cos\left[\pi\alpha/2\right]\sin\omega t \\
&= \frac{\mathcal{R}}{\Gamma\left(1-\alpha\right)}\left(\frac{t}{\tau}\right)^{-\alpha}.
\end{aligned}
\tag{7.50}
$$

Thus, the non-Debye relaxation has the form of the Nutting law, that is, an inverse power law [18].

Complex Creep Compliance   We can also use the fact that the complex compliance is the inverse of the complex modulus, so that using (7.42)

$$
\hat{J}\left(\omega\right) = \frac{1}{\mathcal{R}\left(i\omega\tau\right)^{\alpha}}
\tag{7.51}
$$

and the real part of the creep compliance is given by

$$
\hat{J}_{r}\left(\omega\right) = \frac{\omega^{-\alpha}}{\tau^{\alpha}\mathcal{R}}\cos\left[\alpha\pi/2\right].
\tag{7.52}
$$

Following Schiessel et al. [18] we write for the time rate of change of the creep compliance

$$
\begin{aligned}
\frac{dJ\left(t\right)}{dt} &= \frac{2}{\pi}\int_{0}^{\infty}\frac{\omega^{-\alpha}}{\tau^{\alpha}\mathcal{R}}d\omega\cos\left[\alpha\pi/2\right]\cos\omega t \\
&= \frac{t^{\alpha-1}}{\tau^{\alpha}\mathcal{R}\Gamma\left(\alpha\right)}.
\end{aligned}
\tag{7.53}
$$

Integrating (7.53) and using the initial condition $J\left(0\right) = 1/G\left(0\right) = 0$ we have

$$
J\left(t\right) = \frac{1}{\mathcal{R}\Gamma\left(\alpha\right)}\left(\frac{t}{\tau}\right)^{\alpha}
\tag{7.54}
$$

so that the creep compliance is a power law in time, whose index is given by the order of the fractional derivative. Thus, the creep compliance is slower than linear in time.

7.2   Fractional Relaxation

Solutions to Relaxation Equations   The form of differential equations solved in the last chapter using the method of Laplace transforms consisted of both ordinary differential equations and fractional-differential equations with rational indices using a method perhaps less opaque than the resolvent operator technique of a previous lecture. We now wish to extend those considerations to physical phenomena that do not have such rational indices. For the moment we follow Nonnenmacher and Metzler [14]. Let us examine the simple relaxation process described by the rate equation

$$\frac{d\Phi(t)}{dt} + \frac{1}{\tau_c}\Phi(t) = 0, \tag{7.55}$$

where $t > 0$ and the relaxation time $\tau_c > 0$ determines how quickly the perturbed process returns to its initial state. The solution to (7.55) is, of course, given by

$$\Phi(t) = \Phi(0)e^{-t/\tau_c} \tag{7.56}$$

which is unique and the function has the initial value $\Phi(0)$. Initial attempts at generalizing (7.55) to the fractional calculus simply replace the first-order time derivative with a fractional derivative. Such replacements do not constitute a proper initial value problem, however.

Generalize to Fractional Relaxation Equations The fractional initial value problem was addressed by Nonnenmacher and Metzler by first replacing (7.55) with the integrated form of the equation

$$\Phi(t) - \Phi_0 = -\frac{1}{\tau_c}D_t^{-1}\Phi(t), \tag{7.57}$$

where the initial value is given by $\Phi_0 = \Phi(0)$ and D_t^{-1} is the integral operator. Here Equations (7.57) and (7.55) are completely equivalent because the order of the integral is one. We now generalize (7.57) by replacing the antiderivative operator with a fractional integral operator

$$\Phi(t) - \Phi_0 = -\frac{1}{\tau_c^\alpha}D_t^{-\alpha}\Phi(t), \tag{7.58}$$

where $0 < \alpha < 1$, the time constant has been raised to the αth power to maintain the correct dimensionality, and the dynamical equation incorporates the initial value. Applying the fractional-derivative operator to (7.58) we obtain the fractional-differential equation

$$D_t^\alpha\Phi(t) - \frac{\Phi_0 t^{-\alpha}}{\Gamma(1-\alpha)} = -\frac{1}{\tau_c^\alpha}\Phi(t), \tag{7.59}$$

where we have applied the fractional derivative to the constant initial condition to obtain the time-dependent inhomogeneous term.

Equations of the form (7.59) are mathematically well defined, and strategies for solving them have been developed by a number of investigators. As Nonnenmacher and Metzler point out, Miller and Ross [?] devote their book almost exclusively to finding solutions to equations of this form for the case of rational α. Here we do not restrict α to rational values, since the phenomena in which we are interested do not do so.

Solve by Laplace Transforms To solve the fractional-differential equation (7.59) or its integral equivalent (7.58) we consider the Laplace transform of (7.58)

$$\widetilde{\Phi}(s) - \frac{\Phi_0}{s} = -\frac{1}{\tau_c^\alpha} \mathcal{LT}\left\{D_t^{-\alpha}\Phi(t) \; ; s\right\} \tag{7.60}$$

and using the Laplace transform of the fractional integral

$$
\begin{aligned}
\mathcal{LT}\left\{D_t^{-\alpha}\Phi(t) \; ; s\right\} &= \frac{1}{\Gamma(\alpha)}\mathcal{LT}\left\{t^{\alpha-1} * \Phi(t) \; ; s\right\} \\
&= \frac{1}{\Gamma(\alpha)}\mathcal{LT}\left\{t^{\alpha-1} \; ; s\right\}\mathcal{LT}\left\{\Phi(t) \; ; s\right\} = \frac{\widetilde{\Phi}(s)}{s^\alpha} \tag{7.61}
\end{aligned}
$$

Thus, after some algebra, we obtain for the Laplace transform of the solution to the fractional differential equation

$$\widetilde{\Phi}(s) = \frac{\Phi_0}{s}\frac{(\tau_c s)^\alpha}{1 + (\tau_c s)^\alpha}. \tag{7.62}$$

The solution to the fractional initial value problem is then given by the inverse Laplace transform

$$\Phi(t) = \mathcal{LT}^{-1}\left\{\frac{\Phi_0}{s}\frac{(\tau_c s)^\alpha}{1 + (\tau_c s)^\alpha} \; ; t\right\}. \tag{7.63}$$

However, there is some difficulty involved in carrying out the contour integration implied by (7.63) and so we turn to a less familiar, but more straightforward, procedure for evaluating the solution to the fractional initial value problem, one that does not involve taking the inverse Laplace transform. Subsequently we also solve equations of this kind using fractional Green's functions.

7.2.1   Using Fox Functions

When Things Are Rational It is worthwhile to point out that when α in (7.63) is rational, the expression $(1 + s^\alpha)^{-1}$ can be rationalized to arrive at an expression of the form

$$\frac{1}{1 + s^\alpha} = \frac{1}{1 \pm s^n}\sum_{j=0}^{M} s^{\rho_j},$$

where we have decomposed the denominator into n roots. Therefore we have a sum of expressions of the form

$$\frac{s^\rho}{s + a}$$

so that we can say for $\alpha \in C$, the inverse Laplace transform from (7.63),

$$LT^{-1}\left\{\frac{1}{s}\frac{s^\alpha}{1+s^\alpha};t\right\}$$

is a linear combination of generalized exponentials. Here we consider a different approach that does not require α to be rational.

Mellin Transforms Can Be Useful Consider the Mellin transform (see Appendices for a brief description of Mellin transforms) of a function indicated by

$$\widehat{\Phi}(p) = MT\{\Phi(t);p\} = \int_0^\infty t^{p-1}\Phi(t)\,dt. \tag{7.64}$$

If we now introduce the Laplace transform of the function $\Phi(t)$ into (7.64) we obtain

$$
\begin{aligned}
\widehat{\Phi}(p) &= \int_0^\infty t^{p-1}\frac{1}{2\pi i}\int_{C-i\infty}^{C+i\infty}\widetilde{\Phi}(s)\,e^{st}\,ds\,dt \\
&= \frac{1}{\Gamma(1-p)}\int_0^\infty s^{-p}\widetilde{\Phi}(s)\,ds,
\end{aligned}
\tag{7.65}
$$

where we scaled the variables $x = -st$ and used the definition of the gamma function in the complex plane; see ([7], page 935). Thus, we have the relation

$$MT\{\Phi(t);p\} = \frac{1}{\Gamma(1-p)}MT\{LT\{\Phi(t);s\};1-p\} \tag{7.66}$$

between the Mellin and Laplace transforms involving the change in indices. Inserting the Laplace transform of the solution to the initial value problem (7.62) into (7.65) gives us

$$
\begin{aligned}
\widehat{\Phi}(p) &= \frac{\Phi_0}{\Gamma(1-p)}\int_0^\infty s^{-p-1}\frac{(\tau_c s)^\alpha}{1+(\tau_c s)^\alpha}\,ds \\
&= \frac{\tau_c^p\Phi_0}{\Gamma(1-p)}\int_0^\infty \frac{x^{\alpha-p-1}}{1+x^\alpha}\,dx.
\end{aligned}
\tag{7.67}
$$

Using ([7], page 292) we have the integral for the Beta function

$$\int_0^\infty \frac{x^{\alpha-p-1}}{1+x^\alpha}\,dx = \frac{B(1-p/\alpha,p/\alpha)}{\alpha} \tag{7.68}$$

so that expressing the Beta function in terms of gamma functions we obtain for (7.67),

$$\widehat{\Phi}(p) = \frac{\Gamma(1-p/\alpha)\Gamma(p/\alpha)}{\Gamma(1-p)}\frac{\Phi_0}{\alpha}\tau_c^p. \tag{7.69}$$

The inverse Mellin transform of (7.69) involves the use of Fox functions (see Appendices for a brief discussion of Fox functions). Here we note that the inverse Mellin transform obtained from (7.69) has the form (10.88) and therefore with the appropriate choice of parameters it can be written as a Fox function.

Solution in Terms of Fox Functions The inverse of the Mellin transform of (7.69) is given by

$$\Phi(t) = \frac{\Phi_0}{2\pi i \alpha} \int_C \frac{\Gamma(1+p/\alpha)\,\Gamma(-p/\alpha)}{\Gamma(1+p)} \left(\frac{t}{\tau_c}\right)^p dp \tag{7.70}$$

which is obtained by substituting $-p$ for p in (7.69). The C on the integral denotes the contour integral as in the Fox function (10.87). Comparing (7.70) with (10.87) we obtain for the parameter values

$$\begin{aligned}
\Gamma(1 - a_j + \alpha_j p) &\Rightarrow & j = 1,\ a_1 = 0,\ \alpha_1 = 1/\alpha \\
\Gamma(b_j - \beta_j p) &\Rightarrow & j = 1,\ b_1 = 0,\ \beta_1 = 1/\alpha \\
\Gamma(1 - b_j + \beta_j p) &\Rightarrow & j = 2,\ b_2 = 0,\ \beta_2 = 1 \\
\Gamma(a_j - \alpha_j p) &\Rightarrow & j = 1,\ a_1 = 1,\ \alpha_j = 0
\end{aligned}$$

$$\tag{7.71}$$

so that the solution to the fractional relaxation equation can be expressed in terms of the Fox function

$$\Phi(t) = \frac{\Phi_0}{\alpha} H_{12}^{11} \left(\frac{t}{\tau_c} \middle| \begin{matrix} (0, 1/\alpha) \\ (0, 1/\alpha),(0,1) \end{matrix}\right). \tag{7.72}$$

Series Form of Solution In our particular case (7.72), the Fox function reduces to the series

$$\Phi(t) = \Phi_0 \sum_{k=0}^{\infty} \frac{(-1)^k}{\Gamma(1+k\alpha)} \left(\frac{t}{\tau_c}\right)^{k\alpha},\ t \ge 0 \tag{7.73}$$

where (7.73) is the series expansion of the standard Mittag-Leffler function

$$E_\alpha\left(-\left(\frac{t}{\tau_c}\right)^\alpha\right) \equiv \sum_{k=0}^{\infty} \frac{(-1)^k}{\Gamma(1+k\alpha)} \left(\frac{t}{\tau_c}\right)^{k\alpha}. \tag{7.74}$$

In the limit $\alpha \to 1$ the series expansion becomes that for an exponential function so that

$$\lim_{\alpha \to 1} E_\alpha\left(-\left(\frac{t}{\tau_c}\right)^\alpha\right) = \exp\left[-\frac{t}{\tau_c}\right]. \tag{7.75}$$

In general, however, the Mittag-Leffler function is very different from the exponential.

Asymptotic Forms for the Solution In Figure 7.3 is depicted a log-log plot of the Mittag-Leffler function (7.74) for $\alpha = 0.6$ (solid line). The light line indicates the stretched exponential (Kohlrausch-Williams-Watts Law) function

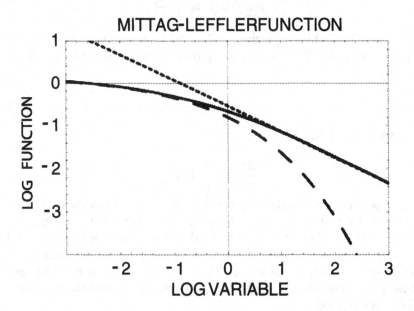

Figure 7.3: *The Mittag-Leffler function is graphed (heavy solid line) and compared with the stretched exponential (curves below) at early times and the inverse power law (extends above) at late times. We see that the Mittag-Leffler function extrapolates smoothly between the Kohlrausch-Williams-Watts Law and the Nutting Law.*

$$\lim_{t \to 0} \Phi_{KWW}(t) \approx \Phi_0 \, e^{-(t/\tau_c)^\alpha}. \tag{7.76}$$

In 1863 Kohlrausch proposed a power law to relate residual strain in slow recovery to time. He considered that if a wire is held twisted and the relaxing torque measured, the torque after a time t being $Y(t)$ and asymptotically $Y(\infty)$, would be determined by the equation

$$\frac{dY(t)}{dt} = -\lambda \left(Y(t) - Y(\infty)\right) t^{-\beta} \tag{7.77}$$

with the solution

$$Y(t) = Y(\infty) + Ce^{-at^\alpha}, \tag{7.78}$$

where $\alpha = \beta - 1$, $a = \lambda/\alpha$, and C is an integration constant. As noted by Scott Blair et al. [19] these early theories almost always applied only to recoverable phenomena.

Nutting's Law In an Inverse Power Law   The straight line in Figure 7.3 represents the inverse power law (Nutting Law)

$$\lim_{t \to \infty} \Phi_N(t) \approx \frac{\tau_c^\alpha \, \Phi_0}{t^\alpha}. \tag{7.79}$$

Note that (7.79) has been previously obtained from our discussion of fractional memory; see (7.50). We note in passing that Nutting first wrote his law in the form [15]

$$\sigma \propto t^\alpha \, \varepsilon^\beta \tag{7.80}$$

and later in the form [16],

$$\frac{d\sigma}{\sigma} = \alpha \frac{dt}{t} + \beta \frac{d\varepsilon}{\varepsilon}. \tag{7.81}$$

Nutting pointed out that some materials stiffen and others soften with increasing stress. He found that $\alpha < 1$ in all the materials he studied, but $\alpha \to 1$ for soft materials, whereas β could be either greater or less than one. A high value of β indicates a rubbery texture, and a low value of this parameter indicates properties such as quicksand [20]. It should also be mentioned that the inverse power-law relation (7.80) was discovered independently by Scott Blair and Coppen in 1939.

Mittag-Leffler Function Is the Full Solution   The heavy curve in Figure 7.3 denotes the asymmetric form of the Mittag-Leffler function in time. The inverse power law indicates that the fractional-differential relaxation process described by (7.59) has self-similar behavior asymptotically, that is, after an initial transient period in which the material relaxes as a stretched exponential, it relaxes as an inverse power law. The parameter τ_c is the time scale characteristic of the transition from the Kohlrausch-Williams-Watts Law to the Nutting Law behavior and is one of the two fitting parameters in the Mittag-Leffler function. The other fitting parameter, the inverse power-law index, is obtained by fitting the slope of the data asymptotically in the doubly logarithmic plot.

7.2.2   A Fractional Theory of Viscoelasticity

Standard Model   The fractional calculus that we have been discussing was successfully used by Glöckle and Nonnenmacher [6] to describe the linear viscoelastic behavior of solids and liquids. They generalized the Zener model, the standard model discussed earlier, to a fractional differential form, thereby incorporating the early observations of Scott Blair into a truly dynamical model. The Zener model of the stress-strain relationship is given by the first-order differential equation

$$\sigma + \tau_0 \, \dot{\sigma}(t) = (G_m + G_e) \tau_0 \, \dot{\varepsilon}(t) + G_e \varepsilon(t), \tag{7.82}$$

where the relaxation time of the Maxwell element τ_0 is given by the ratio of the viscosity η_m and the spring constant G_m, and G_e is the constant of the spring parallel to the Maxwell unit in Figure 7.2. The Zener model is a combination of Hookean springs $[\sigma(t) = G\varepsilon(t)]$ and a Newtonian dash pot $\left[\sigma(t) = \eta\,\dot\varepsilon(t)\right]$ with the appropriate parameters. We have used the notation of Glöckle and Nonnenmacher [6] here to facilitate reference to their papers. It is a simple matter to reexpress their parameter in terms of those we used earlier in the standard model.

Generalization of Standard Model to Fractional Derivatives To generalize the standard model using the relaxation arguments from earlier lectures we rewrite (7.82) as

$$\frac{1}{\tau_0^\beta} D_t^{-\beta}\sigma(t) + \sigma(t) - \sigma_0 = \frac{G_e}{\tau_0^\mu} D_t^{-\mu}\varepsilon(t) + G_0\left[\sigma(t) - \sigma_0\right], \qquad (7.83)$$

here σ_0 and ε_0 are the initial stress and strain, respectively, and $G_0 = [G_m + G_e]$. Recall that the strategy is to first rewrite (7.82) as a first-order integral equation and then to replace the first-order integral operators by fractional integral operators. Glöckle and Nonnenmacher generalize this idea even further by using different fractional integrals on the right- and left-hand sides of the constitutive equation. For reference we record the fractional-differential constitutive equation corresponding to (7.83) by multiplying through the equation with the fractional-derivative operator D_t^β, even though it is the integral form of the equation that we solve using Laplace transforms:

$$
\begin{aligned}
& D_t^\beta\sigma(t) - \frac{\sigma_0 t^{-\beta}}{\Gamma(1-\beta)} + \frac{1}{\tau_0^\beta}\sigma(t) \\
= {}& \frac{G_e}{\tau_0^\mu} D_t^{\beta-\mu}\varepsilon(t) + G_0 D_t^\beta\varepsilon(t) - \frac{G_0\varepsilon_0}{\Gamma(1-\beta)}t^{-\beta}.
\end{aligned}
\qquad (7.84)
$$

Interpreting the Generalized Model We can identify on the left-hand side of (7.84) the dynamical term given by the fractional derivative of the stress, the dissipation term that is linear in the stress, and the time-dependent term that ensures the proper behavior of the stress at the initial time $t = 0$. There are comparable strain terms on the right-hand side of (7.84). If the right-hand side of (7.84) is set to zero we would have the fractional-relaxation model discussed in the previous section. Therefore we expect the solution to the generalized standard model to have some aspects of the Mittag-Leffler function, but to be modified by the influence of the strain.

Laplace Transforms and the Transfer Function Laplace transforming (7.83) allows us to write the linear transfer relation between the Laplace transforms of the stress and the strain

$$\widetilde{\sigma}\left(s\right) = \widetilde{T}\left(s\right)\,\widetilde{\varepsilon}\left(s\right),\tag{7.85}$$

where the transfer function $T\left(t\right)$ contains all the dynamical information we have available about the viscoelastic system. To obtain the Laplace transform of the transfer function $\widetilde{T}\left(s\right)$, we Laplace transform (7.83) yielding

$$\frac{\widetilde{\sigma}\left(s\right)}{\left(s\tau_0\right)^{\beta}} + \widetilde{\sigma}\left(s\right) - \frac{\sigma_0}{s} = \frac{G_e\widetilde{\varepsilon}\left(s\right)}{\left(s\tau_0\right)^{\mu}} + G_0\left[\widetilde{\varepsilon}\left(s\right) - \frac{\varepsilon_0}{s}\right]$$

which combines to give for the ratio of $\widetilde{\sigma}\left(s\right)$ to $\widetilde{\varepsilon}\left(s\right)$,

$$\widetilde{T}\left(s\right) = \frac{G_0 + G_e\left(s\tau_0\right)^{-\mu}}{1 + \left(s\tau_0\right)^{-\beta}},\tag{7.86}$$

where the initial stress and strain are chosen such that

$$\sigma_0 = G_0\varepsilon_0.\tag{7.87}$$

A Solution for a Particular Experiment   To discuss stress relaxation experiments, where the strain is imposed at time $t = 0$ and is held constant thereafter, the strain, given by

$$\varepsilon\left(t\right) = \varepsilon_0\,\Theta\left(t\right),\tag{7.88}$$

is substituted into (7.83) where $\Theta\left(t\right)$ is the Heaviside unit step function. In this way the Laplace transform of the strain, given by

$$\widetilde{\varepsilon}\left(s\right) = \frac{\varepsilon_0}{s},\tag{7.89}$$

may be substituted into the Laplace transform for the stress (7.85) and using the Laplace transform of the transfer function (7.86) yields

$$\widetilde{\sigma}\left(s\right) = \frac{\varepsilon_0}{1 + \left(s\tau_0\right)^{-\beta}}\left[G_0 s^{-1} + G_e s^{-\mu-1}\right].\tag{7.90}$$

Recall that we have defined the stress relaxation function by

$$G\left(t\right) = \frac{\sigma\left(t\right)}{\varepsilon_0},\tag{7.91}$$

so that using the Laplace transform given by (7.90), we obtain

$$\widetilde{G}\left(s\right) = \frac{1}{1 + \left(s\tau_0\right)^{-\beta}}\left[G_0 s^{-1} + G_e s^{-\mu-1}\right].\tag{7.92}$$

We can now take the Mellin transform of the Laplace transform of the stress relaxation function to put it in a form where the solution can be expressed in terms of a Fox function. Inserting $\widetilde{G}\left(s\right)$ into the right-hand side of (7.67) enables us to write the Mellin transform of the stress relaxation function

$$\hat{G}(p) = \frac{\tau_0^p}{\beta\Gamma(1-p)}\left[\Gamma\left(\frac{p}{\beta}\right)\Gamma\left(1-\frac{p}{\beta}\right) + \tau_0^\mu\Gamma\left(\frac{p+\mu}{\beta}\right)\Gamma\left(1-\frac{p+\mu}{\beta}\right)\right].$$

$$(7.93)$$

The inverse Mellin transform of (7.93) is given by, using (10.79),

$$G(t) = \frac{1}{2\pi i}\int_{c-i\infty}^{c+i\infty}\hat{G}(p)\,t^{-p}dp,$$

$$(7.94)$$

so that using the general term from the Appendix we obtain for the stress relaxation function in terms of Fox functions

$$G(t) = \frac{G_0}{\beta}H_{12}^{11}\left(\frac{t}{\tau_0}\left|\begin{array}{c}(0,1/\beta)\\(0,1/\beta),(0,1)\end{array}\right.\right)$$
$$+ \frac{G_e}{\beta}\left(\frac{t}{\tau_0}\right)^\mu H_{12}^{11}\left(\frac{t}{\tau_0}\left|\begin{array}{c}(0,1/\beta)\\(0,1/\beta),(-\mu,1)\end{array}\right.\right).$$

$$(7.95)$$

We can now use the general properties of Fox functions to understand the relaxation behavior of the stress, using the stress relaxation function (7.95).

Asymptotic Forms for the Relaxation Function The asymptotic time behavior of the stress relaxation function (7.95) is given by

$$G(t) \propto t^{\mu-\beta}$$

$$(7.96)$$

in the case where the elastic modulus is nonzero, $G_e \neq 0$, and

$$G(t) \propto t^{-\beta}$$

$$(7.97)$$

in the case where the elastic modulus vanishes, since the second term in (7.95) contributes a factor of t^μ to the solution. For asymptotic convergence of the stress relaxation function we therefore require the fractional-derivative indices to satisfy the relation $\mu \leq \beta$, so that,

$$\lim_{t\to\infty}G(t) = 0.$$

$$(7.98)$$

The Complete Time-Dependent Solution In the short-time limit the stress relaxation is determined by the Mittag-Leffler series solution for the Fox function, just as we found in the preceding section:

$$G(t) = G_0\sum_{k=0}^\infty\frac{(-1)^k}{\Gamma(1+k\beta)}\left(\frac{t}{\tau_0}\right)^{k\beta}$$
$$+ G_0\sum_{k=0}^\infty\frac{(-1)^k}{\Gamma(1+\mu+k\beta)}\left(\frac{t}{\tau_c}\right)^{k\beta}.$$

$$(7.99)$$

For small t the relaxation function decreases with time only if $\mu \geq \beta$, otherwise the second series dominates with increasing time. Thus, the relaxation function is a strongly monotonically decreasing function for all $t > 0$ only if $\mu = \beta$, in which case (7.99) can be written as

$$G(t) = \frac{G_0 - G_e}{\beta} H_{12}^{11} \left(\frac{t}{\tau_0} \middle| \begin{array}{c} (0, 1/\beta) \\ (0, 1/\beta), (0, 1) \end{array} \right) + G_e. \tag{7.100}$$

In Figure 7.4 data from stress relaxation experiments are compared with the Glöckle-Nonnenmacher model with $G_e = 0$, which is the simplest version of the fractional rheology theory. The glassy modulus G_0 or the initial stress $\sigma_0 = G_0 \varepsilon_0$ is obtained by extrapolation to the initial time. The relaxation parameter τ_0 is determined to be 350 seconds in this experiment, so it takes almost six minutes for the transition from the stretched exponential relaxation to the inverse power law decay with $\beta = 0.6$. The parameter τ_0 determines the transition time $t > \tau_0$ from the stretched exponential to the inverse power-law form of the stress relaxation. The slope of the curve yields $-\beta$ in the long-time region of the log-log plot of the data, as we mentioned previously.

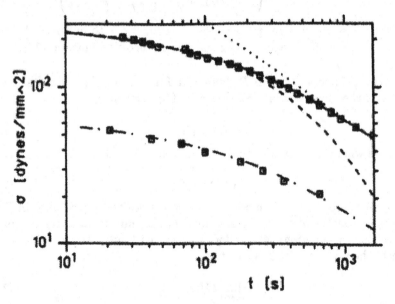

Figure 7.4: *Stress relaxation at constant strain for two different initial conditions: datapoint from ([19]), (———) stretched exponential law with $\sigma_0 = 250$ dynes/mm^2, $\tau_0 = 350$ sec, and $\beta = 0.6$; (....) inverse power law with an overall coefficient of 120 and $\beta = 0.6$; (____) Fox function with $\sigma_0 = 250$ dynes/mm^2, $\tau_0 = 350$ sec, $G_e = 0$, and $\beta = 0.6$; Fox function with $\sigma_0 = 65$ dynes/mm^2, $\tau_0 = 350$ sec, $G_e = 0$, and $\beta = 0.6$. Taken from Glöckle and Nonnenmacher [6] with permission.*

The Fit to Experimental Data Is Extremely Good We see from Figure 7.4 that the fit of the theoretical relations to the data, obtained using the theory developed by Glöckle and Nonnenmacher, is extremely good throughout the total time domain. Glöckle and Nonnenmacher have also compared their theoretical results with experimental data sets obtained by stress-strain experiments carried out on polysobutylene and natural rubber, and have found agreement over more than 10 orders of magnitude in the stress relaxation time. The fractional calculus approach represented by the fractional relaxation has been applied successfully to model self-similar protein dynamics in myoglobin [4] and to formulate slow diffusion processes in biological tissue [9].

7.3 Path Integrals

Macromolecules and Polymers Up to this point we have by and large avoided discussion of the microscopic properties of complex materials, focusing instead on the sometimes bizarre characteristics of the macroscopic equations of motion. When we have discussed the microscopic features of a system it was usually in terms of the statistical physics of ball-like particles, nothing very complicated. Now we wish to discuss the behavior of materials made up of macromolecules, very long chainlike molecules, the basic material of living matter. These macromolecules are strings of small groups of atoms (monomers), repeating units, that interact strongly with each other. Such structures are also called polymers. The dynamics of these complicated structures have been successfully described by means of path integrals and so we now examine how such descriptions relate to the ideas we have been developing.

A Little Background In the physics literature there are a number of ways of introducing a path integral, all of which are intended to enable us to systematically describe the evolution of systems with many degrees of freedom, see, for example, Wiegel [24]. For our purposes the story begins with Wiener's introduction of a measure into function space in order to define an integral over a space of functions. A simple discussion of this concept is given by Wiener [25], but presenting that argument here would lead us too far from our main argument. Essentially what is required is an extension of the notion of integration from functions to functionals. We approach the matter from a different perspective using some of the ideas we have already introduced, rather than using the language of measure theory.

The BCSK Chain Condition and the Path Integral We begin with the BCSK chain condition for the conditional probability density for a particle starting at x_0 at the initial time t_0 to be in the interval $(x, x + dx)$ at time t,

$$P(x, t | x_0, t_0) = \int P(x, t | z, t - \tau) P(z, t - \tau | x_0, t_0) \, dz. \qquad (7.101)$$

The question is how to go from this chain condition to the path integrals of Wiener and others. To answer this question assume that we can partition the total interval between (x_0, t_0) and (x, t) into successively smaller intervals lying between these two endpoints. To facilitate the construction we introduce the notation

$$P(x,t\,|x_0,t_0) = \int P(x,t\,|x_1,t_1)\,P(x_1,t_1\,|x_0,t_0)\,dx_1,$$

where $t \geq t_1 \geq t_0$ so that

$$P(x,t\,|x_0,t_0) = \int P(x,t\,|x_1,t_1)\,dx_1 \int P(x_1,t_1\,|x_2,t_2)\,P(x_2,t_2\,|x_0,t_0)\,dx_2,$$

where $t \geq t_2 \geq t_1 \geq t_0$. Repeating this operation N times we obtain

$$P(x,t\,|x_0,t_0) = \prod_{j=0}^{N-1} \int P(x_{j+1},t_{j+1}\,|x_j,t_j)\,dx_{j+1}, \qquad (7.102)$$

where $x_N = x$. The product of the N conditional probabilities yields the probability that a trajectory starting from (x_0, t_0) will pass through the window (x_1, t_1) in the first time interval, that it will pass through (x_2, t_2) in the second time interval, and so on to (x_N, t_N). This is the behavior of the particle trajectory depicted in Figure 7.5.

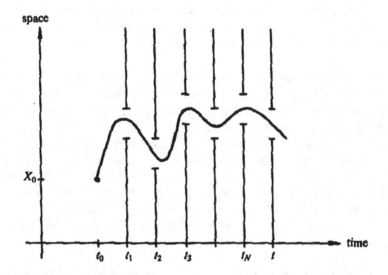

Figure 7.5 : *Trajectory of a Brownian particle which starts at (x_0, t_0) and passes through each of the turnstiles $dx_1, dx_2, \cdots, dx_N$ at times $t_1, t_2, \cdots, t_N$, at equally spaced time intervals*

7.3.1   Wiener Path Integral

Gaussian Probability Densities   For a Wiener process we know that the conditional probability densities in (7.102) are Gaussian

$$P\left(x_{j+1}, t_{j+1} \mid x_j, t_j\right) = \left(4\pi D\left[t_{j+1} - t_j\right]\right)^{-1/2} \exp\left[-\frac{\left(x_{j+1} - x_j\right)^2}{2D\left[t_{j+1} - t_j\right]}\right] \quad (7.103)$$

and this description was successfully used to describe the motion of a Brownian particle. Using the Gaussian conditional probability densities the chain condition (7.102) can be written

$$P\left(x, t \mid x_0, t_0\right) = \left(4\pi D\Delta t\right)^{-(N+1)/2} \prod_{j=0}^{N-1} \int dx_j \exp\left[-\frac{\left(x_{j+1} - x_j\right)^2}{2D\Delta t}\right], \quad (7.104)$$

where we have used equally spaced time intervals, $t_{j+1} - t_j = \Delta t$ for all j. The Wiener path integral is obtained by replacing the finite-sized windows (Δt) with infinitesimally small time intervals such that the dynamical variable threads each of the windows [24],

$$
\begin{aligned}
X(t_0) &= x_0 , \\
x_1 &< X(t_1) < x_1 + dx_1 , \\
x_2 &< X(t_2) < x_2 + dx_2 , \\
&\vdots \\
x_N &< X(t_N) < x_N + dx_N , \\
X(t) &= x.
\end{aligned}
$$

The process of the trajectory weaving its way through each of the differential windows is encapsulated in the notation

$$\mathcal{D}\left[x\left(\tau\right)\right] \Leftrightarrow \prod_{j=1}^{N-1} dx_j$$

such that the integrals in (7.104) become

$$\int_{(x_0, t_0)}^{(x, t)} \mathcal{D}\left[x\left(\tau\right)\right] \Leftrightarrow \frac{1}{\left(4\pi D\Delta t\right)^{(N+1)/2}} \int dx_1 \cdots \int dx_{N-1}.$$

In the same way, the limit of the integrand in (7.104) yields the continuum form

$$\sum_{j=0}^{N} \frac{\left(x_{j+1} - x_j\right)^2}{\Delta t} \Leftrightarrow \int_{t_0}^{t} \left(\frac{dx}{d\tau}\right)^2 d\tau$$

such that we obtain the Wiener path integral

$$P\left(x, t \mid x_0, t_0\right) = \int_{(x_0, t_0)}^{(x, t)} \mathcal{D}\left[x\left(\tau\right)\right] \exp\left[-\frac{1}{2D} \int_{t_0}^{t} \left(\frac{dx}{d\tau}\right)^2 d\tau\right]. \quad (7.105)$$

However, we know that (7.105) is just a symbolic representation of (7.104) so that using the properties of Gaussian integrals we obtain for the conditional probability density, the well-known result

$$P\left(x,t\,|x_0,t_0\right) = \frac{\exp\left[-\frac{(x-x_0)^2}{2D(t-t_0)}\right]}{\sqrt{4\pi D\left(t-t_0\right)}}. \qquad (7.106)$$

This is the conditional probability density for a free Brownian particle.

Trajectories Are Not Differentiable   For Brownian motion we know that the spatial interval increases as the square root of the time interval

$$|x_{j+1} - x_j| = O\left(\sqrt{D\Delta t}\right).$$

Therefore the velocity for the Brownian particle is given by

$$\frac{|x_{j+1} - x_j|}{\Delta t} = O\left(\sqrt{\frac{D}{\Delta t}}\right)$$

which clearly diverges as $\Delta t \to 0$. We obtained this result somewhat earlier, but now it is clear that this divergence indicates that the trajectories depicted in Figure 7.3, although continuous, are not differentiable.

Solution to the Diffusion Equation   We know that the conditional probability density given by (7.106) is the solution to the inhomogeneous diffusion equation, with a point source at the initial time at the initial position

$$\left(\frac{\partial}{\partial t} - D\frac{\partial^2}{\partial x^2}\right) P_0\left(x,t\,|x_0,t_0\right) = \delta\left(x - x_0\right)\delta\left(t - t_0\right). \qquad (7.107)$$

This is the equation of motion for a free Brownian particle. Note that we specify this solution with a zero subscript in order to distinguish it from that of a Brownian particle in a potential $V\left(x\right)$:

$$\left(\frac{\partial}{\partial t} - D\frac{\partial^2}{\partial x^2} - V\left(x\right)\right) P\left(x,t\,|x_0,t_0\right) = \delta\left(x - x_0\right)\delta\left(t - t_0\right). \qquad (7.108)$$

The solution to (7.108) may be expressed in terms of the solution to (7.107) by the integral equation

$$P\left(x,t\,|x_0,t_0\right) = P\left(x,t\,|x_0,t_0\right)$$
$$+ \int P\left(x,t\,|x',t'\right) V\left(x',t'\right) P\left(x',t'\,|x_0,t_0\right) dx'\,dt' \qquad (7.109)$$

The validity of (7.109) is determined by operating on the equation with

$$\left(\frac{\partial}{\partial t} - D\frac{\partial^2}{\partial x^2}\right),$$

using (7.107) and integrating over the intermediate variables. The resulting equation is (7.108), as we suggested.

Wiener Path Integral with a Potential   Let us now follow the tried and true method for solving difficult problems, that being, assume the form of the solution and then verifying that the assumed form does indeed satisfy the equation of motion. Assume a potential such that at the discrete time points $\{t_j\}$ the potential can be written $V_j \equiv V(x(t_j)) = V(x_j, t_j)$. The proposed generalization of the path integral for the free Brownian particle (7.105) is obtained from the discrete version of the equation

$$P(x,t\,|x_0,t_0) = \frac{1}{(4\pi D\Delta t)^{(N+1)/2}} \prod_{j=0}^{N-1} \int dx_{j+1} \exp\left[-\frac{(x_{j+1}-x_j)^2}{2D\Delta t} - V_j\right].$$
(7.110)

We wish to take the limit of (7.110) as $N \to \infty$ and $\Delta t \to 0$ such that $N\Delta t = t - t_0$, as before. It is convenient to expand the second factor in the integrand as follows

$$\exp\left[-\Delta t \sum_{j=0}^{N} V_j\right] = 1 - \Delta t \sum_{j=0}^{N} V_j + \frac{\Delta t^2}{2!} \sum_{j,k=0}^{N} V_j V_k - \cdots$$
(7.111)

which when inserted into (7.110) and collecting terms yields the infinite series of perturbation terms (in an obvious notation)

$$\begin{aligned}
P(x,t\,|x_0,t_0) &= P_0(x,t\,|x_0,t_0) - \Delta t \sum_{j=0}^{N} \int dx_j P(x,t\,|x_j,t_j) V_j P_{j0} \\
&+ \frac{\Delta t^2}{2!} \sum_{j,k=0}^{N} \int dx_j \int dx_k P(x,t\,|x_j,t_j) V_j P_{jk} V_k P_{k0} - \cdots.
\end{aligned}$$
(7.112)

In the limit $\Delta t \to 0$ the sums over discrete times in (7.112) are replaced with time integrals. Ordering the times such that $t_0 < t_1 < \cdots < t_N$ we can absorb the combinatorics factor $N!$ using the fact that $P(x,t\,|x_0,t_0) = 0$ when $t < t_0$. In this way the continuum form of expression (7.112) can be written

$$P(x,t\,|x_0,t_0) = P_0(x,t\,|x_0,t_0) - \int_0^t dt_1 \int dx_1 P(x,t\,|x_1,t_1) V_1 P_{10}$$

$$+ \int_0^t dt_1 \int_0^t dt_2 \int dx_1 \int dx_2 P\left(x,t \,|\, x_1,t_1\right) V_1 P_{12} V_2 P_{20}$$
$$- \cdots . \tag{7.113}$$

Equation (7.113) is, of course, the iterated form of the integral equation for a Brownian particle in a potential given by (7.109). Therefore our guess as the appropriate Wiener integral for a Brownian particle in a potential is given by the continuum limit of (7.110) to be

$$P\left(x,t \,|\, x_0,t_0\right) = \int_{(x_0,t_0)}^{(x,t)} \mathcal{D}\left[x\left(\tau\right)\right] \exp\left[-\frac{1}{2D} \int_{t_0}^t \left(\frac{dx}{d\tau}\right)^2 d\tau - \int_{t_0}^t V\left(x\left[\tau\right]\right) d\tau\right]. \tag{7.114}$$

An alternative way to write the Wiener path integral is

$$P\left(x,t \,|\, x_0,t_0\right) = \int_{(x_0,t_0)}^{(x,t)} \mathcal{D}_W\left[x\left(\tau\right)\right] \exp\left[-\int_{t_0}^t V\left(x\left[\tau\right]\right) d\tau\right], \tag{7.115}$$

where the subscript denotes the weighting of the differential paths with respect to the factor

$$\exp\left[-\frac{1}{2D} \int_{t_0}^t \left(\frac{dx}{d\tau}\right)^2 d\tau\right].$$

This notation is convenient when we extend the arguments to include Lévy stable processes. These arguments interrelating the Wiener path integral, the partial differential equations of motion, and the integral equations were made rigorous by Marc Kac [8]. A very readable discussion of the applications of these ideas to polymer science is given by Wiegel [24].

7.3.2 Lévy Path Integral

Other Kinds of Path Integrals A number of investigators have recognized that the Wiener path integral is only one of an infinite number of possible choices of measure in function space. One of the most notable of the other choices is, of course, Feynman's treatment of propagators in quantum mechanics using path integrals. It was Kac [8] who showed the equivalence between the Wiener and the Feynman path integrals; the former leading to the classical diffusion equation, as we have shown, and the latter leading to the Schrödinger equation, which we do not show. We are interested in another result that was rigorously shown by Kac and that is the existence of a path integral for Lévy stable processes. There has been some recent interest in this formalism, especially as applied to quantum mechanics [13, 23, 11].

The Infinitely Divisible Let us consider an alternate way of writing the Wiener path integral for the free Brownian particle using the characteristic function for the Gaussian distribution

$$\int_{(x_0,t_0)}^{(x,t)} \mathcal{D}_W[x(\tau)] = \int_{-\infty}^{\infty} \frac{dK}{2\pi} \exp[iK(x-x_0)]$$
$$\times \exp\left[-(t-t_0)2DK^2\right]. \tag{7.116}$$

The characteristic function used in (7.116) is a consequence of the BCSK chain condition. However, we know that the most general solution to the chain condition is that of a Lévy stable distribution, so we may replace the path integral in (7.116) by

$$\int_{(x_0,t_0)}^{(x,t)} \mathcal{D}_L[x(\tau)] = \int_{-\infty}^{\infty} \frac{dK}{2\pi} \exp[iK(x-x_0)]$$
$$\times \exp\left[-(t-t_0)b|K|^{\alpha}\right] \tag{7.117}$$

with impunity for $2 \geq \alpha > 0$. Of course, it is now necessary to interpret what we mean by the path integral with the new subscript. It is clear that what we intend to do is repeat the argument presented in the previous section, replacing the Gaussian distribution with the Lévy distribution. There is no problem with this because both distributions are infinitely divisible and therefore can be used to thread the windows in Figure 7.3.

The Path Integral Itself Repeating the arguments from the last section yields the path integral expression for the Lévy distribution

$$P_L(x,t|x_0,t_0) = \int_{(x_0,t_0)}^{(x,t)} \mathcal{D}_L[x(\tau)]$$
$$= \lim_{N\to\infty,\Delta t\to 0} \int_{(x_0,t_0)}^{(x,t)} \prod_{j=0}^{N-1} dx_j \int_{-\infty}^{\infty} \frac{dK}{2\pi} \exp[iK(x_{j+1}-x_j)]$$
$$\times \exp\left[-\Delta t b|K|^{\alpha}\right] \tag{7.118}$$

where it is clear that the trajectory threading the windows has Lévy statistics. An equivalent expression in terms of the probability density at each point in time is

$$P_L(x,t|x_0,t_0) = \lim_{N\to\infty,\Delta t\to 0} \int_{(x_0,t_0)}^{(x,t)} \prod_{j=0}^{N-1} dx_j \int_{-\infty}^{\infty} \frac{dK}{2\pi}$$
$$\times \frac{\exp[iK(x_{j+1}-x_j)]}{(b\Delta t)^{N/\alpha}} P_L\left(\frac{x_{j+1}-x_j}{(b\Delta t)^{1/\alpha}}\right), \tag{7.119}$$

where we explicitly note the scaling in the Lévy distribution in the integrand.

Equation of Motion It is clear from (7.119) that the equation of motion for the free Lévy distribution, obtained by taking the time derivative of the conditional probability density in (7.117), is

$$\frac{\partial P_L\left(x,t\,|x_0,t_0\right)}{\partial t} = -b\int_{-\infty}^{\infty}\frac{dK}{2\pi}\,|K|^{\alpha}\exp\left[iK\left(x-x_0\right)\right]\exp\left[-\left(t-t_0\right)b\,|K|^{\alpha}\right].$$

(7.120)

It was shown by West and Seshadri that the Fourier transform on the right-hand side of (7.120) yields the fractional diffusion equation

$$\frac{\partial P_L\left(x,t\,|x_0,t_0\right)}{\partial t} = \frac{b}{\pi}\Gamma\left(\alpha+1\right)\sin\left[\pi\alpha/2\right]$$
$$\times\int_{-\infty}^{\infty}\frac{P_L\left(x',t\,|x_0,t_0\right)dx'}{|x'-x|^{\alpha+1}},$$

(7.121)

where the integral is the Reisz fractional derivative which we denote by R_x^{α}. We therefore write the fractional diffusion response of the system to a point source at the initial space-time point

$$\left(\frac{\partial}{\partial t}-R_x^{\alpha}\right)P_L\left(x,t\,|x_0,t_0\right)=\delta\left(x-x_0\right)\delta\left(t-t_0\right).$$

(7.122)

Solution to the Fractional Diffusion Equation We can again use the same approach and construct the generalization to the Lévy path integral by introducing a potential. Kac also made this mathematically rigorous and determined that the conditional probability density in such a case is given by

$$P_L\left(x,t\,|x_0,t_0\right)=\int_{(x_0,t_0)}^{(x,t)}\mathcal{D}_L\left[x\left(\tau\right)\right]\exp\left[-\int_{t_0}^{t}V\left(x\left[\tau\right]\right)d\tau\right],$$

(7.123)

where we now know that $\mathcal{D}_L\left[x\left(\tau\right)\right]$ is a functional measure for a Lévy stable process. Correspondingly we can write the fractional diffusion equation

$$\left(\frac{\partial}{\partial t}-R_x^{\alpha}-V\left(x\right)\right)P_L\left(x,t\,|x_0,t_0\right)=\delta\left(x-x_0\right)\delta\left(t-t_0\right)$$

(7.124)

and the integral equation

$$P\left(x,t\,|x_0,t_0\right)=P_L\left(x,t\,|x_0,t_0\right)+\int P_L\left(x,t\,|x',t'\right)$$
$$\times V\left(x',t'\right)P\left(x',t'\,|x_0,t_0\right)dx'dt'.$$

(7.125)

So we see that the inclusion of the fractional derivative in the equation of motion does not change the general form of the solution, but we might argue that the convergence of (7.124) is even slower than in the Wiener case because of the long tails in the Lévy distribution.

7.4 Commentary

Where the Techniques Meet Data The reason we have devoted so much time to the application of fractional operators to materials is that rheology is the most completely developed area of research in which such operators are more than mere curiosities. The fact that polymers, rubber, and plastics are used in so many areas of engineering forced the relatively early development of techniques that could describe phenomena for which ordinary and partial differential equations of evolution could not explain the results of experiment. The same may be said for the recent investigations into the properties of biomedical phenomena, for example, the morphogenesis of physiological structures such as the mammalian lung and the branching systems of veins in the eye. These latter phenomena are only now being modeled as fractional dynamical systems as the techniques are being developed in biophysics; see, for example, West [22].

Things also Work for Nonlinear Phenomena Our focus here has been on the linear stress-strain properties of materials, but as we mentioned, we are also interested in the nonlinear aspects of material dynamics. We have elsewhere examined the generalization of one nonlinear dynamical system to fractional form and that is the Riccati equation. Riccati-type equations have historically played a prominent role in the mathematical modeling of nonlinear phenomena, in large part because the Riccati equation is one of the few nonlinear dynamical equations that we know how to solve exactly in closed form. Such equations arise, for example, in the context of ecological systems and chemical reactions, because the number of pairings of species for reproduction or chemical reaction is quadratic. Furthermore, many nonlinear partial differential equations can be mapped, via similarity transformations, onto Riccati-type equations and therefore their solutions can be expressed in terms of the inverse transforms of the solutions to the Riccati equation. Because of its simple structure and widespread applications, the Riccati equation represents one of the most often studied ordinary differential equations.

Solution to the Riccati Equation The Riccati equation also describes the balance between the hydrodynamic drag of the air on a free-falling body, being quadratic in the speed of the falling body, and the attractive force of gravity, see, for example, Davis [1]. Metzler et al. [12] studied bungee cord jumping rather than free fall to determine whether the cord would exceed its elastic limit and creep to an irreversible increased length during the jump. They used a phenomenological argument to generalize the Riccati equation in terms of the fractional calculus. However, they did not solve the nonlinear fractional differential equation. Instead, they generalized, to fractional order, the ordinary linear differential equation onto which the Riccati equation can be mapped, and solved it. The solution is found to be inverse power law asymptotically in agreement with the Nutting Law. This is the first time that, starting from a standard Riccati equation, a generalized equation for the fall of a heavy body in a fractionally resisting medium has been proposed, discussed, and solved.

The Path Integrals   We mention that although we have only briefly discussed path integrals, the connection between Lévy stable processes and path integrals appears to be quite important. This is particularly true because in the last decade, the understanding of complex physical phenomena has more and more required the application of infinitely divisible statistical distributions such as those of Lévy. As the name implies one can partition processes described by such distributions into smaller and smaller parts without changing the statistical character of the phenomenon. In the physical sciences this property is called scaling. Diffusion, whether ordinary or anomalous, has such a scaling property. It is interesting to investigate the consequences of such a partitioning when applied to quantum phenomena. In this regard the Lévy-Khintchine formula has been applied to the investigation of relativistic quantum mechanics [3, 13, 23, 11]. In addition, the quantum dynamics of the reduced density matrix for certain complex systems has been shown to be Lévy stable [10].

Bibliography

[1] H. Davis, *Introduction to Nonlinear Differential and Integral Equations*, Dover, New York (1962).

[2] W. N. Findley, J. S. Lai and K. Onaran, *Creep and Relaxation of Nonlinear Viscoelastic Materials*, Dover, New York (1976).

[3] P. Garbaczewski, in *Chaos-The Interplay between Stochastic and Deterministic Behavior*, P. Garbaczewski, M. Wolf and A. Weron, eds., Springer-Verlag, Berlin (1995).

[4] W. G. Glöckle and T. F. Nonnenmacher, A fractional Calculus approach to self-similar protein dynamics, *Biophys. J.* **68**, 46-53 (1995).

[5] W. G. Glöckle and T. F. Nonnenmacher, Fox function representation of non-Debye relaxation processes, *J. Stat. Phys.* **71** (1993) 741.

[6] W. G. Glöckle and T. F. Nonnenmacher, Fractional relaxation and the time-temperature superposition principle, *Rheol. Acta* **33** (1994) 337.

[7] I. S. Gradshteyn and I. M. Ryzhik, *Table of Integrals, Series, and Products*, corrected and enlarged edition, Academic, New York (1980).

[8] M. Kac, *Probability and Related Topics in Physical Sciences*, Interscience, New York (1959).

[9] M. Köpf, R. Metzler, O. Haferkamp and T. F. Nonnenmacher, NMR Studier of Anomalous Diffusion in Biological Tissue: Experimental Observations of Lévy Stable Processes, in *Fractals in Biology and Medicine, vol. II*, G. A. Losa, D. Merlini, T. R. Nonnenmacher and E. R. Weibel, eds., Birkhäuser, Basel (1998).

[10] D. Kusnezov, A. Bulgac and G. D. Dang, *Phys. Rev. Lett.* **82**, 1136 (1999).

[11] N. Laskin, *Fractional Quantum Mechanics and Lévy Path Integrals*, preprint.

[12] R. Metzler, W. G. Glöckle, T. F. Nonnenmacher and B. J. West, Fractional tuning of the Riccati equation, *Fractals* **5**, 597 (1997).

[13] E. Montroll, *On the quantum analogue of the Lévy distribution*, in *Physical Reality and Mathematical Description*, C. Mehra, M. Reidel, eds.,Dordrecht, The Netherlands, 501-508 (1974).

[14] T. F. Nonnenmacher and R. Metzler, *Fractals* **3**, 557 (1995).

[15] P. G. Nutting, *J. Franklin Inst.* **191**, 679 (1921).

[16] P. G. Nutting, *Proc. Am. Soc. Test. Mater.* **21**, 1162 (1921).

[17] Yu. N. Rabotnov, *Elements of Hereditary Solid Mechanics*, MIR , Moscow (1980).

[18] H. Schiessel, R. Metzler, A. Blumen and T. F. Nonnenmacher, Generalized viscoelastic models: their fractional equations with solutions, *J. Phys.: Math. Gen.* **28**, 6567-84 (1995).

[19] G. W. Scott Blair, B. C. Veinoglou and J. E. Caffyn, Limitations of the Newtonian time scale in relation to non-equilibrium rheological states and a theory of quasi-properties, *Proc. Roy. Soc. Ser. A* **187**, 69 (1947).

[20] G. W. Scott Blair, *A Survey of General and Applied Rheology*, Pitman, London (1949).

[21] R. K. Schofield and W. G. Blair, *Proc. R. Soc. A* **138**, 707 (1932).

[22] B. J. West, *Physiology, Promiscuity and Prophecy at the Millennium : A Tale of Tails*, Studies of Nonlinear Phenomena in the Life Sciences vol. **7**, World Scientific, Singapore (1999).

[23] B. J. West, Quantum Lévy Propagators, *J. Phys. Chem. B* **104**, 3830 (2000).

[24] F. W. Wiegel, *Introduction to Path-Integral Methods in Physics and Polymer Science*, World Scientific, Singapore (1986).

[25] N. Wiener, *Nonlinear Problems in Random Theory*, MIT Press, Cambridge, MA (1958).

Chapter 8

Fractional Stochastics

Review of Fractional Random Walk Results The modeling of complex phenomena using random walks was discussed earlier. In that discussion we outlined how a simple random walk model of a normal diffusion process leads to Gaussian statistics and a mean-square displacement that increases linearly with time. We also examined how inverse power-law memory in the random fluctuations, that is, in the steps of the walker, can produce a system response that is anomalous in that the mean-square displacement is proportional to t^{2H}, where $0 < H < 1$. We saw that such time series are random fractals with fractal dimension given by $D = 2 - H$. The most complex phenomena studied earlier involved the limit of fractional differences becoming fractional derivatives, so that a stochastic process with long-term memory can be generated by taking the fractional derivative of a Wiener process. We learned that such processes have Gaussian statistics, but they also have inverse power-law spectra.

Review of Phase Space Equations We also learned that the BCSK chain condition is the starting point for studying the evolution of the probability density for most physical scientists. If the dynamical variable $X(t)$ yields for the average increment $\langle \Delta X(t, \Delta t) \rangle$, where the increment is defined $\Delta X(t, \Delta t) = X(t + \Delta t) - X(t)$, and a second moment $\left\langle \Delta X(t, \Delta t)^2 \right\rangle$, that increases no faster than Δt, the chain condition reduces to the Fokker-Planck equation. When these central moments diverge, the Fokker-Planck equation is replaced by the fractional-differential equation of West and Seshadri [33] whose simplest solution is the Lévy distribution. However, in both these cases the statistics of the dynamical variable are Markovian. In the next few lectures one of the things we discuss is how a statistical process can be Gaussian, but because it has a long-time memory, can be non-Markovian. This is the situation for fractional Brownian motion (fBm) first discussed by Mandelbrot and van Ness [19]; fBm is Gaussian but non-Markovian. In the past few years fBm has been used to model a wide range of phenomena, from the biology of DNA sequences [4] and the dynamics of nerve growth [23], to the physics of polymers [7], crack profiles

[14], and non-Fikian diffusion [2], to image analysis [8].

Fractional Langevin Equations In the next few lectures, although we discuss the probability density for non-Markovian processes, we focus most of our discussion on fractional stochastic dynamical equations. In particular we emphasize how the solutions to these equations differ from those of the familiar Langevin equation, when the usual differentials are replaced with fractional differentials. We define these fractional stochastic dynamical equations as fractional Langevin equations. The Mittag-Leffler function replaces the exponential function in the analysis of the solutions to these latter equations and can be used as a nonunitary evolution operator to formally describe the evolution of the dynamical variables. An application of these ideas is also made to transport processes in fluctuating media using fractional eigenvalue equations.

8.1   Fractional Stochastic Equations

The Langevin Equation We now want to shift our focus from deterministic to stochastic phenomena. We have the technical apparatus necessary to generalize one of the standard approaches for modeling complex, statistical, physical phenomena, that being the Langevin equation. We previously considered a number of the problems associated with the derivation and interpretation of the Langevin equation for simple Brownian motion, which in one dimension we write as

$$\frac{dv\left(t\right)}{dt} + \lambda v\left(t\right) = f\left(t\right). \tag{8.1}$$

This is often referred to as an Ornstein-Uhlenbeck process, due to its dependence on the dissipation parameter λ, and the fact that these two scientists gave the first complete mathematical description of the solution to this equation [29]. Physically the dissipation parameter is a consequence of the Stokes drag on Brown's pollen mote, by the lighter particles of water, or on Ingen-Housz's charcoal, by the lighter particles of alcohol. As we also discussed, the proper interpretation of (8.1) is not as a differential equation, but as an integral equation of the form

$$dv\left(t\right) + \lambda v\left(t\right) dt = dB\left(t\right), \tag{8.2}$$

where $dB\left(t\right)$ is a differential Wiener process.

The Fractional Langevin Equation We now want to generalize (8.2) to account for nonlocal influences, that is, for the kind of relaxation that occurs in polymers and the viscoelastic materials discussed in the previous lectures. Thus, if we wanted to model a polymer in contact with a thermal heat bath we would adapt the arguments of the preceding lectures regarding the fractional relaxation in the stress-strain relations. Note that the Langevin equation (8.2)

is phenomenological in character, so that it is reasonable, in the case of a viscoelastic material, to replace Newton's force law with a fractional derivative of
the velocity. Physically, this replacement means that the force is only defined
on a fractal set of points. To ensure the physical reasonability of this model,
the fractional force law ought to include a dependence on the initial velocity in
order to provide a proper interpretation of the initial value problem. In addition
the dissipation parameter should have the appropriate scaled units. Thus, we
write the fractional-dynamical equation of motion for the velocity as

$$D_t^\alpha \left[v\left(t\right) \right] - v_0 \frac{t^{-\alpha}}{\Gamma\left(1-\alpha\right)} = -\lambda^\alpha v\left(t\right) + f\left(t\right), \qquad (8.3)$$

where $f\left(t\right)$ is a Wiener process. Equation (8.3) is clearly a fractional Langevin
equation. The question is: Is (8.3) a reasonable generalization of the usual
Langevin equation given by (8.1) to describe the thermodynamics of viscoelastic
materials?

8.1.1 A Simpler Equation

Nondissipative Stochastic Process Before we work on solving the full fractional Langevin equation, (8.3), let us look at a somewhat simpler version of
this equation, one without dissipation

$$D_t^\alpha \left[x\left(t\right) \right] - x_0 \frac{t^{-\alpha}}{\Gamma\left(1-\alpha\right)} = f\left(t\right). \qquad (8.4)$$

Here we do not give $x\left(t\right)$ a physical interpretation, since we are only interested
in the formal properties of the solution to (8.4). The solution to (8.4), as we
now know, can be written in terms of a fractional integral operator

$$x\left(t\right) - x_0 = D_t^{-\alpha} \left[f\left(t\right) \right]. \qquad (8.5)$$

We also learned that the statistics of the solution to this equation are Gaussian
when $f\left(t\right)$ is a Wiener process and the spectrum of the solution is an inverse
power law. We did not, however, explicitly calculate the two-point correlation
function. We avoided that calculation, in part because we had not yet mastered
the tools. But now we wish to evaluate the two-point correlation function using
the formal properties of (8.5) so that

$$\langle \left[x\left(t_1\right) - x_0\right] \left[x\left(t_2\right) - x_0\right] \rangle = \frac{1}{\Gamma\left(\alpha\right)^2} \int_0^{t_1} dt_1' \int_0^{t_2} dt_2'$$
$$\times \frac{\langle f\left(t_1'\right) f\left(t_2'\right) \rangle}{\left(t_1 - t_1'\right)^{1-\alpha} \left(t_2 - t_2'\right)^{1-\alpha}}, \qquad (8.6)$$

where the fluctuations are assumed to have Gaussian statistics and to be delta
correlated in time

$$\langle f(t_1') f(t_2') \rangle = 2D\delta(t_1' - t_2'). \tag{8.7}$$

Of course, D is the strength of the fluctuations.

Formal Expression for the Two-Time Correlation Function The integral (8.6) is completely symmetric in the times t_1 and t_2, but we know that the delta function will restrict the integration to the earlier of the two times, since this is where both variables can be equal. Therefore, we introduce the notation $t_>$ for the greater time and $t_<$ for the lesser time, and implementing the delta function (8.7) under the integral (8.6) yields

$$\langle [x(t_>) - x_0][x(t_<) - x_0] \rangle = \frac{4D}{\Gamma(\alpha)^2} \int_0^{t_<} dt\, (t_> - t)^{\alpha-1}(t_< - t)^{\alpha-1}. \tag{8.8}$$

Introducing the normalized variable $\xi = t/t_<$ we obtain, after some algebra,

$$\langle [x(t_>) - x_0][x(t_<) - x_0] \rangle = \frac{4D t_>^{\alpha-1} t_<^{\alpha}}{\Gamma(\alpha)^2} \int_0^1 d\xi \left(1 - \frac{t_<}{t_>}\xi\right)^{\alpha-1}(1 - \xi)^{\alpha-1}. \tag{8.9}$$

Using the following integral representation of the *hypergeometric function*, see, for example, Miller and Ross ([20], page 304) or Abramowitz and Stegun [1],

$$_2F_1(a;b;c:z) = \frac{\Gamma(c)}{\Gamma(a)\Gamma(c-a)} \int_0^1 d\xi\, \xi^{a-1}(1-\xi)^{c-a-1}(1-\xi z)^{-b}, \tag{8.10}$$

where $Rec > Rea > 0$, and equating coefficients with the terms in (8.9) we obtain

$$\langle [x(t_>) - x_0][x(t_<) - x_0] \rangle = \frac{4D t_>^{\alpha-1} t_<^{\alpha}}{\alpha\Gamma(\alpha)^2} F\left(1; 1-\alpha; 1+\alpha : \frac{t_<}{t_>}\right), \tag{8.11}$$

where we have suppressed the suffixes on the hypergeometric function. Note that although the statistics of the solution to (8.5) are Gaussian, they are also nonstationary, since the correlation function depends on $t_>$ and $t_<$ separately, and not just on the difference $t_> - t_<$.

The Time-Dependent Variance Of course, we can also use (8.11) to write the second moment at time $t = t_> = t_<$,

$$\left\langle [x(t) - x_0]^2 \right\rangle = \frac{4D t^{2\alpha-1}}{\alpha\Gamma(\alpha)^2} F(1; 1-\alpha; 1+\alpha : 1) \tag{8.12}$$

and using [1],

$$F(a; b; c : 1) = \frac{\Gamma(c)\Gamma(c - a - b)}{\Gamma(c - a)\Gamma(c - b)}, \tag{8.13}$$

provided $\operatorname{Re} c > \operatorname{Re}[a + b]$ and c is not a nonpositive integer, we obtain

$$\left\langle [x(t) - x_0]^2 \right\rangle = \frac{4D}{(2\alpha - 1)\Gamma(\alpha)^2} t^{2\alpha - 1}. \tag{8.14}$$

This result, (8.14), agrees with that obtained for anomalous diffusion, if we make the identification $H = \alpha - 1/2$, where, in general, the parameter α can be in the range $1/2 \leq \alpha \leq 3/2$ so that $1 \geq H \geq 0$. However, if $\alpha < 1$ as we hypothesized, then $1/2 \geq H \geq 0$ requires $1/2 \leq \alpha \leq 1$, and the process is therefore antipersistent.

8.1.2 Non-Markovian Statistics

Statistics Are Gaussian We know, from the linear nature of the differentiation procedure, that the statistics of the fractional-dynamical process described by (8.5) is Gaussian, since the random force is assumed to be Gaussian, but the variance, given by (8.14) is certainly not linear. Therefore the distribution function is

$$P(x - x_0, t) = \left(4\pi\sigma^2 t^{2H}\right)^{-1/2} \exp\left[-\frac{(x - x_0)^2}{4\sigma^2 t^{2H}}\right], \tag{8.15}$$

where the initial time has been set to zero, and the second moment parameter is defined by

$$\sigma^2 = \frac{D}{H\Gamma(H + 1/2)^2}. \tag{8.16}$$

Since we know that the Gaussian distribution is a solution to the Fokker-Planck equation (FPE), and the FPE arises from an expansion, the BCSK chain condition, one might expect that the underlying process described by (8.15) is Markovian. However, this turns out not to be the case. To show that the distribution (8.15) does not describe a Markov process we follow Jumarie [16] and assume that (8.15) satisfies the chain condition and show that this assumption leads to a contradiction.

Not a Fokker-Planck Equation For ease of reference let us rewrite the BCSK chain condition here,

$$P(x, t + \Delta t \,|x_0) = \int dz P(x, t + \Delta t \,|z, t) P(z, t \,|x_0) \tag{8.17}$$

so that using (8.15) the transition probability density can be written

$$P(x, t + \Delta t \mid z, t) = \left(4\pi\sigma^2 \Delta t^{2H}\right)^{-1/2} \exp\left[-\frac{(x-z)^2}{4\sigma^2 \Delta t^{2H}}\right]. \tag{8.18}$$

Now one would expect that for $t \gg \Delta t$ we can obtain the FPE from (8.17) by expanding the transition probability in the vicinity of $\Delta t = 0$ as follows,

$$P(x, t + \Delta t \mid z, t) \approx P(x, t \mid z, t) + \Delta t \, \dot{P}(x, t + \Delta t \mid z, t) \mid_{\Delta t = 0}, \tag{8.19}$$

where the dot denotes the partial derivative with respect to Δt. We can see from the form of the transition probability density that the first term on the right-hand side of (8.19) is the delta function

$$P(x, t \mid z, t) = \delta(x - z). \tag{8.20}$$

Furthermore, taking the time derivative of (8.18) we obtain

$$\dot{P}(x, t + \Delta t \mid z, t) = 2H\sigma^2 \Delta t^{2H-1} \frac{\partial^2 P(x, t + \Delta t \mid z, t)}{\partial z^2} \tag{8.21}$$

so the second term on the right-hand side of (8.19) is given by

$$\dot{P}(x, t + \Delta t \mid z, t) \mid_{\Delta t = 0} = 2H\sigma^2 \Delta t^{2H-1} \delta''(x - z), \tag{8.22}$$

where the prime denotes the derivative with respect to z. Thus, inserting (8.20) and (8.22) into (8.19) yields

$$P(x, t + \Delta t \mid z, t) \approx \delta(x - z) + 2H\sigma^2 \Delta t^{2H} \delta''(x - z) \tag{8.23}$$

so that putting (8.23) into (8.17) and integrating yields

$$P(x, t + \Delta t \mid x_0) \approx P(x, t \mid x_0) + 2H\sigma^2 \Delta t^{2H} \frac{\partial^2 P(x, t \mid x_0)}{\partial x^2}. \tag{8.24}$$

Reordering the factors in (8.24) we obtain

$$\frac{P(x, t + \Delta t \mid x_0) - P(x, t \mid x_0)}{\Delta t^{2H}} \approx 2H\sigma^2 \frac{\partial^2 P(x, t \mid x_0)}{\partial x^2} \tag{8.25}$$

which in the limit $\Delta t \to 0$ is clearly not the FPE for general H.

Fractional Diffusion Equation In the case $H = 1/2$ we see that in the limit $\Delta t \to 0$ the left-hand side of (8.25) becomes a partial derivative with respect to time and we obtain the FPE for a simple diffusion process

$$\frac{\partial P(x, t \mid x_0)}{\partial t} = \sigma^2 \frac{\partial^2 P(x, t \mid x_0)}{\partial x^2}. \tag{8.26}$$

The solution to (8.26) is clearly a Gaussian distribution in this case. However, we do not obtain this equation for $H \neq 1/2$, instead we have a fractional diffusion equation

$$
\begin{aligned}
D_t^{2H} \left[P\left(x,t \,|x_0\right) \right] &\equiv \lim_{\Delta t \to 0} \frac{P\left(x,t+\Delta t \,|x_0\right) - P\left(x,t \,|x_0\right)}{\Delta t^{2H}} \\
&= 2H\sigma^2 \frac{\partial^2 P\left(x,t \,|x_0\right)}{\partial x^2}.
\end{aligned}
\tag{8.27}
$$

Note that unlike the fractional diffusion equation obtained earlier, here we have a fractional derivative in time rather than in space. We discuss such equations more completely in later lectures. Equation (8.27) is not necessarily of the correct form to describe the evolution of the probability density away from a given initial state. However, the fractional-derivative form of the equation in time is sufficient to show that the underlying process is not Markovian.

8.2 Memory Kernels

The Master Equation Earlier we emphasized that a Fokker-Planck equation (FPE) can often be used as a basis for the discussion of stochastic processes involving continuous variables. We also discussed complex phenomena for which the FPE could not be used, such as the non-Markovian process discussed in the last lecture, but for the moment let us focus on the simpler processes. The discrete analogue of the FPE is the master equation (so named by Nordseick et al. [22] in 1940):

$$
\frac{dP\left(l,t\right)}{dt} = \lambda \sum_{l'} \left[p\left(l,l'\right) P\left(l',t\right) - p\left(l',l\right) P\left(l,t\right) \right].
\tag{8.28}
$$

Here the indices l and l' represent possible states for the system of interest, with $\lambda p\left(l',l\right)$ being the probability per unit time of a transition from l to l'. Equation (8.28) should be considered as a gain-loss equation with the positive term measuring the rate of transition into state l and the negative terms the rate of transition out of state l. Since the transition probabilities are normalized by

$$
\sum_{l'} p\left(l',l\right) = 1
\tag{8.29}
$$

the master equation has the alternate form

$$
\frac{dP\left(l,t\right)}{dt} = -\lambda P\left(l,t\right) + \lambda \sum_{l'} p\left(l,l'\right) P\left(l',t\right).
\tag{8.30}
$$

Oppenheim et al. [24] reviewed the properties of the master equation, and critiqued and reprinted many of the basic papers in the field.

The Generalized Master Equation   The probabilities $P(l,t)$ of (8.28) are sometimes interpreted to be diagonal elements of the quantum mechanical density matrix. With that interpretation the equation is called the Pauli equation, since it was derived by Pauli in 1928 [25] in a Sommerfeld Festschrift volume. Subsequently, a set of equivalent generalized master equations was derived independently by several authors [24] from basic quantum mechanical equations under the assumption that at time $t = 0$ the density matrix is diagonal. The form of the generalized master equation is

$$\frac{dP(l,t)}{dt} = \int_0^t d\tau \sum_{l'} [K_{ll'}(t-\tau) P(l',\tau) - K_{l'l}(t-\tau) P(l,\tau)]. \qquad (8.31)$$

The rules for the calculation of the elements of the kernel $K_{ll'}$ from first principles are given in the cited publications [24] and are, unfortunately, rather complicated involving the projection of the probabilities onto different spaces. We are not concerned with that program here, as it would lead us too far afield, but merely consider (8.31) to characterize certain stochastic processes.

Continuous Time Random Walks and the Telegrapher's Equation Here we merely note the connection between special cases of (8.31) and the CTRW formalism of Montroll and Weiss previously discussed. In the special lattice walk, considered in that earlier section, all lattice points are equivalent and the lattice has translational invariance. The appropriate form of (8.31) would have the factorized kernel functions

$$K_{ll'}(t) = p(l - l') K(t), \qquad (8.32)$$

where $p(l)$ is just the transition probability that appears in (8.28). The generalized master equation then reduces to

$$\begin{aligned}
\frac{dP(l,t)}{dt} &= \int_0^t K(t-\tau) d\tau \sum_{l'} [p(l-l') P(l',\tau) - p(l'-l) P(l,\tau)] \\
&= -\int_0^t d\tau K(t-\tau) P(l,\tau) + \int_0^t d\tau \sum_{l'} p(l-l') K(t-\tau) P(l',\tau),
\end{aligned}$$
$$(8.33)$$

where again we have used the normalization condition on the transition probabilities. Note that with the choice of kernel that is a Dirac delta function in time $K(t-\tau) = \lambda \delta(t-\tau)$, the generalized master equation (8.33) reduces to (8.28). On the other hand, with the choice $K(t) = \lambda e^{-t/\tau_c}$, we obtain from (8.33),

$$\frac{d^2 P(l,t)}{dt^2} + \frac{1}{\tau_c} \frac{dP(l,t)}{dt} = -\lambda P(l,t) + \lambda \sum_{l'} p(l,l') P(l',t), \qquad (8.34)$$

which is the discrete version of the telegrapher's equation when the transition probabilities only allow for nearest neighbor steps; that is, the right-hand side of (8.34) is a discrete approximation to the second derivative in space. It is known that at early times an initial pulse described by (8.34) propagates as a wave, and at later times it propagates as a diffusive packet. This phenomenon was observed in the early days of telegraphy. Signal diffusion reduced the data rate in long cables such as the early trans-Atlantic cable. Applications of (8.34) were also made to the propagation of impulses in nerves and to exciton transport in photosynthetic units during the 1970s. Recall the discussion of the telegrapher's equation from Chapter 4.

Correlations of a Normalized Dynamical Variable The same memory kernel that appears in the generalized master equation (8.33), was shown by Zwanzig [36] to determine the evolution of the correlation function for any dynamical variable, such as the end-to-end distance of a molecular chain or the dipole moment of a molecule, within the limits of linear response theory in a physical system. So let us consider the phase space for a complex phenomenon, consisting of the canonical coordinates $(\mathbf{q}, \mathbf{p})$ for all the degrees of freedom of a system in thermal equilibrium. A dynamical variable $X(t)$ has the equilibrium average value

$$\langle X(t) \rangle_{eq} \equiv \int d\mathbf{q} d\mathbf{p} X(\mathbf{q}, \mathbf{p}, t) \rho_{eq}(\mathbf{q}, \mathbf{p}), \qquad (8.35)$$

where we explicitly indicate the dependence of the dynamical variable on the phase space variables and $\rho_{eq}(\mathbf{q}, \mathbf{p})$ is the equilibrium phase space density function. Here we take the system to be classical, and define the normalized variable

$$\xi(t) = \frac{X(t) - \langle X(t) \rangle_{eq}}{\sqrt{\left(X(t) - \langle X(t) \rangle_{eq}\right)^2}} \qquad (8.36)$$

such that $\langle \xi(t) \rangle_{eq} = 0$. The correlation function

$$\phi(t) \equiv \langle \xi(t) \xi(0) \rangle_{eq} \qquad (8.37)$$

is defined to be the normalized relaxation function so that initially $\phi(0) = 1$. Using the projection operator technique Zwanzig showed that the evolution of the correlation function is given by

$$\frac{d\phi(t)}{dt} = -\int_0^t d\tau K(t - \tau) \phi(\tau), \qquad (8.38)$$

where the detailed structure of the memory kernel is determined by the Hamiltonian for the underlying phase space dynamics. Glöckle and Nonnenmacher [11] give some examples of the memory kernel and discuss the properties of the physical systems that would give rise to such kernels. For example, a Dirac delta function kernel such as we used above, $K(t - \tau) = \lambda \delta(t - \tau)$, gives rise to an exponential correlation function. This implies that the traditional exponential relaxation occurs in a system that is Markovian, and what occurs at a given time is completely independent of what precedes it in time.

The Correlation Function as a Dissipative Harmonic Oscillator   A second example is the exponential kernel we used above, $K(t) = \lambda e^{-t/\tau_c}$, which, after taking a derivative of (8.38) and rearranging terms, yields

$$\frac{d^2\phi(t)}{dt^2} + \frac{1}{\tau_c}\frac{d\phi(t)}{dt} + \lambda\phi(t) = 0, \tag{8.39}$$

the equation of evolution for a dissipative harmonic oscillator. The solution to the initial value problem (8.39) for the relaxation function is given by

$$\phi(t) = e^{-t/2\tau_c}\left\{\cos\omega t + \frac{1}{2\omega\tau_c}\sin\omega t\right\} \tag{8.40}$$

where we used the initial conditions $\phi(0) = 1$ and

$$\left.\frac{d\phi(t)}{dt}\right|_{t=0} = 0.$$

The decay of the memory kernel determines the dissipation rate of the relaxation function and along with the coupling strength determines the frequency of oscillation,

$$\omega = \sqrt{\lambda^2 - \frac{1}{4\tau_c^2}}.$$

Motivation for the Inverse Power Law   From our early lectures we formed the impression that the inverse power law is ubiquitous, but we have not provided much motivation for its form. In particular, the above arguments provide no rationale for an inverse power-law memory kernel. Let us now turn to this problem. A familiar motivation is that used in many physico-chemical phenomena involving a hopping particle (random walker) traversing a distribution of random activation barriers. Among the phenomena that are described using this model are the $1/f$-noise in semiconductors [35], the propagation of dislocations due to impact [34], the structural or dielectric relaxation in glassy materials [12], and the transport phenomena in disordered material [15, 32], to name a few.

Model for the Inverse Power Law Waiting Time Distribution   Following Vlad [31] we outline the simplest theoretical description of the above hopping processes in terms of a mean field approximation, the random activation energy model [RAEM]. This model refers to the activated hopping of a particle over an energy barrier having a random height. It is assumed that the jump rate λ has the Arrhenius form

$$\lambda(\Delta E) = \nu \exp\left[-\frac{\Delta E}{kT}\right], \qquad (8.41)$$

for an energy barrier of height ΔE. The waiting time distribution $\psi(\Delta E, t)$ with a given activation energy ΔE has the exponential form

$$\psi(\Delta E, t) = \lambda(\Delta E) e^{-\lambda(\Delta E)t} \qquad (8.42)$$

and is properly normalized

$$\int_0^\infty \psi(\Delta E, t)\, dt = 1. \qquad (8.43)$$

Here we assume that the activation energy ΔE is a random variable whose statistics are determined by a probability law $p(\Delta E)$, so that $p(\Delta E)\, d\Delta E$ is the probability that the activation energy is in the interval $(\Delta E + d\Delta E, \Delta E)$. Thus, a literal waiting time distribution function $\psi(t)$ can be constructed as an average of $\psi(\Delta E, t)$ over all possible activation energies:

$$
\begin{aligned}
\psi(t) &= \int_0^\infty p(\Delta E)\,\psi(\Delta E, t)\, d\Delta E \\
&= \int_0^\infty \lambda(\Delta E) e^{-\lambda(\Delta E)t} p(\Delta E)\, d\Delta E. \qquad (8.44)
\end{aligned}
$$

The Inverse Power Law The most commonly used distribution of activation energies also has the Arrenheus form

$$p(\Delta E) = \frac{1}{kT_0} \exp\left[-\frac{\Delta E}{kT_0}\right] \qquad (8.45)$$

where T_0 is a fixed temperature. Inserting (8.45) into (8.44) yields

$$\psi(t) = \int_0^\infty \lambda(\Delta E)\ e^{-\lambda(\Delta E)t} \exp\left[-\frac{\Delta E}{kT_0}\right] \frac{d\Delta E}{kT_0}$$

which after some algebraic manipulation and substitution from (8.42) gives

$$\psi(t) = \frac{1}{\nu^\beta t^{1+\beta}} \int_0^{\nu t} z^\beta e^{-z}\, dz, \qquad (8.46)$$

where $\beta = T/T_0$. Asymptotically in time, the integral in (8.46) becomes the gamma function and the waiting time distribution has the inverse power-law form

$$\psi(t) = \frac{\Gamma(1+\beta)}{\nu^\beta t^{1+\beta}}. \qquad (8.47)$$

If $\beta < 1$ all the moments of the waiting time distribution diverge and there is no characteristic scale between jumps in the random walk process. Consequently, the asymptotic behavior of this process is described by a random fractal. Vlad [31] also generalizes the above argument to include the internal structure of the random walk process. He shows that the averaged waiting time distribution in the generalized RAEM is nothing more than that given by a CTRW between the internal states.

The Memory Kernel from CTRW   The waiting time distribution function can be used to determine the memory kernel using the CTRW theory

$$K\left(t\right) = \mathcal{L}T^{-1}\left\{\frac{s\widetilde{\psi}\left(s\right)}{1 - \widetilde{\psi}\left(s\right)};t\right\},\tag{8.48}$$

where using (8.47) we obtain the memory kernel

$$K\left(t\right) \approx \frac{\nu^{\beta}}{\Gamma\left(1+\beta\right)}t^{\beta-2}.\tag{8.49}$$

Introducing the time constant τ_c we write the equation for the correlation function as

$$\frac{d\phi\left(t\right)}{dt} = -\frac{1}{\tau_c^{\beta}\Gamma\left(\beta\right)}\int_0^t d\tau\left(t-\tau\right)^{\beta-2}\phi\left(\tau\right),\tag{8.50}$$

where τ_c is a real phenomenological relaxation time, and we have

$$\tau_c^{\beta} = \frac{\Gamma\left(1+\beta\right)}{\Gamma\left(\beta\right)\nu^{\beta}}.\tag{8.51}$$

The same results have been obtained by Nigmatullin in 1992 [21] and again by Stanislavsky [28]. Thus, (8.50) reduces to

$$\frac{d\phi\left(t\right)}{dt} = -\frac{1}{\tau_c^{\beta}}D_t^{1-\beta}\left[\phi\left(t\right)\right].\tag{8.52}$$

By applying the antiderivative operator to both sides of (8.52) and then applying the fractional derivative operator D_t^{β}, the resulting equation is

$$D_t^{\beta}\left[\phi\left(t\right)\right] - \frac{t^{-\beta}}{\Gamma\left(1-\beta\right)}\phi\left(0\right) = -\frac{1}{\tau_c^{\beta}}\phi\left(t\right).\tag{8.53}$$

Thus, a memory function of the form of an inverse power law leads to a resulting fractional relaxation equation.

An Inconsistent Fractional Derivative Equation   Conversely, a relaxation equation of the form

$$D_t^{\beta}\left[\phi\left(t\right)\right] = -\frac{1}{\tau_c^{\beta}}\phi\left(t\right)$$

resulting from the standard relaxation equation by formally replacing d/dt with D_t^{β}, is not compatible with the Zwanzig form of (8.38) [11].

The Time-Temperature Superposition Principle Is there perhaps a more penetrating physical argument that we can use to obtain the fractional relaxation equation from (8.38)? Glöckle and Nonnenmacher [11] show that the time-temperature superposition principle in combination with a separation assumption leads to a memory kernel of the inverse power-law form. In order to incorporate thermodynamical properties into the description, they assumed that $K(t)$ is not only a function of time but also a function of the temperature of the heat bath T : $K(t, T)$. Note that the temperature enters (8.49) through $\beta = T/T_0$. The time-temperature superposition principle applied to the relaxation function $\phi(t)$ states that the functional form at two different temperatures is the same apart from a change of time scale; that is,

$$\phi(t, T_1) = \phi(\lambda t, T_2) \tag{8.54}$$

with the time scale factor being a function of both temperatures $\lambda(T_1, T_2)$. As Findley et al. [9] point out, the time-temperature superposition principle states that the effect of temperature on the time-dependent mechanical behavior of a material is equivalent to stretching (or shrinking) the real time for temperatures above (or below) the reference temperature. In other words, the behavior of materials at high temperature and high strain rate is similar to the behavior of the material at low temperature and low strain rate. Such materials are called thermorheologically simple.

Scaling in the Correlation Transport Equation The evolution of the relaxation function given by (8.38) for thermorheologically simple materials is

$$\frac{d\phi(t, T_1)}{dt} = -\int_0^t K(t - \tau, T_1)\,\phi(\tau, T_1)$$

so that introducing the scaling (8.54) yields

$$\frac{d\phi(\lambda t, T_2)}{d(\lambda t)} = -\frac{1}{\lambda^2}\int_0^t K(t - \tau, T_1)\,\phi(\lambda\tau, T_2)\,d(\lambda\tau)$$

from which we obtain the scaling relation for the memory kernel

$$K(t, T_2) = \frac{K(t/\lambda, T_2)}{\lambda^2}. \tag{8.55}$$

Introducing a reference temperature T_r such that $K_r(t) = K(t, T_r)$ and $\lambda_r(T) = \lambda(T_r, T)$ the memory function (8.55) can be written as

$$K(t, T) = \frac{K_r[t/\lambda_r(T)]}{\lambda_r^2(T)}. \tag{8.56}$$

An Additional Assumption Glöckle and Nonnenmacher [11] go on to argue that in addition to the relation (8.56) one needs a further assumption to obtain the power-law form of the memory kernel, that being the separation of variables given by

$$K(t,T) = g(T) f(t). \tag{8.57}$$

Combining the scaling relation (8.56) and the separation of variables (8.57) one obtains

$$K_r(t) = h(\lambda_r) f(\lambda_r t) \tag{8.58}$$

by expressing the temperature dependence through the scaling parameter

$$h(\lambda_r) = \lambda_r^2 g(T).$$

Differentiating (8.58) with respect to the scaling parameter λ_r yields

$$\frac{dK_r(t)}{d\lambda_r} = h'(\lambda_r) f(\lambda_r t) + t h(\lambda_r) f'(\lambda_r t) = 0,$$

where the prime denotes the derivative with respect to the argument of the function. Thus, we have

$$\frac{\lambda_r h'(\lambda_r)}{h(\lambda_r)} = -\frac{\tilde{t} f'(\tilde{t})}{f(\tilde{t})} , \tilde{t} \equiv \lambda_r t, \tag{8.59}$$

where we see that both sides of the equations are functions of different variables and therefore both are equal to the same constant. Integrating (8.59) gives the desired power law for the memory kernel, $h(\lambda_r) \propto \lambda_r^\gamma$, $f(\tilde{t}) \propto \tilde{t}^\gamma$, so that

$$K(t,T) = \frac{C}{\lambda_r^{\gamma+2}(T)} t^\gamma, \tag{8.60}$$

where the power-law index is a phenomenological parameter. Hence, a combination of the time-temperature principle and the assumption of separation of variables is sufficient to yield the fractional relaxation equation with $\alpha = 2 + \gamma$, implying that fractional relaxation is a special type of a non-Markovian process. For further discussion of this point see [11] and the original literature.

Back to the Telegrapher's Equation Mathematically, the diffusion equation is a parabolic partial differential equation, whereas the telegrapher's equation is a hyperbolic partial differential equation. The first to have noticed this difference seems to have been Cattaneo [6] in 1948, wherein he introduced modified constitutive equations that produced a relaxation of the flux. Note the discussion regarding Maxwell and the telegrapher's equation in Chapter 4. One way to write such an equation is the integrodifferential equation

$$\frac{\partial \rho(x,t)}{\partial t} = \int_0^\infty K(t-t') \frac{\partial^2}{\partial x^2} \rho(x,t') dt'. \tag{8.61}$$

When the memory kernel is of the exponential form

$$K(t - t') = \lambda D e^{-\lambda(t-t')} \tag{8.62}$$

we obtain the telegrapher's equation by taking the derivative of (8.61) with respect to the time t:

$$\frac{\partial^2 \rho(x,t)}{\partial t^2} + \frac{1}{\tau_c} \frac{\partial \rho(x,t)}{\partial t} = D \frac{\partial^2 \rho(x,t)}{\partial x^2}. \tag{8.63}$$

Equation (8.63) was also obtained using the CTRW formalism reviewed earlier, when the waiting time distribution function is an exponential. However, it is also quite possible to generalize this result using more exotic forms for the waiting time distribution function and the memory kernel, such as the inverse power law. To do this we need to generalize the master equation to the continuum.

8.3 The Continuous Master Equation

The Memory Kernel   As we have mentioned more than once, an evolving phenomenon can be represented in two distinct ways: using the dynamical equation approach of Langevin and/or the phase space equation approach of Fokker and Planck. [1] We have examined how the dynamical equations can be generalized to include the ideas of the fractional calculus. Now let us see what can be done for the description of the evolution of the probability density. To address this issue let us go back to the Markov master equation introduced earlier, but using its continuum form in one spatial dimension:

$$\frac{\partial P(x,t)}{\partial t} = \int_{-\infty}^{\infty} K(x - x') P(x',t) \, dx', \tag{8.64}$$

where, for simplicity, we have assumed the kernel is homogeneous in space and independent of time. If we denote the probability per unit time that the random walker takes a step of length in the interval $(x, x + dx)$ as $\Pi(x)$ then the memory kernel can be written

$$K(x) = \Pi(x) - \delta(x) \int_{-\infty}^{\infty} dx' \Pi(x'). \tag{8.65}$$

Note that the memory kernel is the difference between the number of walkers entering the interval $(x, x + dx)$ and those leaving that interval per unit time.

Two-State Random Process   Let us consider the case where the underlying process is the velocity of the walker and can only exist in the two states $\pm W$. The step randomly shifts back and forth between these two values. This means that a transition of length $|x|$ implies a time $t = |x|/W$, which ties the space

---

[1] This section was taken from Bologna et al. [5] and suitably modified for the present discussion.

and time scales together. Shlesinger et al. [27] considered such a transition probability in their study of turbulence:

$$\pi(x,t) = \psi(t)\,\delta\left(|x| - Wt\right), \tag{8.66}$$

where $\psi(t)$ is the waiting time distribution function. We can then construct the transition probability per unit time using the Markov property [5] and integrating $\pi(x,t)$ over time

$$\Pi(x) = \int_0^\infty \pi(x,t)\,dt, \tag{8.67}$$

we obtain the time-independent transition rate in terms of the form of the waiting time distribution function

$$\Pi(x) = \frac{1}{W}\psi\left(\frac{|x|}{W}\right). \tag{8.68}$$

Inverse Power-Law Correlation Function   Now if the correlation function of the velocity fluctuations is an inverse power law

$$\Phi(t) = \frac{\overline{T}^\beta}{\left|\overline{T} + t\right|^\beta}, \tag{8.69a}$$

using the relation between the correlation function and the waiting time distribution function developed by Geisel and Thomas [10],

$$\Phi(t) = \frac{\mu - 2}{\overline{T}}\int_t^\infty (t' - t)\,\psi(t')\,dt' \tag{8.70}$$

which has the equivalent relation

$$\psi(t) = \frac{\overline{T}}{\mu - 2}\frac{d^2\Phi(t)}{dt^2}, \tag{8.71}$$

we obtain by substituting (8.69a) into (8.71) and setting $\mu = \beta + 2$,

$$\psi(t) = \frac{(\mu - 1)\overline{T}^{\mu-1}}{\left|\overline{T} + t\right|^\mu}. \tag{8.72}$$

Thus, (8.68) can be written

$$\Pi(x) = \frac{(\mu - 1)\left(W\overline{T}\right)^{\mu-1}}{\left(W\overline{T} + |x|\right)^\mu}. \tag{8.73}$$

Substituting (8.73) into the memory kernel (8.65) yields

$$\frac{\partial P\left(x,t\right)}{\partial t} = \frac{(\mu-1)\left(W\overline{T}\right)^{\mu-1}}{\overline{T}}\left[\int_{-\infty}^{\infty}\frac{P\left(x',t\right)dx'}{\left(W\overline{T}+|x-x'|\right)^{\mu}}\right.$$
$$\left.-P\left(x,t\right)\int_{-\infty}^{\infty}\frac{dx'}{\left(W\overline{T}+|x'|\right)^{\mu}}\right], \tag{8.74}$$

where $0 < \beta < 1$.

Divergencies Cancel in the Evolution Equation To evaluate the integral terms in (8.74) we use the convolution theorem for Fourier transforms to write the equation for the characteristic function

$$\frac{\partial \phi\left(k,t\right)}{\partial t} = (\mu-1)\left(W\overline{T}\right)^{\mu-1}\left[\mathcal{FT}\left\{\frac{1}{\left(W\overline{T}+|x|\right)^{\mu}};k\right\}\right.$$
$$\left.-\mathcal{FT}\left\{\frac{1}{\left(W\overline{T}+|x|\right)^{\mu}};0\right\}\phi\left(k,t\right)\right]. \tag{8.75}$$

But we have previously determined the value of the integral

$$\mathcal{FT}\left\{\frac{1}{\left(a+|x|\right)^{\mu}};k\right\} = 2\int_{0}^{\infty}\frac{\cos kx}{\left(a+x\right)^{\mu}} = g\left(k\right)|k|^{\mu-1}, \tag{8.76}$$

where the coefficient is

$$g\left(k\right) = 2\Gamma\left(1-\mu\right)\left[\sin\left[\mu\pi/2+|ka|\right] - \widetilde{\sin}_{\mu-1}\left[\mu\pi/2+|ka|\right]\right]. \tag{8.77}$$

To use the latter integral, (8.76), in the evaluation of the former, (8.75), we use $a = W\overline{T}$, and note that we are interested in the asymptotic in time solution of the fractional partial differential equation, that is, the solution in the limit $ka \to 0$. Thus, using the series expansion for the generalized sine and generalized cosine

$$\sin_{\mu-1}|ka| = \sum_{n=0}^{\infty}|ka|^{n+1-\mu}\frac{\sin\left[(n+1-\mu)\pi/2\right]}{\Gamma\left(n+2-\mu\right)}$$

$$\cos_{\mu-1}|ka| = \sum_{n=0}^{\infty}|ka|^{n+1-\mu}\frac{\cos\left[(n+1-\mu)\pi/2\right]}{\Gamma\left(n+2-\mu\right)}$$

in the expression

$$\widetilde{\sin}_{\mu-1}\left[\mu\pi/2+|ka|\right] = \cos\left(\mu\pi/2\right)\sin_{\mu-1}|ka| + \sin\left(\mu\pi/2\right)\cos_{\mu-1}|ka|$$

and retaining the lowest order diverging items in (8.77) we obtain

$$\lim_{ka \to 0} \mathcal{FT}\left\{ \frac{1}{(a+|x|)^{\mu}} ; k \right\} \approx 2\Gamma(1-\mu)\left[\sin(\mu\pi/2) - |ka|\cos(\mu\pi/2) \right.$$

$$\left. - \frac{|ka|^{1-\mu}}{\Gamma(2-\mu)} + \frac{|ka|^{3-\mu}}{\Gamma(4-\mu)} \right] |k|^{\mu-1} \qquad (8.78)$$

and the k-independent term in this expansion cancels against the first term in (8.75).

The Evolution Is Described by an Old Friend Thus, (8.75) reduces to

$$\frac{\partial \phi(k,t)}{\partial t} = -b|k|^{\mu} \phi(k,t), \qquad (8.79)$$

where the parameter b is given by

$$b = 2\cos[\mu\pi/2]\,\Gamma(1-\mu)\,\frac{(W\overline{T})^{\mu}}{\overline{T}} \qquad (8.80)$$

and therefore in the limiting case of $\overline{T} \to 0$ the term within the square brackets of (8.74) becomes the regularized form of the Riesz fractional derivative as shown in (6.165). In this sense the above result coincides with the expression found by West and Seshadri [33]. As pointed out by Bologna et al. [5], keeping $\overline{T} > 0$ makes it possible to cross the critical condition $\beta = 1$ without meeting the divergence corresponding to the Lévy prescription, namely, the divergence of b in (8.79) at $\mu = 1$.

Lévy Basin of Attraction The region of interest for the application of the fractional calculus is where the mean sojourn time

$$\overline{T} = \int_0^{\infty} t\psi(t)\,dt \qquad (8.81)$$

is finite (Kac's Theorem), but the second moment of the sojourn time distribution diverges

$$\int_0^{\infty} t^2\psi(t)\,dt \to \infty. \qquad (8.82)$$

The latter condition keeps the process away from the basin of attraction of the Gaussian distribution. Since $\beta = \mu - 2$ we restrict the analysis to the interval $0 < \beta < 1$. Note that the relation between the correlation function and the distribution of sojourn times is exact if the time interval between the transition from one to the other velocity state is instantaneous.

8.4 Back to Langevin

Homogeneous Fractional Differential Equation Let us examine the solution to the homogeneous Langevin equation and then consider the inhomogeneous case. The homogeneous case is one that we have previously solved, but it is good practice to do it again. The homogeneous equation for the fractional Brownian particle is given by the fractional Langevin equation without the thermal fluctuations

$$D_t^\alpha [v(t)] - v_0 \frac{t^{-\alpha}}{\Gamma(1-\alpha)} = -\lambda^\alpha v(t). \tag{8.83}$$

Equation (8.83) is mathematically well defined, but what does it mean physically? From statistical physics we know that the fluctuations in the equation of motion are intimately related to the dissipation, and that in fact they have the same source. This is what gives rise to the fluctuation-dissipation relation, relating the strength of the fluctuations to the ratio of the temperature to the dissipation parameter. However, in (8.83) we have a dissipation without a corresponding set of fluctuations. Since all the operators in (8.83) are linear we could interpret this equation in terms of the average velocity.

Solve with Laplace Transforms For now we treat (8.83) as a mathematical expression with the initial velocity given by v_0; the time-dependence is included so as to have a well-defined initial value problem, and the dissipation parameter is appropriately scaled to have the units corresponding to the order of the fractional derivative. The solution to this equation is obtained from the corresponding fractional integral equation

$$v(t) - v_0 = -\lambda^\alpha D_t^{-\alpha} [v(t)]; \tag{8.84}$$

by taking the Laplace transform we obtain after some algebra

$$\tilde{v}(s) = \frac{v_0/s}{1 + \left(\frac{\lambda}{s}\right)^\alpha}. \tag{8.85}$$

We calculated the inverse Laplace transform of (8.85) earlier using Fox functions to give

$$v(t) = v_0 \sum_{k=0}^{\infty} \frac{(-1)^k}{\Gamma(1+k\alpha)} (\lambda t)^{k\alpha} \tag{8.86}$$

so that the solution to the homogeneous equation is the Mittag-Leffler function

$$E_\alpha(-(\lambda t)^\alpha) = \sum_{k=0}^{\infty} \frac{(-1)^k}{\Gamma(1+k\alpha)} (\lambda t)^{k\alpha}. \tag{8.87}$$

Mittag-Leffler Replaces the Exponential   Thus, the fundamental process is not that of exponential relaxation as it is for the Ornstein-Uhlenbeck process, rather it is the short-time stretched exponential relaxation and long-time inverse power law decay of a Mittag-Leffler process, with the transition time between the two relaxation domains determined by the dissipation parameter λ.

8.4.1   Inhomogeneous Fractional Solution

Solve with the Fluctuations Back In   Let us now look at the solution to the complete fractional Langevin equation. Again we begin by replacing this equation, (8.3), with the equivalent fractional integral equation

$$v\left(t\right) - v_0 = -\lambda^{\alpha} D_t^{-\alpha}\left[v\left(t\right)\right] + D_t^{-\alpha}\left[f\left(t\right)\right]. \tag{8.88}$$

The Laplace transform of this equation yields

$$\widetilde{v}\left(s\right) - \frac{v_0}{s} = -\left(\frac{\lambda}{s}\right)^{\alpha} \widetilde{v}\left(s\right) + \frac{\widetilde{f}\left(s\right)}{s^{\alpha}}, \tag{8.89}$$

so that after some algebra the Laplace transform of the solution to the fractional Langevin equation is given by

$$\widetilde{v}\left(s\right) = \frac{v_0 s^{\alpha-1}}{\lambda^{\alpha} + s^{\alpha}} + \frac{\widetilde{f}\left(s\right)}{\lambda^{\alpha} + s^{\alpha}}. \tag{8.90}$$

We note the difference in the s-dependence of the two coefficients on the right-hand side of (8.90). The inverse Laplace transform of the first term on the right-hand side of (8.90) is the Mittag-Leffler function that we found in the homogeneous case. The inverse Laplace transform of the second term is the convolution of the random fluctuations and a stationary kernel. The kernel is given in terms of a Fox function

$$\mathcal{LT}^{-1}\left\{\frac{1}{\lambda^{\alpha} + s^{\alpha}}; t\right\} = \frac{\left(\lambda t\right)^{\alpha-1}}{\alpha} H_{12}^{11}\left(\lambda t \left|\begin{matrix}\left(0, 1/\alpha\right)\\ \left(0, 1/\alpha\right), \left(1 - \alpha, 1\right)\end{matrix}\right.\right) \tag{8.91}$$

as we obtained for the second term in (7.95). The series expansion for this Fox function can be written as

$$\frac{1}{\alpha} H_{12}^{11}\left(\lambda t \left|\begin{matrix}\left(0, 1/\alpha\right)\\ \left(0, 1/\alpha\right), \left(1 - \alpha, 1\right)\end{matrix}\right.\right) = \sum_{k=0}^{\infty} \frac{\left(-1\right)^k}{\Gamma\left(\alpha + k\alpha\right)}\left(\lambda t\right)^{k\alpha}, \tag{8.92}$$

where the series is a representation of the generalized Mittag-Leffler function, defined by

$$E_{\alpha,\beta}\left(z\right) \equiv \sum_{k=0}^{\infty} \frac{z^k}{\Gamma\left(k\alpha + \beta\right)}, \quad \alpha > 0, \beta > 0. \tag{8.93}$$

The generalized Mittag-Leffler function reduces to the familiar form for $\beta = 1$,

$$E_{\alpha,1}(z) = \sum_{k=0}^{\infty} \frac{z^k}{\Gamma(k\alpha + 1)} = E_\alpha(z), \tag{8.94}$$

so that both the homogeneous and inhomogeneous terms in the solution to (8.88) can be expressed in terms of Mittag-Leffler functions.

Solution in Terms of a Green's Function We now write the general solution to the fractional Langevin equation, using the inverse Laplace transform of (8.90), as

$$
\begin{aligned}
v(t) &= v_0 E_\alpha\left(-(\lambda t)^\alpha\right) + \int_0^t (t - t')^{\alpha-1} \\
&\quad \times E_{\alpha,\alpha}\left(-(\lambda[t - t'])^\alpha\right) f(t')\, dt'.
\end{aligned} \tag{8.95}
$$

In the case $\alpha = 1$, the Mittag-Leffler function becomes an exponential, so that the solution to the fractional Langevin equation becomes identical to that for an Ornstein-Uhlenbeck process

$$v(t) = v_0 e^{-\lambda t} + \int_0^t e^{-\lambda(t-t')} f(t')\, dt' \tag{8.96}$$

as it should. These results were obtained by Kobelev and Romanov [17] using standard techniques for solving Volterra integral equations.

8.4.2 Velocity Autocorrelation Function

Velocity Correlation at Two Times The traditional quantities calculated from the velocity time series are the velocity autocorrelation function and the mean-square velocity and displacement of the Brownian particle. We can also calculate these quantities using the solution to the fractional Langevin equation (8.95). The velocity autocorrelation function is constructed using (8.95) to be

$$
\begin{aligned}
\langle v(t_1) v(t_2) \rangle &= v_0^2 E_\alpha\left(-(\lambda t_1)^\alpha\right) E_\alpha\left(-(\lambda t_2)^\alpha\right) \\
&\quad + \int_0^{t_1} dt_1' (t_1 - t_1')^{\alpha-1} \int_0^{t_2} dt_2' (t_2 - t_2')^{\alpha-1} \langle f(t_1') f(t_2') \rangle \\
&\quad \times E_{\alpha,\alpha}\left(-\lambda^\alpha (t_1 - t_1')^\alpha\right) E_{\alpha,\alpha}\left(-\lambda^\alpha (t_2 - t_2')^\alpha\right)
\end{aligned} \tag{8.97}
$$

where the meaning of this equation is tied to the statistics of the random force driving the system. The traditional assumption is that the random fluctuations have Gaussian statistics and no memory; that is, they are delta correlated in time

$$\langle f(t_1') f(t_2') \rangle = 2D\delta(t_1' - t_2') \qquad (8.98)$$

and D is the strength of the fluctuations.

Evaluating the Integral Term Here again we observe that the correlation integral is completely symmetric in the times t_1 and t_2, so that introducing the greater and lesser times, $t_>$ and $t_<$, and implementing the delta function, the integral term in (8.97) reduces to

$$
\begin{aligned}
I &= 2\int_0^{t_<} dt\, (t_> - t)^{\alpha-1} (t_< - t)^{\alpha-1} E_{\alpha,\alpha}\left(-\lambda^\alpha (t_> - t)^\alpha\right) \\
&\quad \times E_{\alpha,\alpha}\left(-\lambda^\alpha (t_> - t)^\alpha\right).
\end{aligned}
\qquad (8.99)
$$

Making use of the series expression for the generalized Mittag-Leffler function in (8.99) and changing the initial value on the sums yields

$$I = 2\sum_{k=1}^\infty \sum_{l=1}^\infty \frac{(-\lambda^\alpha)^{k+l}}{\Gamma(k\alpha)\,\Gamma(l\alpha)} I_{kl}, \qquad (8.100)$$

where we have introduced the integral

$$I_{kl} = \int_0^{t_<} dt\, (t_> - t)^{k\alpha-1} (t_< - t)^{l\alpha-1}. \qquad (8.101)$$

Factoring the times $t_>$ and $t_<$ out of the integral and introducing the scaled variable $\xi = t/t_<$ allows us to write

$$I_{kl} = t_>^{k\alpha-1} t_<^{l\alpha} \int_0^1 dt\, \left(1 - \xi\frac{t_<}{t_>}\right)^{k\alpha-1} (1-\xi)^{l\alpha-1}$$

so that we can again use the integral representation of the hypergeometric function (8.10) to evaluate this integral as

$$I_{kl} = t_>^{k\alpha-1} t_<^{l\alpha} \frac{\Gamma(l\alpha)}{\Gamma(l\alpha+1)} F\left(1; 1 - k\alpha; 1 + l\alpha : \frac{t_<}{t_>}\right). \qquad (8.102)$$

Thus, the integral term in the velocity autocorrelation function becomes

$$I = 2\sum_{k=1}^\infty \sum_{l=1}^\infty \frac{(-\lambda^\alpha)^{k+l}\, t_>^{k\alpha-1} t_<^{l\alpha}}{\Gamma(k\alpha)\,\Gamma(l\alpha+1)} F\left(1; 1 - k\alpha; 1 + l\alpha : \frac{t_<}{t_>}\right) \qquad (8.103)$$

and the entire autocorrelation function is

$$\langle v(t_>)v(t_<)\rangle = v_0^2 E_\alpha\left(-(\lambda t_>)^\alpha\right)E_\alpha\left(-(\lambda t_<)^\alpha\right)$$
$$+4D\sum_{k=1}^{\infty}\sum_{l=1}^{\infty}\frac{(-\lambda^\alpha)^{k+l}t_>^{k\alpha-1}t_<^{l\alpha}}{\Gamma(k\alpha)\,\Gamma(l\alpha+1)}F\left(1;1-k\alpha;1+l\alpha:\frac{t_<}{t_>}\right).$$

$$(8.104)$$

Clearly, this is a non-stationary result, because the result depends on both times separately and not on their difference. There is not much more that we can do analytically with (8.104) due to its generality; let us therefore simplify the expression somewhat.

The Time-Dependence of the Second Moment of the Velocity The second moment of the velocity is obtained by setting $t_> = t_< = t$ in (8.104) to yield

$$\left\langle v(t)^2\right\rangle = v_0^2\left[E_\alpha\left(-(\lambda t)^\alpha\right)\right]^2$$
$$+4D\sum_{k=1}^{\infty}\sum_{l=1}^{\infty}\frac{(-\lambda^\alpha)^{k+l}t^{(k+l)\alpha-1}}{\Gamma(k\alpha)\,\Gamma(l\alpha+1)}F\left(1;1-k\alpha;1+l\alpha:1\right)$$

$$(8.105)$$

where we can use (8.13) to replace the hypergeometric function by ratios of gamma functions. After some cancellation of terms (8.105) reduces to

$$\left\langle v(t)^2\right\rangle = v_0^2\left[E_\alpha\left(-(\lambda t)^\alpha\right)\right]^2+4D\sum_{k=1}^{\infty}\sum_{l=1}^{\infty}\frac{(-\lambda^\alpha)^{k+l}}{\Gamma(k\alpha)\,\Gamma(l\alpha+1)}\frac{l\alpha}{l\alpha+k\alpha-1}t^{(k+l)\alpha-1}$$

where if the second term on the right-hand side of this equation is denoted by $\mathcal{I}$ we can write [17],

$$\frac{d\mathcal{I}}{dt} = \sum_{k=1}^{\infty}\sum_{l=1}^{\infty}\frac{(-\lambda^\alpha)^{k+l}}{\Gamma(k\alpha)\,\Gamma(l\alpha)}t^{(k+l)\alpha-2}$$

which clearly integrates to

$$\mathcal{I} = \int_0^t\left[\sum_{k=1}^{\infty}\frac{(-\lambda^\alpha t'^\alpha)^k}{\Gamma(k\alpha)}\right]^2\frac{dt'}{t'^2}.$$

We can also take the derivative of the Mittag-Leffler function

$$\frac{dE_\alpha\left(-(\lambda t)^\alpha\right)}{dt} = \sum_{k=1}^{\infty}\frac{(-\lambda^\alpha t^\alpha)^k}{\Gamma(k\alpha)}\frac{1}{t},$$

where the $k = 0$ term vanishes due to the pole of the gamma function, so that the second moment of the velocity can be rewritten [17],

$$\left\langle v\left(t\right)^2 \right\rangle = v_0^2 \left[E_\alpha \left(-\left(\lambda t\right)^\alpha\right)\right]^2 + 4D \int_0^t \left[\frac{dE_\alpha \left(-\left(\lambda t'\right)^\alpha\right)}{dt'}\right]^2 dt'. \tag{8.106}$$

The Asymptotic Time-Dependence of the Second Moment of the Velocity   We can determine the asymptotic properties of this second moment, since we know that the Mittag-Leffler function has an inverse power-law form,

$$\lim_{t\to\infty} E_\alpha \left(-\left(\lambda t\right)^\alpha\right) \sim t^{-\alpha} \tag{8.107}$$

so that the integral term in (8.106) is guaranteed to converge as $t \to \infty$ for all $\alpha > 0$, and to decay a factor of t^{-1} faster than the $t^{-2\alpha}$ dependence of the first term. Thus, the leading term in the asymptotic analysis of the second moment in the velocity is

$$\lim_{t\to\infty} \left\langle v\left(t\right)^2 \right\rangle \sim t^{-2\alpha} \tag{8.108}$$

the signature of a long-time memory process.

Another Expression for the Integral Term   Note that we can rewrite the derivative of the Mittag-Leffler function in terms of the generalized Mittag-Leffler function $-\lambda^\alpha t^{\alpha-1} E_{\alpha,\alpha} \left(-\left(\lambda t\right)^\alpha\right)$, so that the mean-square velocity can also be written as

$$\left\langle v\left(t\right)^2 \right\rangle = v_0^2 \left[E_\alpha \left(-\left(\lambda t\right)^\alpha\right)\right]^2$$
$$+ 4D\lambda^{2\alpha} \int_0^t \left[t'^{\alpha-1} E_{\alpha,\alpha} \left(-\left(\lambda t'\right)^\alpha\right)\right]^2 dt' \tag{8.109}$$

which, of course, has all the same properties as (8.106).

The Ornstein-Uhlenbeck Limit   Here again in the limit $\alpha \to 1$ we obtain the Ornstein-Uhlenbeck result from (8.106) or (8.109),

$$\left\langle v\left(t\right)^2 \right\rangle = v_0^2 e^{-2\lambda t} + 2\lambda D \left(1 - e^{-2\lambda t}\right). \tag{8.110}$$

In the asymptotic limit (8.110) gives the fluctuation-dissipation relation between the dissipation parameter and the diffusion coefficient

$$D = \frac{kT}{2\lambda}. \tag{8.111}$$

To obtain (8.111) we have used the energy equipartition theorem from equilibrium statistical physics to replace the kinetic energy $\left\langle v(t)^2 \right\rangle / 2$ of a unit mass particle in one dimension by $kT/2$.

8.4.3 Fractional Mean-Square Displacement

The Relationship Between Position and Velocity in a Fractal Random Process The displacement of the fractional Brownian particle can be expressed as the time integral over the velocity, so that the mean-square displacement can be expressed as

$$\left\langle x(t)^2 \right\rangle = \int_0^t dt_1 \int_0^t dt_2 \left\langle v(t_1) v(t_2) \right\rangle. \tag{8.112}$$

This is, of course, only one of the possible choices of the relationship between velocity and displacement once we have allowed for the introduction of fractional derivatives. Kobelev and Romanov [17] explore two different relations between the velocity and the displacement. The one we choose here, the first derivative of the displacement with respect to time, is the usual definition. A second choice is one in which the displacement is a fractional integral of the velocity, so that the particle's displacement is defined by velocity only on the set of points within a time interval of dimension α. They argue that microscopically the trajectory of a diffusing particle is nondifferentiable, as was pointed out by both Boltzmann and Perrin, but Kobelev and Romanov go on to say that a fractional derivative of such a curve can be taken. They refer to the work of Kolwankar and Gangal [18] as being a justification for the observed motion of the particle being represented by their average motion. Kobelev and Romanov could just as well have referred to Rocco and West [26]. They [17] argue that some of the instantaneous velocities and displacements do not contribute to the resulting macroscopic motion, even in the case of classical Brownian motion. This, they contend is the source of anomalous diffusion. We concur that this is one of the sources of anomalous diffusion.

Doing the Algebra For the moment let us restrict our investigation to the evaluation of the integral in (8.112) when we substitute the solution to the fractional diffusion equation under the integral

$$
\begin{aligned}
\left\langle x(t)^2 \right\rangle &= v_0^2 \int_0^t dt_1 \int_0^t dt_2 E_\alpha\left(-(\lambda t_1)^\alpha\right) E_\alpha\left(-(\lambda t_2)^\alpha\right) \\
&\quad + \int_0^t dt_1 \int_0^{t_1} dt_1' (t_1 - t_1')^{\alpha-1} \int_0^t dt_2 \int_0^{t_2} dt_2' (t_2 - t_2')^{\alpha-1} \\
&\quad \times E_{\alpha,\alpha}\left(-\lambda^\alpha (t_1 - t_1')^\alpha\right) E_{\alpha,\alpha}\left(-\lambda^\alpha (t_2 - t_2')^\alpha\right) \left\langle f(t_1') f(t_2') \right\rangle.
\end{aligned}
\tag{8.113}
$$

The first set of integrals on the right-hand side of this equation is readily evaluated using the series representation for the Mittag-Leffler function

$$\int_0^t dt_1 E_\alpha \left(-(\lambda t_1)^\alpha\right) = \sum_{k=0}^\infty \frac{(-\lambda^\alpha)^k t^{k\alpha+1}}{\Gamma(k\alpha+1)(k\alpha+1)}$$

$$= t \sum_{k=0}^\infty \frac{(-\lambda^\alpha)^k t^{k\alpha}}{\Gamma(k\alpha+2)} = tE_{2,\alpha}\left(-(\lambda t)^\alpha\right) \quad (8.114)$$

to express the integral in terms of a generalized Mittag-Leffler function. The second term on the right-hand side of (8.113) is of a form we have encountered twice before and therefore the procedure for its evaluation ought to be familiar by now. Using (8.113) we therefore write the expression for the mean-square displacement as

$$\left\langle x(t)^2 \right\rangle = v_0^2 \left[tE_{2,\alpha}\left(-(\lambda t)^\alpha\right)\right]^2 + 2D\mathcal{I}. \quad (8.115)$$

The integral in (8.115) is given by

$$\mathcal{I} = 4 \sum_{k=1}^\infty \sum_{l=1}^\infty \frac{(-\lambda^\alpha)^{k+l}}{\Gamma(k\alpha)\Gamma(l\alpha+1)} \int_0^t dt_> t_>^{k\alpha-1} \int_0^{t_>} dt_< t_<^{l\alpha}$$

$$\times F\left(1; 1-k\alpha; 1+l\alpha : \frac{t_<}{t_>}\right), \quad (8.116)$$

where to evaluate the primed time integrals we have substituted (8.103) into (8.113). Two of the four time integrations remain, and the additional factor of two comes from symmetry.

A Few More Tricks   To evaluate the integral in (8.116) we again scale out the greater time as we did in (8.102) and reduce things to the form

$$I_{kl} = \int_0^t dt_> t_>^{k\alpha+l\alpha} \int_0^1 d\xi \xi^{l\alpha} F\left(1; 1-k\alpha; 1+l\alpha : \xi\right)$$

$$= \frac{t^{k\alpha+l\alpha+1}}{1+k\alpha+l\alpha} \int_0^1 d\xi \xi^{l\alpha} F\left(1; 1-k\alpha; 1+l\alpha : \xi\right). \quad (8.117)$$

We now use the integral representation of the hypergeometric function (8.10) to write after some simplification

$$\int_0^1 \xi^d F\left(a; b; c : \xi\right) d\xi$$

$$= \frac{\Gamma(c)}{\Gamma(a)\Gamma(c-a)} \int_0^1 \eta^{a-1}(1-\eta)^{c-a-1} d\eta \int_0^1 \xi^d (1-\eta\xi)^{-b} d\xi$$

$$= \frac{\Gamma(c)}{(d+1)\Gamma(a)\Gamma(c-a)} \int_0^1 z^{a-1}(1-z)^{c-a-1} F(d+1;b;d+2:z) dz$$

(8.118)

and comparing (8.118) with the integrals of Gradshteyn and Ryzhik ([13], page 849) we obtain for (8.116),

$$\int_0^1 \xi^d F(a;b;c:\xi) d\xi = \frac{\Gamma(c)\Gamma(d+2)\Gamma(a)}{(d+1)\Gamma(a)\Gamma(c-a)} \frac{\Gamma(d+1-a)\Gamma(d+2-a-b)}{\Gamma(c)\Gamma(d+2-b)\Gamma(d+2-a)}$$

$$= \frac{1}{(d+1-b)} = \frac{1}{(l\alpha+k\alpha)},$$

where we have used $a=1, b=1-k\alpha, c=1+l\alpha$, and $d=l\alpha$. Thus, we obtain

$$\mathcal{I} = 4\sum_{k=1}^{\infty}\sum_{l=1}^{\infty} \frac{(-\lambda^\alpha)^{k+l}}{\Gamma(k\alpha)\Gamma(l\alpha+1)} \frac{t^{k\alpha+l\alpha+1}}{(1+l\alpha+k\alpha)(l\alpha+k\alpha)}$$

and following Kobelev and Romanov [17] we can write the second derivative

$$\frac{d^2\mathcal{I}}{dt^2} = 4\sum_{k=1}^{\infty}\sum_{l=1}^{\infty} \frac{(-\lambda^\alpha)^{k+l} t^{k\alpha+l\alpha-1}}{\Gamma(k\alpha)\Gamma(l\alpha+1)} = 2\frac{d}{dt}\left[\sum_{k=1}^{\infty} \frac{(-(\lambda t)^\alpha)^k}{\Gamma(k\alpha+1)}\right]^2$$

$$= 2\frac{d}{dt}\left[E_\alpha\left(-(\lambda t)^\alpha\right)-1\right]^2.$$

Integrating this expression twice and taking into account that both $\mathcal{I}$ and its first derivative vanish at $t=0$, we obtain the mean-square displacement (8.115),

$$\left\langle x(t)^2\right\rangle = v_0^2 \left[tE_{2,\alpha}\left(-(\lambda t)^\alpha\right)\right]^2$$

$$+ 2D\left[t - 2tE_{2,\alpha}\left(-(\lambda t)^\alpha\right) + \int_0^\infty \left[E_\alpha\left(-(\lambda t')^\alpha\right)\right]^2 dt'\right].$$

(8.119)

This Is Still Ordinary Ddiffusion Again using the asymptotic form of the Mittag-Leffler function, the inverse power law, we see that the integral converges for $\alpha > 1/2$, and the leading term on the right-hand side of (8.119) is

$$\left\langle x(t)^2\right\rangle \propto t,$$

just as in the case of ordinary diffusion, with the classical Langevin equation. Even in the case where $\alpha < 1/2$ the divergence of the integral would be very slow, so that the asymptotic region would still be dominated by the linear term. Thus, the mean-square displacement is completely insensitive to the exotic dynamics of this process.

8.5 Commentary

Long-Range Interactions and Long-Time Memories In the above discussion we have seen how to construct and solve fractional stochastic equations modeling phenomena that have long-time memory and/or long-range interactions. We have seen that the long-range power-law correlations that characterize fractional Brownian motion result in a non-Markovian description of the underlying process as, for example, in crack propagation. Another non-Markovian dynamical process was modeled using an inverse power-law memory kernel to describe the evolution of the relaxation function. The continuous form of the master equation was shown to give rise to the equation of evolution of the Lévy distribution when the memory kernel could be represented by an inverse power law in space, based on an inverse power-law waiting time distribution function in time. This is a plausible argument if you accept the reasonableness of the waiting time distribution being an inverse power law in time. *Just how reasonable is the inverse power-law form of the waiting time distribution function anyway?*

Ways to Obtain the Inverse Power-Law Waiting Tme Distribution Function The form of the memory kernel of most interest to us is one that is indicative of a discontinuous process in time. Such a memory kernel could arise in a number of ways. We have indicated two of them here. We showed that the random hopping activation model (RAEM) gives rise to an inverse power-law memory kernel in a way that is equivalent to the CTRW. The second way we discussed is through the time-temperature superposition principle in which scaling led to the form of the kernel. One source that we did not discuss is the chaotic solutions to nonintegrable microscopic Hamiltonian systems that generate fluctuations on macroscopic time scales [30]. A complete discussion of this mechanism would again divert us from our main argument, however, we retain cognizance of the influence of this effect as we continue our discussion of fractional derivatives.

Crack Propagation One should not be lulled into thinking that the inverse power law in the above discussion is always in the time. Take, for example, the phenomenon of crack propagation. Fractional Brownian motion has been used to describe crack trajectories in a piece of material and the mean-square deviation of an ensemble of such trajectories is determined to be [3],

$$\left\langle X\left(z\right)^2 \right\rangle = 2K_f \, z^{2H};\qquad(8.120)$$

where z replaces the time t and is the line-of-sight distance of propagation of the cracks. This is, of course, essentially the same as (8.14), except that the self-similarity of the process is spatial rather than temporal. Here H characterizes the roughness of the non-Fickian nature of the spatial diffusion of the crack, and K_f is a phenomenological parameter required for correct scaling. In experimental studies of 16 propagating cracks it was found that the mean values of the parameters were $H = 0.77$ and $K_f = 0.084$. Addison and Ndumu [3] have used fractal geometry to understand crack surface geometry and energy release in the fracturing of concrete.

Bibliography

[1] M. Abramowitz and I. A. Stegun, *Handbook of Mathematical Functions*, US Dept. of Commerce, NBS, Appl. Math. Ser. 55 (1972).

[2] P. S. Addison, *IAHR J. Hydraulic Research* **34**, 5439 (1996).

[3] P. S. Addison and A. S. Ndumu, Engineering applications of fractional Brownian motion: self-affine and self-similar random processes, *Fractals* **7**, 151 (1999).

[4] P. Allegrini, M. Barbi, P. Grigolini and B. J. West, Dynamical Model for DNA sequences, *Phys. Rev. E* **52**, 5281-96 (19995).

[5] M. Bologna, P. Grigolini and J. Rioccardi, The Lévy diffusion as an effect of sporadic randomness, submitted to *Phys. Rev. E*

[6] G. Cattaneo, *Atti. Sem. Mat. Fis. Univ. Modena* **3**, 83 (1948).

[7] N. Chakravarti and K. L. Sevastian, *Chem. Phys. Lett.* **267**, 9 (1997).

[8] E. L. Chen, P. C. Chung, H. M. Tsai and C. I. Cheng, *IEEE Trans. Biomed. Eng.* **45**, 783 (1998).

[9] W. N. Findley, J. S. Lai and K. Onaran, *Creep and Relaxation of Nonlinear Viscoelastic Materials*, Dover, New York (1976).

[10] T. Geisel and S. Thomas, *Phys. Rev. Lett.* **52**, 1936 (1984).

[11] W. G. Glöckle and T. F. Nonnenmacher, Fractional relaxation and the time-temperature superposition principle, *Rheol. Acta* **33** (1994) 337.

[12] D. G. Le Grand, W. V. Olszewski and J. T. Bendler, *J. Pol. Sci. BB* **25**, 1149 (1987); J. T. Bendler and M. F. Shlesinger, *J. Stat. Phys.* **53**, 531 (1988).

[13] I. S. Gradshteyn and I. M. Ryzhik, *Table of Integrals, Series, and Products*, corrected and enlarged edition, Academic, New York (1980).

[14] A. M. Hammad and M. A. Issa, *Adv. Cement Based Mat.* **1**, 169 (1994).

[15] M. Jaroniec, *Reac. Kinet. Catal Lett.* **8**, 425 (1978)

[16] G. Jumarie, Stochastic differential equations with fractional Brownian motion inputs, *Int. J. Systems Sci.* **24**, 1113 (1993).

[17] V. Kobelev and E. Romanov, Fractional Langevin Equation to describe anomalous diffusion, *Prog. Theor. Phys. Supp.* **139**, 470-476 (2000).

[18] K. M. Kolwankar and A. D. Gangal, Fractional differentiability of nowhere differentiable functions and dimensions, *Chaos* **6**, 505 (1996).

[19] B. B. Mandelbrot and J. W. van Ness, Fractional Brownian motions, fractional noise and applications, *SIAM Rev.* **10**, 422 (1968).

[20] K. S. Miller and B. Ross, *An Introduction to the Fractional Calculus and Fractional Differential Equations*, John Wiley, New York (1993).

[21] R. R. Nigmatullin, *Theor. and Math. Phys.* **90**(3), 245 (1992).

[22] A. Nordseick, W. E. Lamb and G. E. Uhlenbeck, *Physica* **7**, 344 (1940).

[23] D. J. Odde, E. M. Tanaka, S. S. Hawkins and J. M. Buettner, *Biotech. and Bioeng.* **50**, 452 (1996).

[24] I. Oppenheim, K. Shuler and G. Weiss, *The Master Equation*, MIT University Press, Cambridge, MA (1977).

[25] W. Pauli, *Festschrift zum60 gebürtstag A. Sommerfeld*, S. Hirzel, Leipzig (1928).

[26] A. Rocco and B. J. West, Fractional calculus and the evolution of fractal phenomena, *Physica A* **265**, 535 (1999).

[27] M. F. Shlesinger, B. J. West and J. Klafter, Lévy dynamics for enhanced diffusion: an application to turbulence, *Phys. Rev. Lett.* **58**, 1100-03 (1987).

[28] A. A. Stanislavsky, Memory effects and macroscopic manifestation of randomness, *Phys. Rev. E* **61**, 4752-4759 (2000).

[29] G. E. Uhlenbeck and L. S. Ornstein, On the theory of the Brownian motion, *Phys. Rev.* **36**, 823 (1930).

[30] G. Tréfan, P. Grigolini and B. J. West, Deterministic Brownian Motion, *Phys. Rev. A* **45**, 1249 (1992).

[31] M. O. Vlad, An inverse scaling approach to multi-state random activation energy model, *Physica A* **184**, 303-324 (1992).

[32] M. O. Vlad, *J. Coll. Interface Sci.* **128**, 388 (1989).

[33] B. J. West and V. Seshadri, Linear systems with Lévy fluctuations, *Physica A* **113**, 203-216 (1982).

[34] B. J. West and M. F. Shlesinger,Random walk of dislocations following a high velocity impact, *J. Stat. Phys.* **30**, 527 (1983);Random walk model of impact phenomena, *Physica* **127 A**, 490 (1984).

[35] A. van der Ziel, *Physica* **10**, 359 (1950).

[36] R. W. Zwanzig, *Physica* **30**, 1109 (1964).

Chapter 9

The Ant in the Gurge Metaphor

Random Walking in Random Media The *ant in the labyrinth problem* posed by de Gennes in 1976 concerned the description of the random movement of an entity (the ant) in a disordered system (the labyrinth) [8] and was a metaphor for the general problem of transport in disordered media. The general physical problem was to represent the evolution of conduction electrons in amorphous materials, phase dislocations in polymer gels, and myriad other phenomena. On the other hand, the ant as metaphor, like every other localized quantity in physics, has its corresponding nonlocal, wavelike aspect. Material properties, such as the distribution of grain sizes in polycrystalline materials, the degree of homogeneity, the existence of microscopic cracks, inclusions, twin boundaries, and dislocations, all affect fracture micromechanics and wave propagation. To describe the motion of this generalized ant through such disordered, but scaling, materials we change de Gennes' image to that of *an ant in a gurge*, that is, an ant in a kind of turbulent flow field. In terms of this modified image we construct an equation that in one limit models fractional diffusion and in another limit models fractional wave propagation. This new equation is the fractional generalization of the telegrapher's equation. But in addition to these physical processes we also use this metaphor to describe the influence of scaling on the observables in other complex phenomena as well.[1]

Anomalous Diffusion De Gennes' image of the localized ant in a labyrinth was first used as a paradigm for anomalous diffusion, where the mean-square displacement of the diffusing quantity in a time t is given by

$$\left\langle X\left(t\right)^2 \right\rangle \propto t^{2\mu} \tag{9.1}$$

---

[1] This section is taken, in part, from West and Nonnenmacher [41], and suitably modified for the present discussion.

and μ is a constant. Sublinear diffusive growth $(2\mu < 1)$ is familiar from disordered fractal materials [14] and was found in the heart rate variability produced by the cardiac control system [27]. Enhanced diffusion $(2\mu > 1)$, which is not part of de Gennes' ant problem, but is part of the modified metaphor, has been known for 20 years to arise in dynamically chaotic systems [6]. More recently enhanced diffusion has been found to occur in the fluctuations in normal human gait [13, 39], as well as in the genetic variability of molecular evolution [3] and in the long-range correlations of DNA sequences [1].

In Terms of Random Walks One limit of the ant in the gurge problem is described by random walk models [23] wherein $\mu = H$, the Hurst exponent confined to the interval $0 \leq H \leq 1$. The interpretation of the various regimes of the parameter H was given earlier in terms of random walks. Normal diffusion has $H = 1/2$, superdiffusion or persistence has $H > 1/2$, and subdiffusion or antipersistence has $H < 1/2$. Thus, one kind of anomalous diffusion has to do with memory in time for the random steps taken on a regular lattice. This model has been used extensively in the interpretation of the fluctuations in physiological time series, as well as in complex physical phenomena [36].

Antipersistent Processes Another kind of anomalous diffusion has to do with taking uncorrelated steps in time, on a random, or fractal lattice. In this second limit of an ant in a gurge one often sees the mean-square displacement with $\mu = 1/d_w$, where d_w is the anomalous diffusion exponent. For a simple random walk on a fractal object only $d_w > 2$ occurs, because geometrical obstacles exist on all length scales and such obstacles inhibit the ant's movement and the process is therefore subdiffusive. Havlin and Ben-Avraham [14] point out that the anomalous exponent d_w is, in fact, the fractal dimension of the path of the random walker on the lattice.

Taking long steps Finally, in addition to the structure of the space on which the transport takes place, and the correlation of the steps in time, there is the distribution in the sizes of the steps taken by the ant. This is the third source of anomaly in transport. The transition probability is given by

$$p(x) \propto 1/|x|^{1+\alpha}, \qquad (9.2)$$

where $|x|$ is the magnitude of the step, so that for $\alpha < 2$ the resulting random walk is Lévy-stable, which is to say that the set of points formed by the steps of the walk form an α-stable Lévy process [23, 24]. If the random walk is changed to a random flight, so that the time taken to complete a given transition depends on the length of the step, anomalous diffusion results. This aspect of Lévy flights was first used to understand turbulent diffusion by Shlesinger et al. [31] yielding $2\mu \approx 3$. This is consistent with Richardson's t^3 enhanced diffusion law [28].

New Modeling Strategies The properties of scaling media are often described by fractal functions and their space-time evolution. However, such functions contain hierarchies of singularities and are typically nondifferentiable. Thus, the understanding of such phenomena as fractional wave propagation and fractional diffusion comes about through the development and implementation of alternate modeling strategies that do not include differential equations of motion. In recent years there have been a number of attempts to model the phase space evolution of anomalous transport processes, all leading to evolution equations with fractional derivatives [15]. However, just as there are a number of different mechanisms leading to the diffusion anomalies [5, 14], so too, there are a number of different fractional evolution equations. Some of these equations have fractional time derivatives, some have fractional phase-space derivatives, and some have both.

The Physical Properties of Lévy Processes We briefly discussed the fractional diffusion equation that is first-order in time and fractional-order in space in an earlier lecture. We found that the normalizable, positive definite, solution to this fractional diffusion equation is the Lévy distribution. We spend some time here reviewing the properties of Lévy processes, in particular, the renormalization group properties of the distribution function. The characteristic function of the Lévy distribution is shown to be the functional fixed point of a renormalization group relation and consequently much of the physics that we know from phase transitions, for example, in magnetic materials, can be used to physically interpret the solution to the fractional diffusion equation.

Fractional Diffusion Equation and Its Solutions We have seen that stochastic dynamical processes with long-term memory, such as fractional Brownian motion, are described by fractional-diffusion equations (FDEs) that are second-order in space and fractional-order in time. We give an ad hoc derivation of this FDE using the first-passage time distribution function, leading to an equation of the form

$$D_t^\alpha \left[u\left(x, t; x_0\right) \right] - \frac{t^{-\alpha}}{\Gamma\left(1 - \alpha\right)} u_0\left(x; x_0\right) = K_\alpha \frac{\partial^2 u\left(x, t; x_0\right)}{\partial x^2}, \qquad (9.3)$$

where $u\left(x, t; x_0\right)$ is interpreted as the conditional probability density; we have set $\alpha = 2H$ and $K_\alpha = \sigma^2/2$. Equation (9.3) is the fractional partial differential equation constructed by Schneider and Wyss [30] to describe both wave propagation and transport in fractal media. When $0 \leq \alpha \leq 1$ this equation describes fractional diffusion with a solution that is normalizable and can be interpreted as a probability density. When $1 \leq \alpha \leq 2$ our ad hoc derivation of this equation is no longer valid and instead of an evolving probability density the equation describes fractional wave propagation. Therefore this equation describes the full range of dynamics of the ant.

Fractional Eigenvalue Problem We discuss how one may construct an eigenfunction expansion of the solution to (9.3), and its generalization to include a spatial inhomogeneity such as a potential. Metzler et al. [22] discussed the formalism to solve the eigenvalue problem obtained in this way. We discuss their method and show that it gives a Boltzmann distribution as the equilibrium state of the FDE. Furthermore, the FDE constructed in this way gives rise to Einstein's fluctuation-dissipation relation generalized to fractional dynamics. In addition, the solution is obtained for a fractional stochastic oscillator, that is, a fractional dynamical equation driven by a simple harmonic potential and Gaussian delta-correlated fluctuations.

A Fractional Propagation-Transport Equation Finally we examine the solution to the FDE that has both a fractional-time and a fractional-space derivative. The form of the general solution is a generalization of the characteristic function for the Lévy distribution. This distribution has a power-law time dependence and is a direct generalization of fractional Brownian motion. We call this process fractional Lévy motion.

9.1   Lévy Statistics and Renormalization

Renormalization Group Model Applied to Lévy Processes We have repeatedly commented on how the statistics change from Gaussian to Lévy as the underlying process changes from simple to complex. The change in the statistics is associated with the increasing range of the interaction between the elements in the phenomena and this leads us to renormalization group theory. The renormalization group (RG) approach to the modeling of phase transitions was invented to enable us to formalize just such changes in physical systems. The prototype of the RG construction is the heuristic argument made by Kadanoff [21] in the 1960s using a system of N Ising spins on an E-dimensional lattice. The spins lie on the vertices of the lattice and interact with their nearest neighbors and an externally applied magnetic field. The spins are the fundamental elements of the system, and the interactions determine whether the material being modeled will become magnetized, that is, change its phase from an unmagnetized to a magnetized piece of metal. When the temperature of the metal is above the critical temperature, thermal motion jostling the spins ensures that each spin takes on two possible values randomly over space, so there is no net magnetization of the material. As the temperature is lowered to the critical value the lattice begins to take on a patchy appearance, with patches of magnetization of a given orientation in finite spatial domains. We can associate a correlation length ξ with these correlated clusters of spin that increase in size as the critical temperature is approached from above. At the critical temperature the piece of metal changes phase, becoming completely magnetized, so that every vertex of the lattice is tied to every other vertex. The challenge to the physical scientist is to solve the equation of motion for this system, so as to explain this phase transition.

Coarse-Graining the Space of Fluctuations It has not been possible to completely solve the equations of motion for the above Ising spin model except in one Euclidean dimension ($E = 1$), so a method of approximation based on the symmetry of the system was developed. The model lattice is partitioned into cells of side L, where the distance between spins on the original lattice is taken to be unity and $L \gg 1$. Thus, there are N/L^E cells, each of which contains L^E spins. Although L is quite large, the size of the cell is chosen to be much smaller than the correlation length, $\xi \gg L$. In this way there are many cells within each of the patches on the lattice, that is, within a cluster of correlated spins. Now we have the physical picture of a lattice made up of these cells of side L. Each cell is much smaller than the correlated cluster of spins and therefore all the spins in a given cell are oriented in essentially the same direction. In this way the cell takes on a life of its own, with its own spin, and the coarse-grained lattice looks like the original Ising lattice, but in terms of the aggregated spins rather than the original spins.

The Renormalization Group Assumption Here is where the flash of brilliance occurs. Although we cannot solve the equations of motion for either of the two lattices above, the original or the coarse-grained, we impose the restriction that the equations of motion for the spins on the two lattices be identical in form. The condition that the equations of motion for the aggregated spins are of the same form as those for the original spins places a restriction on the spin-spin interaction coefficients, in the two representations, as well as on the interaction of the spins with the externally applied magnetic field. This restriction is, in fact, the renormalization group relation on the generator of the motion for the two systems, the generator of the motion being the Hamiltonian. We encountered such thinking earlier in the context of canonical variables and canonical equations of motion.

The Parameters Retain the Coarse-Grained Information This procedure of coarse-graining the interaction process is repeated over and over again, each time smoothing over the fluctuations at the smaller scale. For example, if the cell size is L with the first application of the RG transformation, then the cell size is of order L^2 with the second application of the transformation and so on. Furthermore, as the explicit dependence on the fluctuations at the smaller scales is lost due to coarse-graining, the effects of these fluctuations are implicitly retained in the parameters of the coarse-grained Hamiltonian. Thus, even though the dynamics of the phenomenon become less erratic, the influences of the small scale fluctuations are not lost, but are parameterized in the scaling properties of the interaction coefficients.

Scaling Replaces Mathematical Rigor The fact that such a renormalization group relation can be constructed for the equation of motion suggested that such relations also exist for the thermodynamic functions describing the system. However, to pursue this argument further would require a level of technical detail regarding a specific physical model that would not be appropriate for the present discussion. Therefore, let us redirect our remarks to the statistical properties of complex physical phenomena and see how we might exploit the RG idea in that context. The hierarchical property of complex phenomena, scales within scales within scales, has been used to describe such things as turbulence, anomalous diffusion, conduction in amorphous semiconductors, fractal wave propagation, and phase transitions. The basis of our understanding of these complex phenomena is scaling and in particular the RG symmetry that is a consequence of coarse-graining. However, before the development of these ideas, Lévy was quite successful in establishing the class of infinitely divisible distributions that concerns processes whose statistical properties persist to all levels of aggregation. We now examine how these two threads of argument are interwoven.

Application of the Arguments from the Central Lmit Theorem Let us apply the coarse-graining technique developed for the study and understanding of phase transitions to statistical processes. Consider the discrete data set $\{X_j\}$ consisting of N identically distributed random variables, $j = 1, 2, \cdots, N$. Without loss of generality we assume that the mean value of the data set is zero; of course, we could always subtract from a nonzero mean to define a new data set having a zero mean. The CLT states that if we add all the members of this data set, as we did for the random walk, with the proper weighting factor,

$$Y(N) = X(N)/N^{1/\alpha}, \tag{9.4}$$

then in the limit $N \to \infty$ the distribution for the variate $Y(N)$ approaches that of Lévy with an index α. To see this we write the characteristic function for the normalized variable

$$\phi_N(k) = \left\langle e^{ikY(N)} \right\rangle, \tag{9.5}$$

where the bracket denotes an average over an ensemble of realizations of the N steps. Now inserting the definition of the normalized sum variable into (9.5) we have

$$\phi_N(k) = \left\langle \exp\left[i\frac{k}{N^{1/\alpha}}\sum_{j=1}^{N}X_j\right]\right\rangle = \prod_{j=1}^{N}\left\langle \exp\left[i\frac{k}{N^{1/\alpha}}X_j\right]\right\rangle, \tag{9.6}$$

which, due to the independence of the data elements, is the product of N single-term characteristic functions

$$\phi_N(k) = \left\langle \exp\left[i\frac{k}{N^{1/\alpha}}X_j\right]\right\rangle^N. \tag{9.7}$$

Here the brackets denote an average over an ensemble of realizations of the jth datapoint, and therefore the average can be replaced by a single-term characteristic function

$$\phi_N(k) = \left[\phi\left(k/N^{1/\alpha}\right)\right]^N. \tag{9.8}$$

If we now insert the characteristic function for the Lévy distribution

$$\phi_L(k) = \exp\left[-b(\alpha)|k|^{\alpha}\right]$$

into (9.8) and obtain

$$\phi_N(k) = \phi_L(k), \tag{9.9}$$

the statistics of the overall sum are exactly the same as the statistics of the individual contributions, the results of the CLT.

Aggregating the Data to Determine Scaling Properties We now coarse-grain the data by aggregating the s nearest neighbors to form N_s new datapoints, each one of which is the sum of s of the original datapoints. This is analogous to partitioning the original spin lattice into cells. It is clear that $N = s[N_s]$ where $[\cdot]$ is the integer closest to the term in the brackets. In this way $[N_s] = N_s$ to a good approximation for N sufficiently large. In order for the statistics of the new data set to be the same as those for the original data set (the imposition of the RG condition of the probability density in analogy with the Hamiltonian of the spin system) we must have in analogy with (9.4),

$$Y(N_s) = \frac{1}{N_s^{1/\alpha}} \sum_{m=1}^{N_s} Y_m(s), \tag{9.10}$$

where each of the $Y_m(s)$ are aggregate variables consisting of a sum of s of the original datapoints $\{X_j\}$ and the index m keeps track of the level of aggregation. In order for the statistics of the original and the aggregate variables to be independent of how we group the data points together we require the RG scaling relation

$$Y(N) = \lambda_s Y(N_s). \tag{9.11}$$

In particular, the scaling parameter λ_s must be chosen such that each $Y_m(s)$ is scaled to ensure that the contribution from each aggregation of s datapoints remains statistically the same as the original data set.

The Lévy Characteristic Function Is a Functional Fixed Point of the Renormalization Group Transformation. If the two representations of the data are to be statistically equivalent then their characteristic functions must be identical

$$\phi_N(k) = \phi_{N_s}(\lambda_s k), \tag{9.12}$$

where, of course, we must now choose the scaling parameter λ_s such that (9.12) is indeed true. Inserting the Lévy characteristic function into both sides of this equation we find the resulting constraint on the scaling parameter

$$N = N_s \lambda_s^{-\alpha}. \tag{9.13}$$

Given that the relation between the original number of datapoints N and the number of aggregate datapoints N_s is the number of datapoints in each aggregated point s, we obtain from (9.13)

$$\lambda_s = \frac{1}{s^{1/\alpha}}. \qquad (9.14)$$

Thus, the process of coarse-graining the data set leaves the Lévy form of the characteristic function invariant when the scaling is properly chosen. The coarse-graining, so constructed, defines a RG transformation for the random variable. Furthermore, if we define a RG transformation T_N that consists of taking the Nth power of the single-point characteristic function $\phi(k)$ and rescaling the Fourier variable with $N^{1/\alpha}$ we obtain

$$T_N\left[\phi(k)\right] = \left[\phi\left(k/N^{1/\alpha}\right)\right]^N. \qquad (9.15)$$

If $\phi(k)$ is the Lévy characteristic function then (9.15) becomes

$$T_N\left[\phi_L(k)\right] = \phi_L(k) \qquad (9.16)$$

so that the Lévy characteristic function is a fixed point of the RG transformation T_N, which is to say, that the application of the renormalization group coarse-graining does not change the functional form of the characteristic function. In terms of the probability density this invariance of the functional form means that the statistical structure of the data is the same at every scale independently of the coarse-graining and therefore the process is fractal.

The Domain of Attraction Is Lévy　The consequences of (9.16) are that the collective behavior of our one-dimensional time series is described by a distribution belonging to the class of stable distributions which is parameterized by the exponent α. The possible initial distributions of the time series separate into universality classes according to which domain of attraction they belong as specified by the index α. In other words, the domain of attraction of a given data set defines the universality class to which the data belong. The fact that the Lévy stable distribution is a fixed point of such a RG transformation seems to have been first noted by Jano-Lasino in 1975 [17]. A somewhat related analysis of the relation between scaling and Lévy stable distributions is given by Hilfer [16] and also by West [35].

Clusters Within Clusters Within Clusters　The α-stable Lévy dynamical process is one in which there are clusters of activity interspersed with quiescent intervals. Such a statistical process can also be observed as patches of fluctuations in space, distinct from intermittent fluctuations in time, but each is a kind of clustering of the activity. However, if one of the clusters is examined more closely, it is seen to be made up of smaller clusters separated by quiescent intervals, and so on, and just as in our discussion of the dust ball, we have clusters, within clusters, within clusters. Scaling, self-affinity, and fractal statistics are all present in Lévy dynamical processes. What we need to understand is how dependent this behavior is on the detailed dynamics of a phenomenon. In other words, how universal is the RG behavior that leads to Lévy statistics?

Universality In physics universality is concerned with how phenomena as microscopically different as a fluid and a magnetic solid, can display identical macroscopic behavior; that is, their phase transition behavior is the same. Coarse-graining the two systems leads to the same type of RG description for the two phenomena, such that they are both described by the same critical exponent. This is what universality has come to mean in physics, that the critical exponents in many thermodynamic variables and differing materials are the same. These thermodynamic indices are analogous to the Lévy indices necessary for the Lévy characteristic function to be a fixed point of the RG transformation. Thus, two phenomena as different as anomalous diffusion in a liquid and the beating of the human heart may share the same statistical properties, because they both have the same Lévy index.

9.2 An ad hoc Derivation

Many Derivations of the Fractional Propagation-Transport Equation A number of methods have been used to construct the propagation-transport equation (9.3) and its generalizations, each one emphasizing a different aspect of the general situation. Some derivations have been based on the non-Markov character of the system dynamics, as we discussed earlier and concern the fractal time nature of the equation. Others have been based on the generalized Taylor expansion of Osler [25], and developed in the present context by West and Grigolini [37] and by Zaslavsky et al. [43] to incorporate the fractional spatial derivatives into the general equation. Here we present an entirely different derivation of (9.3) based on the first passage time distribution.

First Passage Time Distribution for a Stationary Process To facilitate the analysis we assume a stationary process whose two-point probability density is given by

$$P(x_1, t_1 | x_2, t_2) = P(x_1 - x_2, t_1 - t_2).$$ (9.17)

Following the arguments of Siegert [32] we construct the convolution equation

$$P(x_1 - x_2, t) = \int_0^t P(x_1 - z, t - \tau) Q(z - x_2, \tau) d\tau,$$ (9.18)

where $Q(z, \tau) d\tau$ is the probability that the walker has made the transition from $0 \to z$ in the time τ for the first time. Thus, $Q(z, \tau)$ is the first passage time probability density. Using the convolution form of (9.18) we can take the Laplace transform and write

$$\tilde{P}(x_1 - x_2, s) = \tilde{P}(x_1 - z, s)\, \tilde{Q}(z - x_2, s) \tag{9.19}$$

so that, in general, the first passage time density is given by the inverse Laplace transform of the ratio of the Laplace transforms of two transition probabilities

$$\tilde{Q}(z - x_2, s) = \frac{\tilde{P}(x_1 - x_2, s)}{\tilde{P}(x_1 - z, s)}. \tag{9.20}$$

First Passage Time Distribution for an Einstein Process　Consider the simple case of a probability density for an Einstein process, that is, a Gaussian distribution, whose second moment increases linearly in time

$$P(x, t) = \frac{\exp\left[-\frac{x^2}{4Dt}\right]}{\sqrt{4\pi Dt}}. \tag{9.21}$$

In terms of a characteristic function (9.21) can be written as the inverse Fourier transform

$$P(x, t) = \int_{-\infty}^{\infty} \frac{dk}{2\pi} e^{-ikx} e^{-Dtk^2}. \tag{9.22}$$

Using (9.22) we can express the Laplace transform of $P(x, t)$ as

$$\tilde{P}(x, s) = \int_{-\infty}^{\infty} \frac{dk}{2\pi} \frac{e^{-ikx}}{s + Dk^2} = \frac{1}{2\sqrt{sD}} \exp\left[-|x|\sqrt{s/D}\right]. \tag{9.23}$$

The Laplace transform of the probability density (9.23) can be inserted into (9.20), so that if $x_1 > x > x_2$ we have

$$\tilde{Q}(x - x_2, s) = \frac{\tilde{P}(x_1 - x_2, s)}{\tilde{P}(x_1 - x, s)} = \exp\left[-(x - x_2)\sqrt{s/D}\right]. \tag{9.24}$$

The inverse Laplace transform of $\tilde{Q}$ then yields

$$Q(x - x_2, t) = \mathcal{LT}^{-1}\left\{\exp\left[-(x - x_2)\sqrt{s/D}\right]; t\right\}$$
$$= \frac{|x - x_2|}{\sqrt{4\pi Dt^3}} \exp\left[-\frac{(x - x_2)^2}{4Dt}\right]. \tag{9.25}$$

Thus, asymptotically in time, the first passage time density for an Einstein process becomes

$$Q(x - x_2, t) \propto \frac{1}{t^{3/2}}, \tag{9.26}$$

an inverse power law in time [23].

The Fractional Propagation-Transport Equation from the First Passage Time Distribution Let us generalize these results by making use of the chain condition for the first passage time density

$$Q(x_1 - x_2, t) = \int_0^t Q(x_1 - z, t - \tau) \times Q(z - x_2, \tau)\, d\tau. \tag{9.27}$$

From the Laplace transform of (9.27) we obtain

$$\tilde{Q}(x_1 - x_2, s) = \tilde{Q}(x_1 - z, s)\tilde{Q}(z - x_2, s) \tag{9.28}$$

so that (9.28) has the same form as that for the characteristic functions in (6.154), but with space and time interchanged. The general solution to (9.28) is given by

$$\tilde{Q}(y, s) = \exp\left[-by|s|^\lambda\right], \quad 0 < \lambda \le 1/2, \tag{9.29}$$

where $\lambda = 1/2$ is the limiting Gaussian case considered above. This is also of the form of the one-sided Lévy distribution given in (6.68). Using the definition of $\tilde{Q}$ in terms of the ratio of $\tilde{P}$s in (9.24) we can write

$$\tilde{P}(x_1 - x_2, s) = \tilde{P}(x_1 - x, s)\,\exp\left[-b(x - x_2)|s|^\lambda\right]. \tag{9.30}$$

Taking the second spatial derivative of $\tilde{P}$ in (9.30) yields

$$\frac{\partial^2 \tilde{P}(x_1 - x_2, s)}{\partial x_2^2} = s^{2\lambda}\,b^2\tilde{P}(x_1 - x, s)\,\exp\left[-b(x - x_2)|s|^\lambda\right] \tag{9.31}$$

so that rewriting the constant $b^2 = K_\lambda^{-1}$, setting $x = x_2$, and inverse Laplace transforming (9.31) yields

$$P(y, t) = P_0(y) + \frac{K_\lambda}{\Gamma(2\lambda)}\int_0^t \frac{dt'}{(t - t')^{1-2\lambda}}\frac{\partial^2 P(y, t')}{\partial y^2}, \tag{9.32}$$

where $P_0(y) = P(y, t = 0)$. But we know that the integral is just the fractional integral in time, so that taking the fractional time derivative from the left of the equation gives us

$$D_t^{2\lambda}[P(y, t)] - \frac{t^{-2\lambda}}{\Gamma(1 - 2\lambda)}P_0(y) = K_\lambda\frac{\partial^2 P(y, t)}{\partial y^2} \tag{9.33}$$

and identifying 2λ with α we obtain the fractional differential equation given by (9.3).

9.3   Fractional Eigenvalue Equations

Another Form of the Fractional Propagation-Transport Equation   The
derivation of the FDE (9.33) imposes certain restrictions on the index λ to retain
the interpretation of $P(y,t)$ as a probability density. Now that we have the equation, however, we observe that it has the same form as the fractional-wave equation of Schneider and Wyss [30] if the index is in the interval $0 < 2\lambda \leq 2$. Another possible generalization of Equation (9.33) can be made using the Fokker-
Planck equation discussed in earlier lectures. In the latter case the second-
derivative term is supplemented by a term dependent on an external potential
and (9.33) is replaced with

$$D_t^\alpha \left[P(x,t) \right] - \frac{t^{-\alpha}}{\Gamma(1-\alpha)} P_0(x) = \left[\frac{\partial}{\partial x} \frac{V'(x)}{m\gamma_\alpha} + K_\alpha \frac{\partial^2}{\partial x^2} \right] P(x,t), \qquad (9.34)$$

where $V(x)$ is an external potential, γ_α is a friction coefficient with dimensions
$\sec^{\alpha-2}$, K_α is the generalized diffusion coefficient, and we have suppressed the
initial position x_0 for notational convenience.

Formal Solution in Terms of a Spatial Operator   Introducing the spatial
derivative operator

$$\mathcal{L} \equiv \frac{\partial}{\partial x} \frac{V'(x)}{m\gamma_\alpha} + K_\alpha \frac{\partial^2}{\partial x^2} \qquad (9.35)$$

into (9.34) we can write

$$D_t^\alpha \left[P(x,t) \right] - \frac{t^{-\alpha}}{\Gamma(1-\alpha)} P_0(x) = \mathcal{L} P(x,t) \qquad (9.36)$$

which is a formal equation we have solved more than once before. The difference
between (9.36) and the equations solved earlier is that rather than having a
dissipation parameter on the right-hand side we now have an operator. In any
case we can, following Metzler et al. [22], construct the formal solution to (9.36)
assuming the natural boundary conditions

$$\lim_{x \to \pm\infty} P(x,t) = 0 \qquad (9.37)$$

so that we obtain in terms of the Mittag-Leffler function

$$P(x,t) = E_\alpha \left(t^\alpha \mathcal{L} \right) P_0(x). \qquad (9.38)$$

However, the solution (9.38) is in terms of the operator given by (9.35), which
for $\alpha = 1$ reduces to the familiar form

$$P(x,t) = e^{t\mathcal{L}} P_0(x), \qquad (9.39)$$

where, of course, $e^{t\mathcal{L}}$ is the evolution operator. For example, in mechanics $\mathcal{L}$ would be the Liouville operator in terms of Poisson brackets and the Hamiltonian

$$\mathcal{L}P = \{H, P\},$$

and in quantum mechanics it would be the commutator with the Hamiltonian operator

$$\mathcal{L}\psi = \frac{i}{\hbar}[\mathcal{H}, \psi].$$

In just the same way the Mittag-Leffler function generalizes the concept of the evolution operator to fractional time, a result also obtained by Hilfer [16].

Solution to the Boundary Value Problem Using Eigenfunction Expansions One form of the solution to a boundary value problem is given by the series method. This technique involves separating the partial differential equation in terms of the independent variables such that the boundary corresponds to one or more of the coordinate surfaces, $\xi = constant$, or in the case of one spatial dimension, the value of the function at a point. The separation constant is adjusted such that the solution takes on a specified value along this surface or at this point. The analytic solution to (9.36) was obtained by Metzler et al. [22] using separation of variables

$$P(x, t) = \sum_n A_n \phi_{n,\alpha}(x) T_{n,\alpha}(t), \tag{9.40}$$

where the index n refers to a given eigenvalue of the operator $\mathcal{L}$, that is, the separation constant. The complete solution is then obtained by summing over the spectrum of eigenvalues that solve (9.36) subject to the boundary conditions (3.74). Introducing (9.40) into (9.36) yields, after some simplification,

$$\frac{1}{T_{n,\alpha}(t)} D_t^\alpha [T_{n,\alpha}(t)] - \frac{t^{-\alpha}}{\Gamma(1-\alpha)} = \frac{1}{\phi_{n,\alpha}(x)} \mathcal{L}\phi_{n,\alpha}(x) = -\lambda_{n,\alpha}, \tag{9.41}$$

where $\lambda_{n,\alpha}$ is a constant independent of x and t. Thus, (9.41) reduces to the pair of eigenvalue equations

$$D_t^\alpha [T_{n,\alpha}(t)] - \frac{t^{-\alpha}}{\Gamma(1-\alpha)} T_{n,\alpha}(t) = -\lambda_{n,\alpha} T_{n,\alpha}(t), \tag{9.42}$$

and

$$\mathcal{L}\phi_{n,\alpha}(x) = -\lambda_{n,\alpha}\phi_{n,\alpha}(x). \tag{9.43}$$

It is now clear, based on our previous analysis, that (9.42) has the solution

$$T_{n,\alpha}(t) = E_\alpha(-t^\alpha \lambda_{n,\alpha}) \tag{9.44}$$

in terms of the monotonically decreasing Mittag-Leffler function. The complete solution is then obtained using (9.40),

$$P_\alpha\left(x,t\,|x_0\right) = \sum_{n=0}^{\infty} A_n \phi_{n,\alpha}\left(x\right) E_\alpha\left(-t^\alpha \lambda_{n,\alpha}\right) \qquad (9.45)$$

and the coefficients are determined by the initial condition, just as for the Fourier series expansion.

Evaluating the Coefficients in the Series Expansion of the Solution
At time $t = 0$, using the fact that the Mittag-Leffler function is unity for zero argument, we have

$$P_\alpha\left(x,t=0\,|x_0\right) = \sum_{n=0}^{\infty} A_n \phi_{n,\alpha}\left(x\right) \qquad (9.46)$$

so that if $\left\{\phi_{n,\alpha}\left(x\right)\right\}$ forms an orthonormal set of functions, the integral of the product over the domain of the solution yields

$$\int_\Omega \phi_{n,\alpha}\left(x\right) \phi_{m,\alpha}\left(x\right) dx = \delta_{m,n}; \qquad (9.47)$$

then multiplying (9.45) by an eigenfunction and integrating over the domain of the solution we obtain

$$A_n = \int_\Omega \phi_{n,\alpha}\left(x\right) P_\alpha\left(x,t=0\,|x_0\right) dx. \qquad (9.48)$$

Thus, if all the initial weight of the function is concentrated at x_0, that is,

$$P_\alpha\left(x,t=0\,|x_0\right) = \delta\left(x-x_0\right),$$

the solution (9.45 becomes

$$P_\alpha\left(x,t\,|x_0\right) = \sum_{n=0}^{\infty} \phi_{n,\alpha}\left(x\right) \phi_{n,\alpha}\left(x_0\right) E_\alpha\left(-t^\alpha \lambda_{n,\alpha}\right). \qquad (9.49)$$

The form of (9.49) should be familiar in the case $\alpha = 1$ where the Mittag-Leffler function becomes an exponential and the solution is the exponentially weighted sum over eigenfunctions

$$P_\alpha\left(x,t\,|x_0\right) = \sum_{n=0}^{\infty} \phi_{n,\alpha}\left(x\right) \phi_{n,\alpha}\left(x_0\right) \exp\left[-t\lambda_{n,\alpha}\right].$$

These Solutions Are Consistent with Equilibrium Statistical Mechanics   We assume that a steady-state (equilibrium) solution to these equations of motion exist and that the eigenvalues are ordered such that

$$0 \le \lambda_{0,\alpha} \le \lambda_{1,\alpha} \le \cdots. \qquad (9.50)$$

The lowest eigenvalue $\lambda_{0,\alpha} = 0$ implies

$$\frac{\partial}{\partial x}\left[\frac{V'(x)}{m\gamma_\alpha} + K_\alpha\frac{\partial}{\partial x}\right]\phi_{0,\alpha}(x) = 0 \tag{9.51}$$

and for the zero-flux equation of motion

$$\left[\frac{V'(x)}{m\gamma_\alpha} + K_\alpha\frac{\partial}{\partial x}\right]\phi_{0,\alpha}(x) = 0. \tag{9.52}$$

The solution to (9.52) is given by

$$\phi_{0,\alpha}(x) = Z^{-1}\exp\left[-\frac{V(x)}{m\gamma_\alpha K_\alpha}\right], \tag{9.53}$$

where Z is the normalization constant. If the equilibrium solution to the system dynamics is given by the Boltzmann distribution

$$\phi_{0,\alpha}(x) \propto \exp\left[-\frac{V(x)}{k_B T}\right],$$

we have the effective diffusion coefficient being given by

$$K_\alpha = \frac{k_B T}{2m\gamma_\alpha} \tag{9.54}$$

a generalization of the fluctuation-dissipation relation of Einstein to fractional dynamics [22]. Thus, the fractional stochastic analysis given by the fractional diffusion equation in this form is consistent with ordinary statistical mechanics.

9.4 Fractional Stochastic Oscillator

Linear Oscillators The linear harmonic oscillator is the backbone of physics, beginning with Newton's estimate of the speed of sound in air using coupled linear oscillators to model a column of air, right up to all the -*ons* in quantum mechanics. The boson, the photon, the phonons, and so on, are linear quantum oscillators modeling microscopic phenomena. Even our statistical mechanical *derivation* of the Langevin equation began with a *heat bath* of harmonic oscillators. Without reiterating the problems associated with that model, we recall that this modeling approach worked better in representing the reversible process of linear wave propagation than it did in modeling irreversible processes.

Fractional Fokker-Planck Equation for the Linear Oscillator We now investigate the transport-propagation equation description of the dynamics of a system described by a harmonic potential

$$V(x) = m\omega^2 x^2/2$$

and driven by random fluctuations. Thus, the equation of motion for the fractional, stochastic, harmonic oscillator is given by

$$\frac{\partial}{\partial x}\left[\frac{\omega^2 x}{\gamma_\alpha}+K_\alpha\frac{\partial}{\partial x}\right]P_\alpha\left(x,t\,|x_0\right)=D_t^\alpha\left[P_\alpha\left(x,t\,|x_0\right)\right]-\frac{t^{-\alpha}}{\Gamma\left(1-\alpha\right)}P_0\left(x\right)$$

$$(9.55)$$

so that again using the separation of variables and assuming the eigenvalue separation of equations we have

$$\frac{\partial}{\partial x}\left[\frac{\omega^2 x}{\gamma_\alpha}+K_\alpha\frac{\partial}{\partial x}\right]\phi_{n,\alpha}\left(x\right)=-\lambda_{n,\alpha}\phi_{n,\alpha}\left(x\right) \qquad (9.56)$$

and the time equation is given by (9.42).

Eigenfunction Expansion of the Solution   We introduce the temperature through (9.54) to obtain the dynamical equation in terms of the scaled variables $y=x\sqrt{m\omega^2/kT}$:

$$\frac{\partial}{\partial y}\left[y+\frac{\partial}{\partial y}\right]\phi_{n,\alpha}\left(y\right)=-\tilde{\lambda}_{n,\alpha}\phi_{n,\alpha}\left(y\right), \qquad (9.57)$$

where the scaled eigenvalue is

$$\tilde{\lambda}_{n,\alpha}=\frac{\gamma_\alpha}{\omega^2}\lambda_{n,\alpha} \qquad (9.58)$$

and the fluctuation strength is given by (9.54). The analysis of (9.57) has been given by Uhlenbeck and Ornstein [33], which after some rearrangement yields the solution

$$P_\alpha\left(y,t\,|y_0\right) = \sqrt{\frac{m\omega^2}{kT}}\sum_{n=0}^{\infty}\frac{H_n\left(y/\sqrt{2}\right)H_n\left(y_0/\sqrt{2}\right)}{2^n\sqrt{2\pi}\Gamma\left(n+1\right)}$$

$$\times \exp\left[-y_0^2/2\right]E_\alpha\left(-t^\alpha\tilde{\lambda}_{n,\alpha}\right) \qquad (9.59)$$

where $H_n\left(y\right)$ is the Hermite polynomial

$$H_n\left(y\right)\equiv\left(-1\right)^n e^{y^2}\frac{d^n}{dy^n}\left[e^{-y^2}\right] \qquad (9.60)$$

and the numerical coefficients in (9.59) are required for normalization.

From Weighted Mittag-Leffler to Weighted Exponential   We see that when the fractional derivative index $\alpha=1$, the Mittag-Leffler function in (9.59) reduces to the exponential in which case the solution to the harmonic oscillator equation becomes

$$P_1(y,t|y_0) = \sqrt{\frac{m\omega^2}{kT}} \sum_{n=0}^{\infty} \frac{H_n(y/\sqrt{2}) H_n(y_0/\sqrt{2})}{2^n \sqrt{2\pi} \Gamma(n+1)}$$
$$\times \exp\left[-y_0^2/2\right] \exp\left(-t\lambda_{n,1}\right) \tag{9.61}$$

and we find that the standard eigenvalues to the harmonic oscillator equation are given by

$$\lambda_{n,1} = \lim_{\alpha \to 1} \lambda_{n,\alpha} \tag{9.62}$$

where $\lambda_{n,1} = n\lambda$; $\lambda = \omega^2/\gamma$, and γ is the linear dissipation parameter.

The Closed Form Solution Although (9.61) is the formally correct series solution to the phase space equation of motion for the linearly damped harmonic oscillator driven by white noise, it may not be in the form with which the reader is familiar. Uhlenbeck and Ornstein used the properties of Hermite polynomials to show that the series in (9.61) could be explicitly summed to yield

$$P_1(x,t|x_0) = \sqrt{\frac{m}{2\pi kT(1-e^{-2\lambda t})}} \exp\left[-\frac{m}{2kT} \frac{(x - x_0 e^{-\lambda t})^2}{(1 - e^{-2\lambda t})}\right]. \tag{9.63}$$

This solution yields for the average displacement

$$\langle x;t \rangle = x_0 e^{-\omega^2 t/\gamma} \tag{9.64}$$

along with the second moment

$$\langle x^2;t \rangle = \frac{kT}{m\omega^2} + \left(x_0^2 - \frac{kT}{m\omega^2}\right) e^{-2\omega^2 t/\gamma}. \tag{9.65}$$

For very small frequencies, essentially the case of a free particle, these two equations reduce to

$$\langle x;t \rangle = x_0 \tag{9.66}$$

and

$$\langle x^2;t \rangle = x_0^2 + \frac{kT}{m\gamma} t \tag{9.67}$$

as they should. These are essentially the results for the overdamped case where the dissipation is very large and the solution is in the time domain $t \gg 1/\gamma$.

Fractional Transport Equation for the Average Value We can also determine the evolution of the average fractional displacement by multiplying (9.55) by x and integrating over the range of the variate. Thus, using $P_0(x) = \delta(x - x_0)$ we obtain

$$D_t^\alpha \left[\langle x; t \rangle \right] - x_0 \frac{t^{-\alpha}}{\Gamma(1-\alpha)} = -\frac{\omega^2}{\gamma_\alpha} \langle x; t \rangle \tag{9.68}$$

where the term on the right-hand side of the equation is obtained from an integration by parts. The solution to (9.68) is, of course, given in terms of the Mittag-Leffler function

$$\langle x; t \rangle = x_0 E_\alpha \left(-\frac{\omega^2}{\gamma_\alpha} t^\alpha \right) \tag{9.69}$$

which is obtained using Laplace transforms, as we did in earlier lectures.

Solution of the Fractional Transport Equation for the Second Moment In the same way, multiplying (9.55) by x^2 and integrating yields an equation for the second moment that can be solved by again using Laplace transforms. The integral obtained in this way is analogous to (8.95) with $v(t)$ replaced by $\langle x^2; t \rangle$ and the random force $f(t)$ replaced with the constant $2K_\alpha$:

$$\langle x^2; t \rangle = \langle x^2; 0 \rangle E_\alpha \left(-\frac{\omega^2}{\gamma_\alpha} t^\alpha \right) + 2K_\alpha \int_0^t \tau^{\alpha-1} E_{\alpha,\alpha} \left(-\frac{\omega^2}{\gamma_\alpha} \tau^\alpha \right) d\tau. \tag{9.70}$$

Inserting the series expansion for the generalized Mittag-Leffler function under the integral allows us to integrate term-by-term to obtain

$$\int_0^t \tau^{\alpha-1} E_{\alpha,\alpha} \left(-\frac{\omega^2}{\gamma_\alpha} \tau^\alpha \right) d\tau = \sum_{k=0}^\infty \frac{\left(-\frac{\omega^2}{\gamma_\alpha} \right)^k}{\Gamma(k\alpha+\alpha)} \frac{t^{(k+1)\alpha}}{k\alpha+\alpha}$$

$$= -\frac{\gamma_\alpha}{\omega^2} \sum_{j=1}^\infty \frac{\left(-\frac{\omega^2}{\gamma_\alpha} \right)^j}{\Gamma(j\alpha+1)} t^{j\alpha}.$$

Thus, we have for the second moment of the displacement

$$\langle x^2; t \rangle = \left[\langle x^2; 0 \rangle - \frac{kT}{m\omega^2} \right] E_\alpha \left(-\frac{\omega^2}{\gamma_\alpha} t^\alpha \right) + \frac{kT}{m\omega^2}, \tag{9.71}$$

where, as we noted above, $kT/m\omega^2$ is the thermal equilibrium value of the mean-square displacement of the linear oscillator.

Anomalous Diffusion at Early Times and Really Anomalous at Late Times   The difference between the solution to the fractional oscillator equations and the standard oscillator equation is clearly the rate at which the higher-order terms in the solution decay. Equation (9.63) shows that the decay of the Uhlenbeck-Ornstein oscillator is dominated by the exponential decay at a rate given by the lowest eigenvalue. On the other hand, the fractional oscillator solution decays as an inverse power law towards the asymptotic thermal equilibrium value. Also, at very short times, the Mittag-Leffler function can be expanded to give

$$E_\alpha \left(-\frac{\omega^2}{\gamma_\alpha} t^\alpha \right) \sim 1 - 2 \frac{\omega^2}{\gamma_\alpha \Gamma(\alpha+1)} t^\alpha$$

so that (9.71) can be written

$$\langle x^2; t \rangle - \langle x^2; 0 \rangle \sim - \left[\langle x^2; 0 \rangle - \frac{kT}{m\omega^2} \right] \frac{2\omega^2}{\gamma_\alpha \Gamma(\alpha+1)} t^\alpha, \qquad (9.72)$$

a typical anomalous diffusion result [22]. The mean-square displacement of the fractional stochastic oscillator at early times evolves as does anomalous diffusion. But it certainly does not behave this way asymptotically.

A Speculation   On another technical point, it might be possible to sum the series in (9.59) using the properties of Hermite polynomials, as Uhlenbeck and Ornstein did for the simple harmonic oscillator, to obtain the closed-form solution for the probability density, but we have not succeeded in doing this as yet for the fractional case.

9.5   Fractional Propagation-Transport Equation

Fractional Derivative in Space   The first successful generalization of a random walk model to anomalous diffusion used a Langevin equation describing a dynamical process $X(t)$ driven by fluctuations $\xi(t)$ with Lévy statistics, that is, [2]

$$dX(t)/dt = \xi(t). \qquad (9.73)$$

The probability density for this process $P(x,t)$ was then described by an equation of evolution that is first-order in time, but whose phase space derivative is fractional [34]:

$$R^\alpha [P(x,t)] = \frac{b}{\pi} \Gamma(\alpha+1) \sin[\pi\alpha/2] \int_{-\infty}^{\infty} \frac{P(x',t)}{|x-x'|^{\alpha+1}} dx',$$

$$(9.74)$$

---

[2] This lecture is taken from West and Nonnenmacher [41] and is suitably modified.

where $R^\alpha [\cdot]$ is a Riesz fractional derivative with $0 < \alpha \leq 2$ [29]. The equation of evolution that replaces the Fokker-Planck equation is

$$\frac{\partial P(x,t)}{\partial t} = R^\alpha [P(x,t)] \qquad (9.75)$$

the solution to which is the Lévy probability density, as we found in an earlier lecture.

Fractional Derivatives in Time A second process in which the fluctuations $\xi(t)$ are not Lévy, but rather are Gaussian with an inverse power-law correlation function

$$\langle \xi(t)\,\xi(t+\tau) \rangle \propto 1/\tau^{1+\beta}, \qquad (9.76)$$

with $0 < \beta \leq 1$, yields an equation of evolution for the probability density that is second-order in the phase space variable x, but is of fractional order β in time:

$$D_t^\beta [P(x,t)] = \frac{1}{\Gamma(-\beta)} \int_0^t \frac{P(x,t')}{|t-t'|^{1+\beta}} dt', \qquad (9.77)$$

where D_t^β is the Riemann-Liouville fractional derivative. The fractional diffusion equation with this fractional time derivative was obtained using a continuous time random walk formalism [7] and independently using a stochastic two-state process with memory [2]. In this case the equation that replaces the Fokker-Planck equation for a freely diffusing particle is

$$D_t^\beta [P(x,t)] = D\frac{\partial^2 P(x,t)}{\partial x^2}, \qquad (9.78)$$

where D is the diffusion coefficient.

Fractional Derivatives in Both Space and Time We include both the long-time memory effects manifest in the fractional time derivative (9.77) and the long-range spatial effects manifest in the fractional phase-space derivative (9.74) in the same fractional evolution equation using the strategy developed for solving the initial value problem developed earlier,

$$u(x,t) - u_0(x) = D_t^{-\beta} [R^\alpha [u(x,t)]], \qquad (9.79)$$

where the initial function is given by $u_0(x) = u(x,t=0)$. Thus, the equation of evolution for the ant in the gurge is obtained by operating on (9.79) with the β-fractional derivative to obtain

$$D_t^\beta [u(x,t)] - \frac{t^{-\beta} u_0(x)}{\Gamma(1-\beta)} = R^\alpha [u(x,t)], \qquad (9.80)$$

where we have used the fact that $u_0(x)$ is a constant function with respect to the time fractional derivative. Equation (9.80) not only describes fractional diffusion when $0 < \beta \leq 1$, but also describes wave propagation in a non-dissipative fractal medium when $1 < \beta \leq 2$. In addition, when $\alpha = 0$ there are no spatial effects in the dynamics, so that the resulting equation is the same as the one to describe shear relaxation in viscoelastic materials [9] through [12].

General Solution to the Propagation-Transport Equation Equation (9.80) contains the effects of long-range steps taken by the ant that are correlated in time. The α-index determines the spatial range and the β-index the time range of the interdependence of the steps in the underlying process. Of course (9.80) would be of little value if it could not be solved. The method of solution involves taking the Laplace-Mellin transform of (9.80) and in the inversion process writing $u(x, t)$ in terms of Fox H-functions [11], as we did earlier. In this way the normalized solution to the ant's FDE can be expressed in terms of the Fourier transform of the solution to be

$$\phi_{\beta,\alpha}(k,t) = \sum_{n=0}^{\infty} \frac{(-1)^n}{\Gamma(1+\beta n)} \left(bt\,|k|^{\alpha/\beta}\right)^{n\beta} = E_\beta\left(-(bt)^\beta\,|k|^\alpha\right), \qquad (9.81)$$

where $E_\beta\left(-(bt)^\beta\,|k|^\alpha\right)$ is the Mittag-Leffler function [11]. The inverse Fourier transform of (9.81) yields the general asymptotic solution to (9.80) given by

$$\begin{aligned}
u_{\beta,\alpha}(x,t) &= \mathcal{FT}^{-1}\left[\phi_{\beta,\alpha}(k,t); x\right] \\
&= \sum_{n=1}^{\infty} \frac{(-1)^{n+1}\,\Gamma(1+n\alpha)\,\sin[n\alpha\pi/2]}{\Gamma(1+n\beta)}\frac{(bt)^{n\beta}}{|x|^{n\alpha+1}},
\end{aligned} \qquad (9.82)$$

when $u_0(x) = \delta(x)$.

Lévy α-Stable Processes Notice that in the $\beta \to 1$ limit that the Mittag-Leffler function sums to an exponential, so that the Fourier transform in (9.81) becomes

$$\phi_{1,\alpha}(k,t) = \exp\left[-bt\,|k|^\alpha\right] \qquad (9.83)$$

which is the characteristic function for the symmetric α-stable Lévy distribution. Another property of the Mittag-Leffler function that we discussed earlier is that for small values of its argument it gives rise to the stretched exponential

$$\phi_{\beta,\alpha}(k,t) \approx e^{-(bt)^\beta\,|k|^\alpha}. \qquad (9.84)$$

If $\alpha = 2$, the inverse Fourier transform of (9.84) is a Gaussian distribution with a variance that increases as a power law in time, that is,

$$\left\langle X(t)^2 \right\rangle \propto t^\beta,$$

corresponding to fractional Brownian motion. Thus, the inverse Fourier transform of (9.84) with $\alpha \neq 2$ corresponds to *fractional Lévy motion*, that is, a generalization of the Lévy process to include other than linear time dependence.

Same Asymptotic Properties in Space   The series (9.82) reduces to the asymptotic expansion for the Lévy stable distribution [23, 40], $P_\alpha(x,t) \equiv u_{1,\alpha}(x,t)$, both having the well-known, lowest-order, inverse power-law behavior

$$u_{\beta,\alpha}(x,t) \propto \frac{(bt)^\beta}{|x|^{1+\alpha}} \tag{9.85}$$

which was considered the signature of Lévy processes. However, we now see that a solution (9.82), more general than the Lévy form, shares this asymptotic property.

Wave Propagation Through a Fractal Medium   Because of the self-similar character of fractal media, the wavelength of waves propagating in such media always correspond to some hierarchical structure of the medium. Berry [4] dubbed such waves *diffractals*, a new wave regime characterized by a short-wave limit in which ever finer levels of structure are explored and geometrical optics is never applicable. The physical arguments of Berry regarding the influence of fractal media on scalar waves were replaced with more formal arguments by Schneider and Wyss [30]. The latter investigators interpreted (9.79) as a fractional diffusion equation when $0 < \beta \leq 1$ and as a fractional wave equation when $1 < \beta \leq 2$. In the limit $\alpha \to 2$ the general solution to the fractional evolution equation, whose Fourier transform is given by (9.81), agrees with the Schneider and Wyss solution, their (2.33) with the initial value function indexed with $k = 0$. This is an exact solution to the propagation of a scalar wave through a fractal medium and is therefore an exact diffractal in Berry's sense.

Properties of the Moments of the Dynamic Variable   The physical observables of the dynamical space-time process described by (9.80) are given by the noninteger moments $\left\langle X(t)^\vartheta \right\rangle_{\beta,\alpha}$. Using the scaling properties of the general solution (9.82) we obtain

$$\left\langle X(t)^\vartheta \right\rangle_{\beta,\alpha} = \int_{-\infty}^{\infty} u_{\beta,\alpha}(x,t) x^\vartheta \, dx = t^{\vartheta\beta/\alpha} \int_{-\infty}^{\infty} u_{\beta,\alpha}(x',1) x'^\vartheta \, dx' \tag{9.86}$$

so that the moments are finite if the exponents satisfy the constraint $\vartheta < \alpha \leq 2$, where the dimensionless integral on the second line of (9.86) is finite. We can see from (9.86) that when $\alpha = \vartheta = 2$ we obtain the result for fractional Brownian motion as we did above, otherwise we have the more general scaling result for fractional Lévy motion. Using (9.86) the norm of $u_{\beta,\alpha}(x,t)$ can be shown to exist by taking the limit $\vartheta \to 0$.

$$\lim_{\vartheta \to 0} \left\langle X(t)^{\vartheta} \right\rangle_{\beta,\alpha} = \int_{-\infty}^{\infty} dx' \int_{-\infty}^{\infty} e^{-ik'x'} E_{\beta} \left(-|k'|^{\alpha} \right) dk' = 1, \qquad (9.87)$$

since the integration over the dimensionless variable x' yields a Dirac delta function in k', whose integration then gives the Mittag-Leffler function with a zero argument and $E_{\beta}(0) = 1$. Thus, the function $u_{\beta,\alpha}(x,t)$ can be interpreted as a probability density for the appropriate values of the parameters.

9.6 Commentary

The Renormalization Group Concept In these last few lectures we have reviewed some history of the renormalization group concept. In particular, when these ideas are applied to statistical fluctuations, the distribution is found to be that of Lévy. The result is that the α-stable Lévy distribution is a functional fixed point of the renormalization transformation operator. Lévy statistics have been used to explain the characteristics of anomalous diffusive processes.

Long-Range Coupling and the Renormalization Group An example of anomalous diffusion is given by the passage of tracer particles along a line of convective rolls aligned in space to form a one-dimensional lattice. The rolls act to trap the tracer particles for a waiting time on any one roll that is an inverse power law and consequently the distribution of an ensemble of such particles along the entire lattice after a long time is Lévy. The tying together of time scales in anomalous diffusion is therefore a consequence of the direct physical coupling of the dynamical fluid and the diffusing particles. On the other hand, the statistics of the intervals between beats in the human heart are found to be Lévy [26], but without the direct physical coupling. The cardiovascular system is made up of a large number of complex subsystems that control the beating of the heart through a network of feedforward and feedback interactions. Each of these subsystems is responsive to a different set of frequencies in the cardiac output. That is to say, the scaling in the heart rate variability ties together the frequencies in the feedback in such a way that the responses of the individual subsystems become coordinated through the RG relation. The coordination in the frequencies of the output is manifest in the coordination of the responses of the subsystems. This has been called an allometric control system [38, 40, 42].

Long-Term Memory The ad hoc derivation of the fractional transport equation (9.33) demonstrates the dependence of the evolution of the process on its long-term memory. This is true, with and without an external potential, as becomes clear with the solutions expressed in terms of eigenfunction expansions. Even with this long-term memory, asymptotically we still obtain the Boltzmann distribution for a physical system, and therefore we still have traditional statistical mechanics.

Other Drivations of the Propagation-Transport Equation   Zaslavsky also obtained a characteristic function of the form (9.84) using what he called a fractional Fokker-Planck-Kolmogorov (FFPK) equation. His derivation of the FFPK equation was based on the scaling properties of the solutions to non-integrable Hamiltonian equations. In such systems the phase space for the system contains islands of stability embedded in a chaotic sea, where the sea arises from the break up of KAM orbits. Trajectories that start on a stable island remain on that island forever. Trajectories that start in the chaotic sea avoid the islands and remain, like the Flying Dutchman, in the chaotic sea without landfall. Trajectories that start at the island/sea interface typically remain stuck at the interface for a long time, but asymptotically they break away. It has been shown by a number of investigators that an ensemble of orbits initiated at the interface decays as an inverse power law in time. If N_0 orbits are randomly distributed in a narrow ribbon along the island/sea interface initially, then at time t, the fraction of orbits remaining on the interface is $N(t)/N_0 \sim t^{-\alpha}$, $\alpha > 0$, an inverse power law in time.

Importance of Weak Strong Chaos   The kinetics of trajectories in the above phase space can be described by a FPK equation, to a good approximation, if chaos is weak. Usually, weak chaos means that most KAM tori in the Hamiltonian system, cantori, break up. The cantori are embedded in the stochastic sea and play the role of barriers riddled with holes having a Cantor set distribution. These holes act as turnstiles for particle transport. The closer the trajectory is to the boundary of chaos the stronger is the cantori's influence on the transport, in which case the evolution of the probability density can no longer be described by FPK. Using a renormalization argument Zaslavsky shows that the FPK equation is replaced by a FFPK equation of the form given by (9.80).

Bibliography

[1] P. Allegrini, M. Buiatti, P. Grigolini and B. J. West, Fractional Brownian motion as a nonstationary process: An alternative paradigm for DNA sequences, *Phys. Rev. E* **57**, 4558 (1998).

[2] P. Allegrini, P. Grigolini and B. J. West, Dynamical approach to Lévy processes, *Phys. Rev. E* **54**, 4760-4767 (1996).

[3] D. R. Bickel and B. J. West, *J. Mol. Evol.* **47**, 551 (1998).

[4] M. Berry, Diffractals, *J. Phys. A: Math. Gen. vol.* **12**, 781-797 (1979).

[5] J. -P. Bouchaud and A. Georges, *Phys. Rept.* **195**, 127 (1990)

[6] B. V. Chirikov, *Phys. Rept.* **52**, 265 (1979).

[7] A. Compte, Stochastic foundations of fractional dynamics, *Phys. Rev. E* **53**, 4191 (1996).

[8] P. G. de Gennes, *La Recherche* **7**, 919 (1976).

[9] W. G. Glöckle and T. F. Nonnenmacher, *Macromolecules* **24**, 6426 (1991).

[10] W. G. Glöckle and T. F. Nonnenmacher, A fractional Calculus approach to self-similar protein dynamics, *Biophys. J.* **68**, 46-53 (1995).

[11] W. G. Glöckle and T. F. Nonnenmacher, Fox function representation of non-Debye relaxation processes, *J. Stat. Phys.* **71** (1993) 741.

[12] W. G. Glöckle and T. F. Nonnenmacher, Fractional relaxation and the time-temperature superposition principle, *Rheol. Acta* **33** (1994) 337.

[13] J. M. Hausdorff, C. -K. Peng, Z. Ladin, J. Y. Wei and A. L. Goldberger, *J. Appl. Physiol.* **78**, 349 (1995).

[14] S. Havlin and D. Ben-Avraham, *Adv. in Phys.* **36**, 695 (1987).

[15] R. Hilfer, Ed., *Applications of Fractional Calculus in Physics*, World Scientific, Singapore (1999).

[16] R. Hilfer, Classification theory for an equilibrium phase transitions, *Phys. Rev. E* **48**, 2466 (1993).

[17] G. Jano-Lasino, The Renormalization Group: A Probabilistic View, *Nuovo Cimento* 29B, 99-119 (1975).

[18] M. Jaroniec, *Reac. Kinet. Catal Lett.* **8**, 425 (1978)

[19] P. Jörgi, D. Sornette and M. Blank, Fine structure and complex exponents in power-law distributions from random maps, *Phys. Rev. E* **57**, 120 (1998).

[20] D. D. Joseph and L. Preziosi, *Rev. Mod. Phys.* **61**, 41 (1989).

[21] L. P. Kadanoff, Fractals: Where's the beef?, *Physics Today*, 6 (Feb.) (1986).

[22] R. Metzler, E. Barkai and J. Klafter, Anomalous diffusion and relaxation close to thermal equilibrium: A fractional Fokker-Planck equation, *Phys. Rev. Lett.* **82**, 3563-3567 (1999); From continuous time random waks to the fractional Fokker-Planck equation, *Phys. Rev. E* **61**, 132-138 (2000).

[23] E.W. Montroll and B.J. West, On an enriched collection of stochastic processes, in *Fluctuation Phenomena*, 61-206, E.W. Montroll and J.L. Lebowitz, eds., second edition, North-Holland Personal Library, North-Holland, Amsterdam (1987); first edition (1979).

[24] E.W. Montroll and M.F. Shlesinger, On the wonderful world of random walks, in *Nonequilibrium Phenomena II: From Stochastics to Hydrodynamics*, 1-121, E.W. Montroll and J.L. Lebowitz, eds., North-Holland, Amsterdam (1983).

[25] T. J. Osler, An integral analogue of Taylor's series and its use in computing Fourier transforms, *Math. Comp.* **26**, 449-460 (1972).

[26] C. K. Peng, S. Buldyrev, A. L. Goldberger, S. Havlin, F. Sciortino, M. Simons, and H. E. Stanley, Long-range correlations in nucleotide sequences, *Nature* **356**, 168 (1992).

[27] C. K. Peng, J. Mietus, J. M. Hausdorff, , S. Havlin, H. G. Stanley and A. L. Goldberger, Long-range anticorrelations and non-Gaussian behavior of the heartbeat, *Phys. Rev. Lett.* **70**, 1343 (1993).

[28] L. F. Richardson, Atmospheric diffusion shown on a distance-neighbour graph, *Proc. Roy. Soc. London A* **110**, 709-737 (1926).

[29] S. G. Samko, A. A. Kilbas and O. I. Marichev, *Fractional Integrals and Derivatives*, Gordon and Breach, New York (1993).

[30] W. R. Schneider and W. Wyss, *J. Math. Phys.* **30**, 134 (1989).

[31] M. F. Shlesinger, B. J. West and J. Klafter, Lévy dynamics for enhanced diffusion: an application to turbulence, *Phys. Rev. Lett.* **58**, 1100-1103 (1987).

[32] A. J. F. Siegert, *Phys. Rev.* **81**, 617 (1951).

[33] G. E. Uhlenbeck and L. S. Ornstein, On the theory of the Brownian motion, *Phys. Rev.* **36**, 823 (1930).

[34] B. J. West and V. Seshadri, Linear systems with Lévy fluctuations, *Physica A* **113**, 203-216 (1982).

[35] B. J. West, Sensing scaled scintillations, *J. Opt. Soc. Am.* **7**, 1074 (1990).

[36] B. J. West and W. Deering, Fractal Physiology for Physicists : Lévy Statistics, *Phys. Repts.* **246**, 1-100 (1994).

[37] B. J. West and P. Grigolini, Fractional differences, derivatives and fractal time series, in *Applications of Fractional Calculus in Physics*, ed. R. Hilfer, World Scientific, Singapore (1998).

[38] B. J. West, R. Zhang, A. W. Sanders, S. Miniyar, J. H. Zucherman and B. D. Levine, Fractal fluctuations in transcranial Doppler signals, *Phys. Rev. E* **59**, 1 (1999).

[39] B. J. West and L. Griffin, Allometric control, inverse power laws and human gait, *Chaos, Solitons & Fractals* **10**, 1519 (1999); Allometric Control of Human Gait, *Fractals* **6**, 101 (1998).

[40] B. J. West, *Physiology, Promiscuity and Prophecy at the Millennium : A Tale of Tails*, Studies of Nonlinear Phenomena in the Life Sciences vol. **7**, World Scientific, Singapore (1999).

[41] B. J. West and T. Nonnenmacher, An ant in a Gurge, *Phys. Lett. A* **278**, 255 (2001).

[42] B. J. West, R. Zhang, A. W. Sanders, S. Miniyar, J. H. Zucherman and B. D. Levine, Fractal fluctuations in Cardiac Time Series, *Physica A* **270**, 522 (1999).

[43] G. M. Zaslavsky, M. Edelman and B. A. Niyazov, Self-similarity, renormalization, and phase space nonuniformity of Hamiltonian chaotic dynamics, *Chaos* **7**, 159 (1997).

Chapter 10

Appendices

10.1 Special Functions

The special functions of mathematical physics are those analytic functions that have been of assistance in understanding a variety of physical phenomena. For example, wave propagation in homogeneous and inhomogeneous media, heat transport, diffusion, conduction, and so on. These functions have, by and large, been solutions to partial differential equations that describe the evolution of the physical phenomena of interest. Here we investigate how to generate these functions using fractal operators, see also Bologna[2].

10.1.1 Jacobi and Laguerre Polynomials

We recall the definitions of Jacobi and Lagueree polynomials in terms of a differential generating function, after which we define their generalizations in terms of fractional derivatives. We take this approach in order to apply the technique to generalizations of the hypergeometric function, the Bessel function, and so on. We define Jacobi's polynomial using

$$P_n^{(\alpha,\beta)}(z) \quad = \quad \frac{(-1)^n}{2n!}(1-z)^{-\alpha}(1+z)^{-\beta}\frac{d^n}{dz^n}\left[(1-z)^{n+\alpha}(1+z)^{n+\beta}\right].$$

$$(10.1)$$

We generalize (10.1) by setting $n \in \mathcal{R}$ (actually $n \in \mathcal{N}$). The reason for doing this becomes clear after we make the substitution of variables:

$$z' \to \frac{z+1}{2} \tag{10.2}$$

in (10.1) yielding (we have dropped the prime for notational convenience)

$$P_n^{(\alpha,\beta)}(z) = \frac{(-1)^n}{2n!}(1-z)^{-\alpha}z^{-\beta}\frac{d^n}{dz^n}\left[(1-z)^{n+\alpha}z^{n+\beta}\right]. \tag{10.3}$$

The change of variables (10.2) does not change the structure of the definition because it is a simple linear transformation, so (10.3) is still Jacobi's polynomial. Now, by replacing the integer derivatives by fractional derivatives, we can define Jacobi's generalized functions by the generator

$$P_\gamma^{(\alpha,\beta)}(z) = \frac{e^{i\gamma\pi}}{\Gamma(\gamma+1)}(1-z)^{-\alpha} z^{-\beta} D_z^\gamma \left[(1-z)^{\gamma+\alpha} z^{\gamma+\beta}\right]. \qquad (10.4)$$

where $\alpha, \beta, \gamma \in \mathcal{R}$.

In a similar way we can generalize the Laguerre polynomials, defined by

$$L_n^\alpha(z) = \frac{z^{-\alpha}}{n!} e^z \frac{d^n}{dz^n}\left[z^{n+\alpha}e^{-z}\right]. \qquad (10.5)$$

The generalization $L_n^\alpha(z) \rightarrow L_{\beta\gamma}^\alpha(z)$ is accomplished by the replacements $n \rightarrow \beta$ ($\beta \in \mathcal{R}$) and $e^z \rightarrow E_\gamma^z$ such that we obtain

$$L_{\beta\gamma}^\alpha(z) = \frac{z^{-\alpha}}{\Gamma(\beta+1)} e^z D_z^\beta \left[z^{n+\alpha} E_\gamma^{-z}\right] \qquad (10.6)$$

from which we define the generalized Laguerre functions

$$L_\beta^\alpha(z) = L_{\beta 0}^\alpha(z) = \frac{z^{-\alpha}}{\Gamma(\beta+1)} e^z D_z^\beta \left[z^{n+\alpha} e^{-z}\right]. \qquad (10.7)$$

10.1.2 Applications of the Jacobi and Laguerre Generalized Functions

The hypergeometric function can be defined in terms of the definite integral

$$F(\alpha, \beta, \gamma, z) \equiv \frac{\Gamma(\gamma)}{\Gamma(\alpha)\Gamma(\gamma-\alpha)} \int_0^1 t^{\alpha-1} \frac{(1-t)^{\gamma-\alpha-1}}{(1-zt)^\beta} dt. \qquad (10.8)$$

The factor in the denominator of the integrand may be expanded using binomial coefficients to obtain

$$F(\alpha, \beta, \gamma, z) \equiv \frac{\Gamma(\gamma)}{\Gamma(\alpha)\Gamma(\gamma-\alpha)} \sum_{n=0}^\infty \binom{-\beta}{n}(-z)^n \int_0^1 t^{n+\alpha-1}(1-t)^{\gamma-\alpha-1} dt. \qquad (10.9)$$

The integral expression for the Beta function is

$$B(x, y) = \int_0^1 t^{x-1}(1-t)^{y-1} dt = \frac{\Gamma(x)\Gamma(y)}{\Gamma(x+y)} \qquad (10.10)$$

and when this expression is substituted into (10.9) we obtain

$$F(\alpha,\beta,\gamma,z) \equiv \frac{\Gamma(\gamma)}{\Gamma(\alpha)} \sum_{n=0}^{\infty} \binom{-\beta}{n} (-z)^n \frac{\Gamma(\alpha+n)}{\Gamma(\gamma+N)}.$$

(10.11)

This series expression for the hypergeometric function may be reexpressed in terms of the fractional derivative D_z^μ as follows,

$$F(\alpha,\beta,\gamma,z) \equiv \frac{\Gamma(\gamma)}{\Gamma(\alpha)} z^{1-\gamma} D_z^{\alpha-\gamma} \left[z^{\alpha-1} (1-z)^{-\beta} \right],$$

(10.12)

and using symmetry we have the alternative expression for the hypergeometric function

$$F(\alpha,\beta,\gamma,z) = F(\beta,\alpha,\gamma,z) = \frac{\Gamma(\gamma)}{\Gamma(\beta)} z^{1-\gamma} D_z^{\beta-\gamma} \left[z^{\beta-1} (1-z)^{-\alpha} \right].$$

(10.13)

The connection between the hypergeometric function and the generalized Jacobi function can be obtained by inspection of (10.4),

$$F(\alpha,\beta,\gamma,z) = \frac{\Gamma(\gamma)\Gamma(\alpha-\gamma+1)}{\Gamma(\beta)} e^{i\pi(\gamma-\alpha)} (1-z)^{\gamma-\alpha-\beta} P_{\alpha-\gamma}^{(\gamma-\alpha-\beta,\gamma-1)}(z)$$

(10.14)

or again by symmetry

$$F(\alpha,\beta,\gamma,z) = \frac{\Gamma(\gamma)\Gamma(\beta-\gamma+1)}{\Gamma(\alpha)} e^{i\pi(\gamma-\beta)} (1-z)^{\gamma-\alpha-\beta} P_{\beta-\gamma}^{(\gamma-\alpha-\beta,\gamma-1)}(z).$$

(10.15)

Now it is possible to extend the definition of the hypergeometric function to the case $Re[\alpha-\gamma] > 0$. In particular for $\alpha - \gamma = n$ with $n \in \mathcal{N}$ we have as the prescription for generating the hypergeometric function

$$F(\alpha,\beta,\gamma,z) = \frac{\Gamma(\gamma)}{\Gamma(\alpha)} z^{1-\gamma} D_z^n \left[z^{\alpha-1} (1-z)^{-\beta} \right].$$

(10.16)

Another special function of interest is the confluent hypergeometric function defined by the integral

$$F\left(\alpha,\gamma,z\right) \equiv \frac{\Gamma\left(\gamma\right)}{\Gamma\left(\alpha\right)\Gamma\left(\gamma-\alpha\right)}\int_0^1 t^{\alpha-1}\left(1-t\right)^{\gamma-\alpha-1}e^{zt}dt.$$

$$(10.17)$$

It is possible to proceed in direct analogy with the hypergeometric equation, but instead we choose an alternate route using

$$F\left(\alpha,\gamma,z\right) = \lim_{\beta\to\infty}F\left(\alpha,\beta,\gamma,\frac{z}{\beta}\right). \qquad (10.18)$$

Substituting the generator of the hypergeometric function (10.12) into (10.18) we obtain

$$
\begin{aligned}
F\left(\alpha,\gamma,z\right) &= \lim_{\beta\to\infty}\frac{\Gamma\left(\gamma\right)}{\Gamma\left(\alpha\right)}\left(\frac{z}{\beta}\right)^{1-\gamma}D_{\frac{z}{\beta}}^{\alpha-\gamma}\left[\left(\frac{z}{\beta}\right)^{\alpha-1}\left(1-\frac{z}{\beta}\right)^{-\beta}\right]\\
&= \lim_{\beta\to\infty}\frac{\Gamma\left(\gamma\right)}{\Gamma\left(\alpha\right)}z^{1-\gamma}D_z^{\alpha-\gamma}\left[z^{\alpha-1}\left(1-\frac{z}{\beta}\right)^{-\beta}\right],
\end{aligned}
$$

$$(10.19)$$

where we have used the operator property $D_{bz}^{\alpha} = b^{-\alpha}D_z^{\alpha}$. Now we take the indicated limit and obtain

$$F\left(\alpha,\gamma,z\right) = \frac{\Gamma\left(\gamma\right)}{\Gamma\left(\alpha\right)}z^{1-\gamma}D_z^{\alpha-\gamma}\left[z^{\alpha-1}e^z\right]. \qquad (10.20)$$

Finally, using the property of the confluent hypergeometric function, as defined by the integral,

$$F\left(\alpha,\gamma,z\right) = e^z F\left(\gamma-\alpha,\gamma,-z\right)$$

allows us to write (10.20) as

$$F\left(\alpha,\gamma,z\right) = \frac{\Gamma\left(\gamma\right)}{\Gamma\left(\gamma-\alpha\right)}z^{1-\gamma}e^z D_z^{\alpha-\gamma}\left[z^{\gamma-\alpha-1}e^{-z}\right]. $$

$$(10.21)$$

By inspection of the generating equation we constructed for the generalized Laguerre polynomial (10.7) we can see that (10.21) can be written as

$$F\left(\alpha;\gamma;z\right) = \frac{\Gamma\left(\gamma\right)\Gamma\left(\alpha-\gamma+1\right)}{\Gamma\left(\gamma-\alpha\right)}e^{-z}L_{\alpha-\gamma}^{\gamma-1}\left(z\right)$$

$$(10.22)$$

or equivalently as

$$F(\alpha; \gamma; z) = \frac{\Gamma(\gamma)\Gamma(1-\alpha)}{\Gamma(\gamma-\alpha)} L_{-\alpha}^{\gamma-1}(z).$$ (10.23)

Therefore we can assert that the hypergeometric function and the confluent hypergeometric function are both members of the class of Jacobi generalized functions, since the Laguerre polynomials can be derived by a limiting procedure on the Jacobi polynomials.

10.1.3 Bessel Functions

The integral definition of a Bessel function is

$$J_\nu(z) \equiv \frac{(z/2)^\nu}{\sqrt{\pi}\Gamma(\nu+1/2)} \int_{-1}^{1} dt\,(1-t^2)^{\nu-1/2}\cos zt; \quad for\ \nu > -1/2;$$ (10.24)

and in series form the Bessel function can be written

$$J_\nu(z) \equiv \sum_{k=0}^{\infty} \frac{(z/2)^{\nu+2k}}{k!\,\Gamma(\nu+k+1)}.$$ (10.25)

Let us restrict our attention to the particular case $\nu = 0$:

$$J_0(z) \equiv \sum_{k=0}^{\infty} \frac{(z/2)^{2k}}{(k!)^2}$$ (10.26)

so that making the substitution of variables $t = (z/2)^2$ and applying the fractal operator $D_t^{-\alpha}$ we obtain

$$D_t^{-\alpha}\left[J_0(t)\right]\big|_{t=(z/2)^2} \equiv \sum_{k=0}^{\infty} \frac{(-1)^k\,t^{\alpha+k}}{k!\,\Gamma(\alpha+k+1)} = \sum_{k=0}^{\infty} \frac{(-1)^k\,(z/2)^{2\alpha+2k}}{k!\,\Gamma(\alpha+k+1)}$$
$$= \left(\frac{z}{2}\right)^\alpha J_0(z).$$ (10.27)

If we consider the zeroth-order Bessel function directly we have

$$D_z^{-\alpha}\left[J_0(2\sqrt{z})\right] = \left(\frac{z}{2}\right)^\alpha J_0(2\sqrt{z})$$ (10.28)

yielding the same kind of relation that is obtained for the integer derivatives of Bessel functions of half-integer index.

Using the result (10.28) with the condition $\alpha > -1$ we can obtain a straightforward relation for Laplace transforms of the fractional derivatives of Bessel functions. Consider the expression

$$\mathcal{LT}\left[\left(\frac{z}{2}\right)^{\alpha}J_0\left(2\sqrt{z}\right);s\right] = \mathcal{LT}\left[D_z^{-\alpha}\left[J_0\left(2\sqrt{z}\right)\right];s\right]$$

$$= \frac{1}{s^{\alpha}}\mathcal{LT}\left[J_0\left(2\sqrt{z}\right);s\right],$$

$$(10.29)$$

but we also know

$$\mathcal{LT}\left[J_0\left(2\sqrt{z}\right);s\right] = \sum_{k=0}^{\infty}\frac{1}{(k!)^2}\mathcal{LT}\left[z^k;s\right]$$

$$= \sum_{k=0}^{\infty}\frac{(-1)^k}{k!}\frac{1}{s^{k+1}} = \frac{1}{s}\exp\left[-\frac{1}{s}\right], \qquad (10.30)$$

and finally for the total Laplace transform

$$\mathcal{LT}\left[\left(\frac{z}{2}\right)^{\alpha}J_0\left(2\sqrt{z}\right);s\right] = \frac{1}{s^{\alpha+1}}\exp\left[-\frac{1}{s}\right]. \qquad (10.31)$$

We can now determine the connection between Bessel functions and Laguerre functions. We start with the modified Bessel function

$$I_{\nu}\left(z\right) \equiv e^{-i\pi\nu/2}J_{\nu}\left(iz\right)$$

$$= \frac{2z^{\nu}}{\sqrt{\pi}\Gamma\left(2\nu+1\right)}e^{-z}\int_0^1 e^{2zt}\left[t\left(1-t\right)\right]^{\nu-1/2}dt \qquad (10.32)$$

or with the confluent hypergeometric function

$$I_{\nu}\left(z\right) = \frac{2z^{\nu}\Gamma\left(\nu+1/2\right)}{\Gamma\left(2\nu+1\right)}F\left(\nu+1/2;2\nu+1;2z\right)$$

$$(10.33)$$

and expressing the confluent hypergeometric function in terms of the generalized Laguerre function

$$F\left(\nu+1/2;2\nu+1;2z\right) = \frac{\Gamma\left(2\nu+1\right)\Gamma\left(1/2-\nu\right)}{\Gamma\left(\nu+1/2\right)}L_{-\nu-1/2}^{2\nu}\left(2z\right) \qquad (10.34)$$

so that

$$I_{\nu}\left(z\right) = \frac{\Gamma\left(1/2-\nu\right)}{\sqrt{\pi}}\left(2z\right)^{\nu}e^{-z}L_{-\nu-1/2}^{2\nu}\left(2z\right). \qquad (10.35)$$

So it is possible to express cylindrical functions (Bessel, Hankel, etc.) as generalized Laguerre functions. Finally, we can say that it is possible to generate all the special functions starting from the generalized Jacobi functions.

10.2 Fractional Derivatives

Here we gather together a number of equations for easy reference:

$$D_t^\alpha \left[t^\beta \right] = \frac{\Gamma(\beta+1)}{\Gamma(\beta+1-\alpha)} t^{\beta-\alpha} \tag{10.36}$$

$$D_t^\alpha \left[\{a+bt\}^{\alpha-1} \right] = \frac{\Gamma(\alpha)\sin\pi\alpha}{\pi(a+bt)} \left(\frac{a}{t}\right)^\alpha \tag{10.37}$$

$$
\begin{aligned}
D_t^\alpha \left[t^\beta \ln t \right] = \; & \frac{1}{t^\alpha} \frac{\ln t}{\Gamma(1-\alpha+\beta)} \\
& - \frac{\Gamma(\alpha+1)}{t^\alpha} \sum_{n=1}^{\infty} \frac{(-1)^n}{n\Gamma(1-n+\alpha)\Gamma(1+n+\beta-\alpha)}
\end{aligned}
\tag{10.38}
$$

$$
\begin{aligned}
D_t^\alpha \left[\ln t \right] = \; & \frac{1}{t^\alpha} \left\{ \frac{\ln t}{\Gamma(1-\alpha)} \right. \\
& \left. - \frac{\Gamma(\alpha+1)\left[\gamma+\Psi(1-\alpha)\right]\sin\pi\alpha}{\pi\alpha} \right\},
\end{aligned}
\tag{10.39}
$$

where the psi function is defined as

$$\Psi(z) = \frac{d\left[\ln\Gamma(z)\right]}{dz} = -\gamma + \sum_{n=1}^{\infty} \frac{1}{n} - \frac{1}{n+z-1} \tag{10.40}$$

$$D_t^\alpha \left[t^{\alpha-1} \ln(t+a) \right] = \Gamma(\alpha) \left[\frac{1}{t} - \frac{1}{t} \left(\frac{a}{t+a}\right)^\alpha \right] \; ; \; \alpha \neq 0. \tag{10.41}$$

10.2.1 1/2-Fractional Derivatives

Here we record a number of fractional derivatives of order 1/2:

$$D_t^{1/2} \left[(a+bt)^{-1/2} \right] = \sqrt{\frac{a}{\pi t}} \frac{1}{a+bt} \tag{10.42}$$

$$D_t^{1/2} \left[(a+bt)^{1/2} \right] = \sqrt{\frac{b}{\pi}} \left[\arcsin\sqrt{\frac{bt}{a+bt}} + \sqrt{\frac{a}{bt}} \right] \qquad \text{if } b > 0 \tag{10.43}$$

$$D_t^{1/2} \left[(a-bt)^{1/2} \right] = \sqrt{\frac{b}{\pi}} \left[\sqrt{\frac{a}{bt}} - \ln\left(\frac{\sqrt{bt}+\sqrt{a}}{\sqrt{a-bt}}\right) \right] \qquad \text{if } b > 0 \tag{10.44}$$

$$D_t^{1/2}\left[(a+bt)^{-1}\right] = \sqrt{\frac{b}{(a+bt)^3}}\left[\sqrt{\frac{a+bt}{bt}}\right.$$

$$\left. - \ln\left(\frac{\sqrt{bt}+\sqrt{a+bt}}{\sqrt{a}}\right)\right] \qquad \text{if } b>0 \quad (10.45)$$

$$D_t^{-1/2}\left[(a+bt)^{-1/2}\right] = \frac{2}{\sqrt{\pi b}}\arcsin\sqrt{\frac{bt}{a+bt}} \qquad (10.46)$$

$$D_t^{-1/2}\left[\ln(a+t)\right] = \frac{1}{\sqrt{\pi b}}\left\{\left(1-\sqrt{\frac{t+a}{t}}\right)\ln a\right.$$

$$\left. + 2\sqrt{\frac{t+a}{t}}\ln\left(\sqrt{t+a}+\sqrt{t}\right)\right\}.$$

$$(10.47)$$

10.2.2   Generalized Trigonometric Functions

Here we record the series expansions for the generalized trigonometric and hypergeometric functions:

$$\sin_\alpha t = \sum_{k=0}^{\infty}\frac{t^{k-\alpha}}{\Gamma(k+1-\alpha)}\sin\left[(k-\alpha)\pi/2\right] \qquad (10.48)$$

$$\cos_\alpha t = \sum_{k=0}^{\infty}\frac{t^{k-\alpha}}{\Gamma(k+1-\alpha)}\cos\left[(k-\alpha)\pi/2\right] \qquad (10.49)$$

$$\tan_\alpha t = -\tan(\alpha\pi/2) + \frac{t}{\cos^2(\alpha\pi/2)} + \cdots \qquad (10.50)$$

$$\sinh_\alpha t = \sum_{k=0}^{\infty}\frac{t^{2k-\alpha}}{\Gamma(2k+2-\alpha)} \qquad (10.51)$$

$$\cosh_\alpha t = \sum_{k=0}^{\infty}\frac{t^{2k-\alpha}}{\Gamma(2k+1-\alpha)}. \qquad (10.52)$$

These equations can be compared with the series expansion for the ordinary trigonometric functions:

$$\sin t = \sum_{k=0}^{\infty}\frac{t^k}{\Gamma(k+1)}\sin\left[k\pi/2\right] \qquad (10.53)$$

$$\cos t = \sum_{k=0}^{\infty} \frac{t^k}{\Gamma(k+1)} \cos\left[k\pi/2\right] \qquad (10.54)$$

$$\sinh t = \sum_{k=0}^{\infty} \frac{t^{2k+1}}{\Gamma(2k+2)} \qquad (10.55)$$

$$\cosh t = \sum_{k=0}^{\infty} \frac{t^{2k+2}}{\Gamma(2k+1)}. \qquad (10.56)$$

10.2.3 Miscellaneous Integrals

We record here a number of useful integrals of generalized functions:

$$
\begin{aligned}
Erf(t) &= \int_0^t e^{-x^2}\,dx = \frac{\sqrt{\pi}}{2} e^{-t^2} \left[E_{1/2}^{t^2} - \frac{1}{t\sqrt{\pi}} \right] \\
&= \frac{\sqrt{\pi}}{2} e^{-t^2} E_{-1/2}^{t^2} \qquad (10.57)
\end{aligned}
$$

$$\int_0^t e^{\pm x^2}\,dx = \frac{1}{\alpha}\Gamma\left(\frac{1}{\alpha}\right) e^{\pm t^2} E_{-1/2}^{\mp t} \qquad (10.58)$$

$$\int_0^t E_\alpha^x\,dx = E_\alpha^t + \frac{1}{\alpha\Gamma(-\alpha)} + C \qquad (10.59)$$

$$\int_0^t \cosh_\alpha x\,dx = \sinh_\alpha t + C \qquad (10.60)$$

$$\int_0^t \sinh_\alpha x\,dx = \cosh_\alpha t + \frac{1}{\alpha\Gamma(-\alpha)\,t^\alpha} + C \qquad (10.61)$$

$$\int_0^t \frac{\cosh_\alpha x}{x^{\alpha+1}}\,dx = \frac{\Gamma(-\alpha)}{2} E_\alpha^t E_\alpha^{-t} + C \qquad (10.62)$$

$$\int_0^t e^{bx} E_\alpha^{ax}\,dx = \frac{e^{bt}}{a+b}\left[E_\alpha^{at} - \left(\frac{b}{a}\right)^\alpha E_\alpha^{-bt} \right] + C \qquad (10.63)$$

$$\int_0^\infty \frac{\cos \omega x}{(x+a)^\alpha} dx = \Gamma(1-\alpha) \omega^{\alpha-1} \quad [\sin(\alpha\pi/2 + \omega a)$$
$$- \widetilde{\sin}_{\alpha-1}(\alpha\pi/2 + \omega a)]$$

(10.64)

where $\alpha > 0$ and we have adopted the notation

$$\widetilde{\sin}_{\alpha-1}(\alpha\pi/2 + \omega a) = \cos(\alpha\pi/2) \sin_{\alpha-1}(\omega a) + \sin(\alpha\pi/2) \cos_{\alpha-1}(\omega a)$$

(10.65)

so that for the similar integral we have

$$\int_0^\infty \frac{\sin \omega x}{(x+a)^\alpha} dx = \Gamma(1-\alpha) \omega^{\alpha-1} \quad [\cos(\alpha\pi/2 + \omega a)$$
$$- \widetilde{\cos}_{\alpha-1}(\alpha\pi/2 + \omega a)], \quad (10.66)$$

where $\alpha > 0$ and we have adopted the notation

$$\widetilde{\cos}_{\alpha-1}(\alpha\pi/2 + \omega a) = \cos(\alpha\pi/2) \cos_{\alpha-1}(\omega a) - \sin(\alpha\pi/2) \sin_{\alpha-1}(\omega a).$$

(10.67)

10.2.4   Complex Integrals

Here we list a number of integrals that are useful for the analyses discussed in the text [2]:

$$\int_0^\infty \frac{x^\alpha dx}{x+a} \cos \omega x = \frac{\pi a^\alpha}{\sin(\pi\alpha)} [\cos_\alpha \omega a - \cos \omega a]$$

(10.68)

$$\int_0^\infty \frac{x^\alpha dx}{x+a} \sin \omega x = -\frac{\pi a^\alpha}{\sin(\pi\alpha)} [\sin_\alpha \omega a - \sin \omega a]$$

(10.69)

$$\int_0^\infty \frac{x^\alpha}{x^2+a^2} e^{-sx} dx = \frac{\pi a^{\alpha-1}}{\sin(\pi\alpha)} [\sin(sa + \alpha\pi/2) - D_{sa}^\alpha [\sin(sa)]]$$

(10.70)

$$\int_0^\infty \frac{x^\alpha}{x+a} e^{-sx} dx = \frac{\pi a^\alpha}{\sin(\pi\alpha)} [E_\alpha^{sa} - e^{sa}] \tag{10.71}$$

$$\frac{1}{\sqrt{\pi t}} \int_0^\infty e^{-\frac{x^2}{4t}} \sinh_\alpha x dx = E_{\frac{\alpha-1}{2}}^t \tag{10.72}$$

$$\frac{1}{\sqrt{\pi t}} \int_0^\infty e^{-\frac{x^2}{4t}} \cosh_\alpha x dx = E_{\frac{\alpha}{2}}^t \tag{10.73}$$

$$\int_0^\infty \frac{\cos_\alpha \omega x}{x^2 + a^2} dx = \frac{\pi}{2a} [\cosh_\alpha \omega a - 2\sinh \omega a + \sinh_\alpha \omega a] \tag{10.74}$$

$$\int_0^\infty \frac{\sin_\alpha \omega x}{x^2 + a^2} dx = \frac{\pi}{2a} [\{\sinh_\alpha \omega a - \sinh \omega a\} \cot(\alpha\pi/2)]$$
$$- \frac{\pi}{2a} [\{\cosh_\alpha \omega a - \sinh \omega a\} \tan(\alpha\pi/2)] \tag{10.75}$$

$$\int_0^\infty \frac{\sin_\alpha \omega x}{x^\beta} dx = \frac{\cos \pi(\alpha+\beta/2)}{\sin \pi(\alpha+\beta)} \frac{\pi\omega^{\beta-1}}{\Gamma(\beta)}; \quad \text{for } \alpha+\beta < 0, \beta > 0 \tag{10.76}$$

$$\int_0^\infty \frac{\cos_\alpha \omega x}{x^\beta} dx = \frac{\sin \pi(\alpha+\beta/2)}{\sin \pi(\alpha+\beta)} \frac{\pi\omega^{\beta-1}}{\Gamma(\beta)}; \quad \text{for } \alpha+\beta < 1, \beta > 0 \tag{10.77}$$

10.3 Mellin Transforms

The transform and inverse formulae of Mellin are

$$\widehat{\Phi}(p) = \int_0^\infty t^{p-1} \Phi(t) \, dt, \tag{10.78}$$

$$\Phi(t) = \frac{1}{2\pi i} \int_{C-i\infty}^{C+i\infty} \widehat{\Phi}(p) \, t^{-p} dp. \tag{10.79}$$

These reciprocal relations first occurred to Riemann in a famous memoir on prime numbers.

Let $\widehat{\phi}(p)$ be a function expressible as a Dirichlet series

$$\widehat{\phi}(p) = \sum_{n=1}^\infty \frac{a_n}{n^p} \tag{10.80}$$

so that using the Euler equation for the gamma function

$$\Gamma(p) = \int_0^\infty x^{p-1} e^{-x} dx$$

in the series (10.80) yields

$$\widehat{\phi}(p) = \sum_{n=1}^\infty \frac{a_n}{\Gamma(p)} \int_0^\infty x^{p-1} e^{-nx} dx.$$

Thus, carrying the sum through the integral we have the Mellin transform

$$\widehat{\Phi}(p) = \int_0^\infty x^{p-1} \Phi(x) dx \tag{10.81}$$

where we have defined the function

$$\Phi(x) = \sum_{n=1}^\infty a_n e^{-nx} \tag{10.82}$$

as well as

$$\widehat{\Phi}(p) = \Gamma(p) \, \widehat{\phi}(p). \tag{10.83}$$

The inverse Mellin transform is constructed using the Dirichlet series

$$\frac{1}{2\pi i} \int_{C-i\infty}^{C+i\infty} \widehat{\Phi}(p) x^{-p} dp = \frac{1}{2\pi i} \int_{C-i\infty}^{C+i\infty} \Gamma(p) \, \widehat{\phi}(p) \, x^{-p} dp$$

$$= \frac{1}{2\pi i} \sum_{n=1}^\infty a_n \int_{C-i\infty}^{C+i\infty} x^{-p} dp \frac{\Gamma(p)}{n^p} \tag{10.84}$$

and making use of the integral expression for the exponential function

$$e^{-x} = \frac{1}{2\pi i} \int_{C-i\infty}^{C+i\infty} \Gamma(p) \, x^{-p} dp$$

we can simplify (10.84) to

$$\frac{1}{2\pi i} \int_{C-i\infty}^{C+i\infty} \widehat{\Phi}(p) x^{-p} dp = \sum_{n=1}^\infty a_n e^{-nx}$$

which using (10.82) is the inverse Mellin transform

$$\Phi(x) = \frac{1}{2\pi i} \int_{C-i\infty}^{C+i\infty} \hat{\Phi}(p) \, x^{-p} dp.$$

Consider the Mellin transformation (10.78) in which we substitute $t = e^{\xi}$ and $p = c + iq$ so that we obtain

$$\hat{\Phi}(c+iq) = \int_0^\infty e^{\xi(c+iq)} \Phi\left(e^{\xi}\right) d\xi. \tag{10.85}$$

Furthermore, we can make the same substitution in the inverse Mellin transform (10.79) to obtain

$$\Phi(t) = \frac{1}{2\pi} \int_{C-i\infty}^{C+i\infty} \hat{\Phi}(c+iq) \, e^{\xi(c+iq)} dq. \tag{10.86}$$

Thus, $\Phi\left(e^{\xi}\right) e^{\xi c}$ and $\hat{\Phi}(c+iq)$ are Fourier transforms of each other.

10.3.1 Fox Functions

The H-function, or as it is also called, the Fox function, arises naturally in the study of a number of phenomena in the physical sciences, engineering and statistics. Nearly all special functions arising in applied mathematics and statistics are special cases of this function. A complete review of the formal properties of these functions is given by Mathai and Saxena [3], which we follow, in part, in this appendix. In an attempt to unify and extend the then existing results on symmetrical Fourier kernels, Fox [1] defined the H-function in terms of a Barnes-type integral by

$$H_{pq}^{mn}(z) = H_{pq}^{mn}\left(z \left| \begin{matrix} (a_1, A_1) \cdots (a_p, A_p) \\ (b_1, B_1) \cdots (b_p, B_p) \end{matrix} \right. \right) \equiv \frac{1}{2\pi i} \int_C h(\xi) z^{\xi} d\xi, \tag{10.87}$$

where $z \neq 0$. Here the function $h(\xi)$ is given by the ratio of the product of gamma functions

$$h(\xi) = \frac{\prod\limits_{j=1}^{n} \Gamma(1 - a_j + A_j\xi) \prod\limits_{j=1}^{m} \Gamma(b_j - B_j\xi)}{\prod\limits_{j=m+1}^{q} \Gamma(1 - b_j + B_j\xi) \prod\limits_{j=n+1}^{p} \Gamma(a_j - A_j\xi)}, \tag{10.88}$$

where p, q, m, and n are integers satisfying $0 \leq n \leq p$ and $1 \leq m \leq q$; A_j $(j = 1, \cdots, p)$, B_j $(j = 1, \cdots, q)$ are positive numbers; a_j $(j = 1, \cdots, p)$, b_j $(j = 1, \cdots, q)$ are complex numbers such that

$$A_j(b_k + \nu) \neq B_k(a_j - \lambda - 1) \tag{10.89}$$

for $\nu, \lambda = 0, 1, 2, \cdots; k = 1, \cdots, m; j = 1, \cdots, n$. Empty products are interpreted as unity.

The C subscript on the integral (10.87) denotes a contour separating the points

$$\xi = \frac{\nu + b_j}{B_j} \quad (j = 1, .., m; \ \nu = 0, 1, ...) \tag{10.90}$$

which are the poles of $\Gamma \left(b_j - B_j \xi \right) \ (j = 1, .., m)$ from the points

$$\xi = \frac{a_j - \nu - 1}{A_j} \quad (j = 1, .., nm; \ \nu = 0, 1, ...) \tag{10.91}$$

which are the poles of $\Gamma \left(1 - a_j + A_j \xi \right) \ (j = 1, .., n)$. In the complex ξ-plane the former poles lie to the right of the contour C and the latter poles lie to the left of the contour C.

The Fox function is an analytic function of z which is reasonable if (1) for every $z \neq 0$ if $\mu > 0$ and (2) for $0 < |z| < \beta^{-1}$ if $\mu = 0$ where

$$\mu = \sum_{j=1}^{q} B_j - \sum_{j=1}^{p} A_j \tag{10.92}$$

and

$$\beta = \Pi_{j=1}^{p} A_j^{A_j} \Pi_{j=1}^{q} B_j^{-B_j}. \tag{10.93}$$

Due to the occurrence of the factor z^ξ in the integrand of (10.87) it is, in general, multivalued but one-valued on the Riemann surface of $\log z$. Apart from symmetry relations in the parameters which can be determined from (10.88), there are a number of important relations among the Fox functions listed by Glöckle and Nonnenmacher [4].

Using the residue theorem the Fox function can also be expressed as

$$H_{pq}^{mn} \left(z \left| \begin{matrix} (a_1, A_1) \cdots (a_p, A_p) \\ (b_1, B_1) \cdots (b_p, B_p) \end{matrix} \right. \right) = - \sum res \left[h \left(p \right) \ z^p \right], \tag{10.94}$$

where the residues are taken at the poles $p_{j\nu} = (b_j + \nu) / B_j$

$(j = 1, 2, \cdot m; \ \nu = 0, 1, \cdot)$. If the poles are simple then in general (10.94) may be written

$$H_{pq}^{mn} (z) = \sum_{h=1}^{m} \sum_{k=0}^{\infty} \frac{\prod\limits_{j=1}^{n}{}' \Gamma \left(1 - a_j + A_j p_{hk} \right) \prod\limits_{j=1}^{m} \Gamma \left(b_j - B_j p_{hk} \right)}{\prod\limits_{j=m+1}^{q} \Gamma \left(1 - b_j + B_j p_{hk} \right) \prod\limits_{j=n+1}^{p} \Gamma \left(a_j - A_j p_{hk} \right)}$$

$$\times \frac{(-1)^k}{k!} \frac{z^{p_{hk}}}{B_k}, \tag{10.95}$$

where the prime on the product symbol denotes the omission of the $j = h$ term. The formula (10.95) can be used to calculate the special values of the Fox function and to derive the asymptotic behavior for $z \to 0$.

In closing we note that the generalized Mittag-Leffler function can be expressed in terms of the following Fox function [4]

$$E_{\alpha,\beta}(-z) = H_{12}^{11}\left(z \,\middle|\, \begin{matrix} (0,1) \\ (0,1)\,(1-\beta,\alpha) \end{matrix}\right). \tag{10.96}$$

Bibliography

[1] C. Fox, *Trans. Am. Math. Soc.* **98**, 395 (1910).

[2] M. Bologna, *Derivata ad Indice Reale*, Ets Editrice, Italy (1990).

[3] A. M. Mathai and R. K. Saxena, *The Fox-Function with Applications in Staistics and Other Disciplines*, Wilry Eastern Limited, New Delhi (1978).

[4] W. G. Glöckle and T. F. Nonnenmacher, Fox function representation of non-Debye relaxation processes, *J. Stat. Phys.* **71** (1993) 741.

Index